How To Make Christmas Crafts from Empty Toilet Paper Rolls:

A Frugal Crafting DIY Book

Andrea Reynolds

SkyShan Publishing, LLC.

ISBN: 978-1-970106-64-0

TABLE OF CONTENTS

Dedication: To my wonderful family who puts up with my insanity. Your support has meant the world to me.

Introduction:

No exotic crafting supplies are needed for these fantastic designs!

I have a confession to make; I love using toilet paper rolls instead of throwing them out. It's weird, but these tiny cardboard tubes are the perfect size to create unique projects. My imagination sparks with potential ideas. Pandemics notwithstanding, toilet paper rolls are in most homes.

I have two words of caution to all you potential readers before beginning. Firstly, the designs in this book are far from perfect. I am not going to sit there with tweezers and obsess over perfection. I believe in making these projects accessible to everyone. To me, perfection is found in imperfection.

Secondly, I love writing these books as I believe in making them fun. Occasionally my warped sense of humor comes out when least expected. If you want dry reading without a few humorous comments, this might not be the book for you. Hence the current subject matter.

If I haven't scared you away, welcome to this fantastic adventure in toilet paper roll crafting! This book will show you step-by-step instructions on how to create the following unique projects:

Holiday Crackers
Toilet Paper Roll Reindeer
Toilet Paper Roll Moose
Napkin Ring Holders – Multiple Ways
Sitting Bear with Vest

The Christmas season occasionally fills me with waves of anxiety, trying to make beautiful memories for my children. Even acknowledging this self-inflicted stress, finding ways to handle your mental wellbeing is vital to a happy holiday.

Crafting reminds us that the journey is important and having a tangible item at the end of the day fills you with accomplishment. Luckily these projects can be made with your youngsters to create unforgettable memories or even by yourself to destress.

Let's get crafting!

Holiday Crackers to Make Yourself

Holiday crackers have been a part of the holidays in the UK since the Victorian era around the mid-1840s. The traditional cracker contains a paper crown, a tacky joke, a snap, and a tiny little trinket. However, the sky is the limit in what you can add to your personalized cracker.

After purchasing the store-made variety for one Christmas, to say we were disappointed is an understatement. None of the snaps worked, and the quality versus the price was insane. In all honesty, with a neurotic dog who flips out during every thunderstorm, I was glad the snaps failed.

While these instructions do not include the snap, crackle, and pop of it all, these sticks are purchased at most craft stores or online retailers. Search for “Christmas Cracker Snap Strips.”

These crackers come together quickly with only paper, ribbon, an empty TP roll, and glue.

Required Materials / Optional Materials (Marked as **)

For the Cracker

1-Empty Roll of Toilet Paper per Cracker

1-Piece of Copy Paper with an image printed in a Landscape orientation **or** Gift Wrap.

2- 12" (30.5 cm) piece Ribbon for Decoration per Cracker

Tools/Tertiary Equipment

Scissors

Glue

Color Printer / Copier to print holiday image onto paper

Computer with a program like Word to size the image

Optional Decorative Items

If you choose to utilize a custom image, you must ensure proper sizing. A picture of the family member on the cracker is a fabulous idea. However, you need to be able to see the image to have the effort appreciated.

The standard circumference of a toilet paper roll is 5.375” (13.7 cm). Please remember that I am assuming (dangerous word alert) the standard is the same in the rest of the world as the US. (If I’m wrong, please let me know in the review section while dropping a 5-star review for this literary masterpiece.) There is no need to break out π; simply wrap a piece of string around the tube, flatten the ribbon out, and measure the length.

To ensure your picture is the correct size, you can utilize your computer to make the process easier. I will be using Microsoft Word, but you can use any program based on your familiarity.

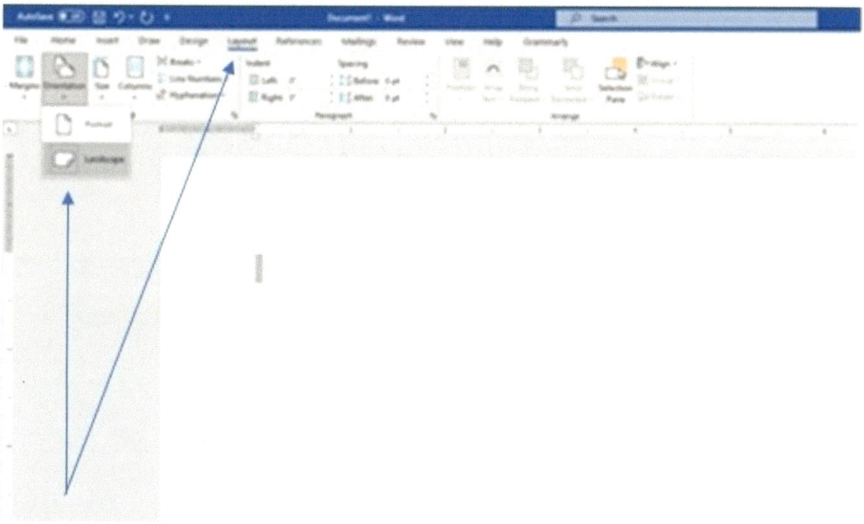

Open a blank document. Click on the “Layout” tab, click on the “Orientation” tab, and select “Landscape.” Insert the picture of your choice.

Click on the “Insert” tab, then click on the “Pictures” icon, and click on “This Device” to choose the image of your choice. Be sure you know the file path of your picture’s location.

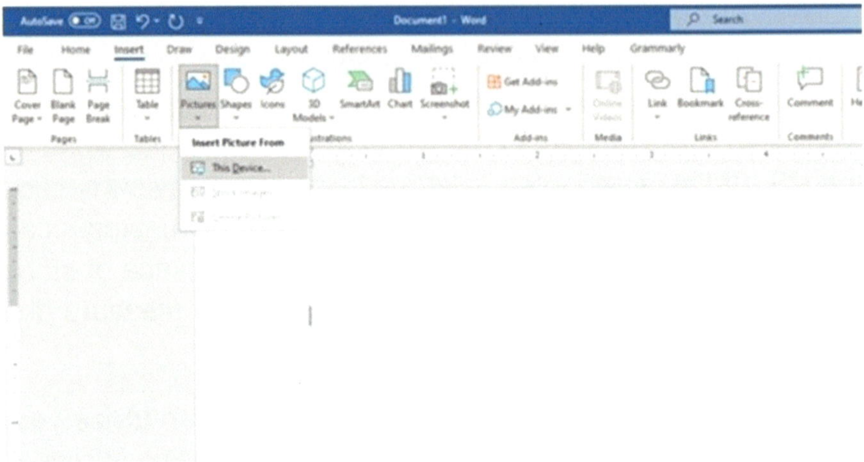

Make sure the picture is active. (Click on the picture, and the “Picture Format” tab will appear as the farthest right tab.) Click on the “Picture Format” tab. In the far right, in the “Size” section, adjust the height to 5.375 inches (13.65 cm). The width must be at least the width of the toilet paper roll. I like at least 2” (5 cm) wider than the cylinder. Print out your image—time to make the Cracker.

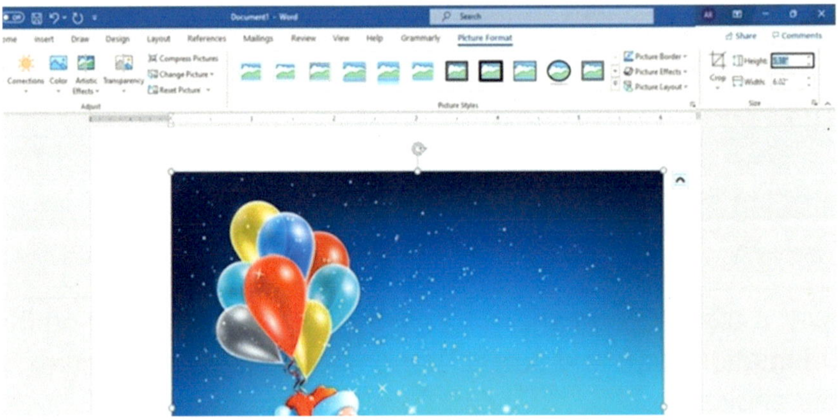

Trim the paper **only** to the proper height. Avoid trimming the width of the paper. (It is easier to trim excess than not have enough paper.) Place a thin layer of glue over the entirety of the outside of the toilet paper roll. Align the paper, centering the image over the cylinder, and cover the cardboard with the picture. Allow the glue to completely dry.

When the glue dries completely, grasp the roll between your thumb and forefinger at the end of the rolls on each side, and gently squeeze.

Trim the sides to be evenly spaced. Next, cut two pieces of ribbon 12” (30.5 cm) in length. Tie one string on one side in a bow. Do not be afraid to pull it tight, but go slowly to prevent ripping the paper.

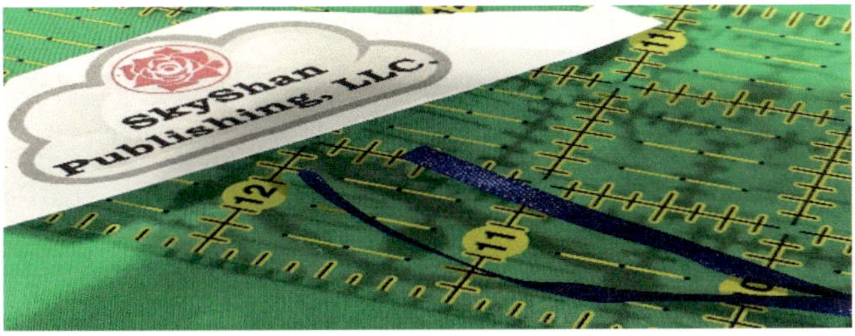

Place your chosen gifts into the open side of the cracker. Some cute ideas for adults could be lottery tickets, an airplane bottle of adult libation, lipstick or nail polish for teens or adults, or individually wrapped high-end chocolates.

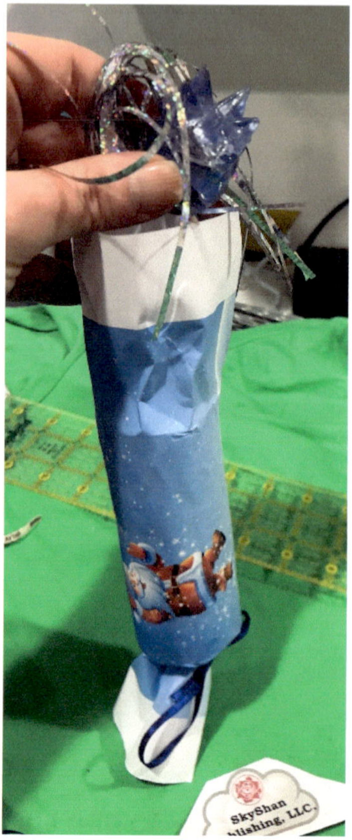

Tie the other end of the tube closed. Place this gift on the place setting. You can even write the recipient's name on the cracker, so there is no confusion. You don't want to accidentally give eight-year-old Bobby eighty-year-old Uncle Buck's bottle of Bourbon. If your penmanship resembles a toddler like my own, you can write the person's name on the interior of one side of the tube.

Your Christmas Cracker is ready to be demolished. The sound of tearing paper will fill your house along with the sound of laughter.

Before and after the joy.

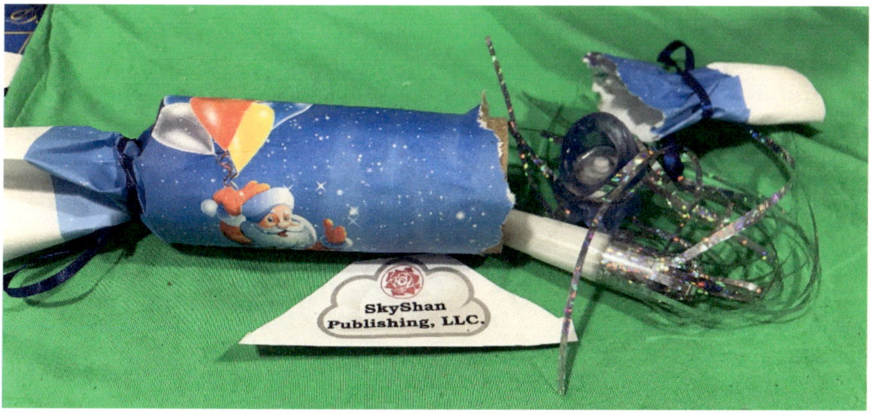

Toilet Paper Roll Reindeer

Sometimes you just run out of crafting supplies. The kids got into the glue and spent hours making fingerprint molds, and left the top open. Never fear for this next project. The bare minimum needed for this project is two empty toilet paper rolls, scissors, and tape.

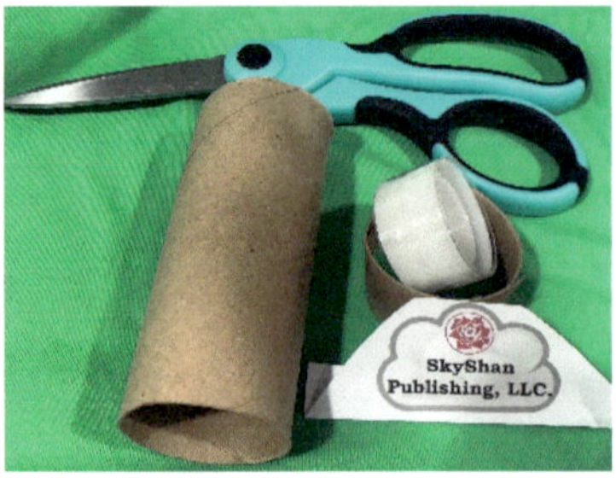

How is it possible to create a beautiful project? Personally, something about the austerity of the reindeer without embellishments appeals to me. While these instructions will include adding felt to the reindeer, the exact instructions apply to the simple design, excluding material addition.

Required Materials / Optional Materials (Marked as **)

Reindeer

2-Empty Roll(s) of Toilet Paper

**1-Piece of Felt (OR Construction Paper) Each– Dark Brown, Light Brown, Black, White, and Red

Tools/Tertiary Equipment

Scissors

**Glue, and paperclip or binder clip

Tape

Measure and cut a piece of brown felt to cover the entire roll of toilet paper. Glue the entirety of the outside of the toilet paper roll, and place the brown felt over the roll. Use paper clips or binder clips to secure the felt. Let dry.

On the second roll, cut a 1” (2.5 cm) wide piece. Push the portion of the tube to create a fold on either side. Cut the cylinder in half at one end.

Take one half of the piece of roll, and using the bottom fold pull the fold toward the top of the cardboard.

Fold the other side of the 1” (2.5 cm) roll.

Measure a piece of dark brown felt to fit the triangle at the bottom of the above picture with a ¼” (0.635 cm) additional height. Cut an extra piece of dark brown felt to fit the triangle. Next, cut 2(two) pieces of light brown felt to cover the 2(two) rectangles above the triangle.

Turn the cardboard around and glue the areas and attach the felt.

Cut out eyes, nose, and glue to the reindeer. Glue the opposite side of the brown triangle. Set aside to dry. Oh, look! My reindeer has a red nose. I wonder what I should call him.

When the reindeer's body is dry, it is time to create the body of our TP buddy. Remove the paper clips and find the top middle at the back of the tube. Measure a ½" (1.25 cm) width and cut 2 (two) ½" (1.25 cm) long incisions into the cylinder. Raise the tab slightly. Your reindeer now has a tail.

Turn the tube over to the other side. In the center of the cardboard, measure 1" (2.5 cm) wide, and cut 1" (2.5 cm) deep on either side.

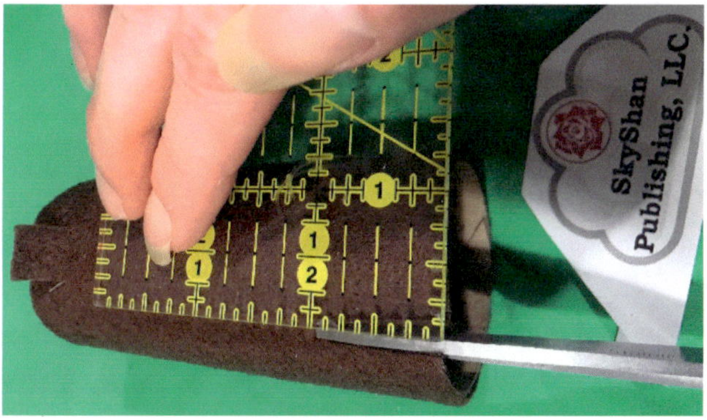

Pull the tab upward and put a bend in the tip of the front about ¼" (0.635cm). This is the neck of the reindeer.

There are some ways to make your life easier when you make your cuts when creating the legs. In the front of your reindeer, cut a 1" (2.5 cm) long strip to mirror the neck above. On the outside of these cuts, make an additional 1" (2.5 cm) long cut ¼" (0.635 cm) apart from the initial incision. (No this is not surgery) Pull these legs down, folding a ¼" (0.635 cm) tab for the feet.

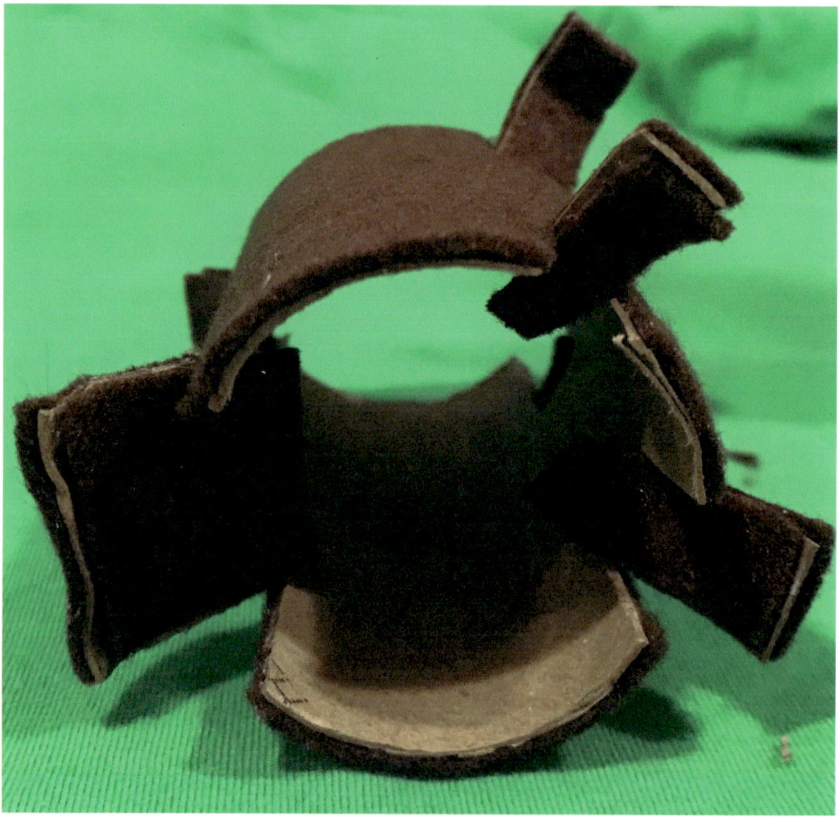

Using a ruler, cut the back legs in the exact placement as the front legs. If you are using felt, cut out felt, glue, and cover the opposite side of the legs, neck, and tail. Allow to dry.

Time to turn your attention back to the head.

Sometimes bigger is not better. If you have a small pair of scissors, I recommend using them. If not, it is okay. You got this. Take your time and slowly cut out antlers. This does not have to be perfect. No two antlers are the same, and it is okay to create an organic look.

Quick Tip: Cut the antlers out slowly, working your way inward. It is okay to have a sharp edge; round it out with small snips.

Wait, I Made a Mistake – I Cut off an Antler

A moment of inattention can lead to disaster. Oopsie!

Relax! This is not the end of the world. First, turn your head over. Get some glue and cover the felt, the cardboard, and the back of the cardboard. Take a piece of tape and cover the glued area; ensure that the glue touches parts of the cardboard that does not have any glue on it. Clamp the damaged area with a binder clip or a paper clip.

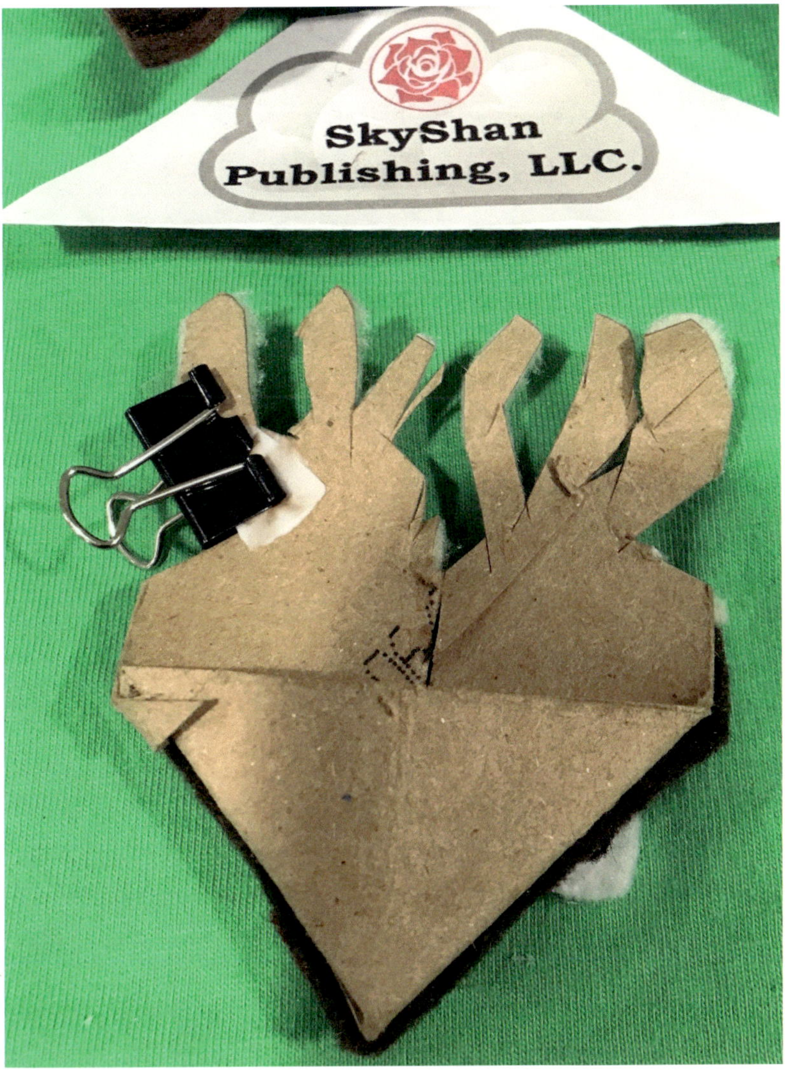

Allow this repair to dry completely. Remove binder clip after drying—time to finish assembling the reindeer.

In the front of the body, push the bottom tab into the body. Next, press each side tab into the body as well.

Repeat this step on the back of the body. Adjust the legs, so the body stands on its four legs.

When attaching the reindeer's head, two options depend upon what you have on hand.

Option one is to glue the head to the neck.

Option two is to use tape to attach the head to the neck. You will need more than these two pieces. I recommend running the adhesive covering the head to the neck.

Construction Paper Reindeer

If you don't have felt on hand, you can easily use construction paper instead. My experience with this is to make sure you use paper clips. Until the glue dries, the paper can slip.

When gluing the paper to the tube, there can be some overlap. There is no need to trim to exact measurements as the paper is thinner than felt. Make sure that you glue the overlap.

My supply of construction paper only has one shade of brown, so I used orange as the antler color. Add eyes and nose to the face.

Once dried, ensure that you cut out the tail, legs, and neck. Attach the head and enjoy your incredible creation.

Please don't disparage the antlers on the left. We can't all have a perfect rack.

Optional Toilet Paper Roll Moose

The reindeer model converts easily to a moose with only a few modifications. This is an excellent idea for moose lovers everywhere. The start of the moose head begins the same as the reindeer until this point.

Traditional depictions of moose heads show their rack of antlers as more horizontal from the head and not vertical. Therefore, the head of your animal needs an additional fold. Glue it down and hold it in place with paper clips until dry.

Cut out two pieces of dark brown felt triangles to fit the face of your moose. Cut out eyes and nose (White and black felt). Glue the face and the eyes and nose. Light brown felt should be sized and cut out.

Once the glue has dried, cut out the moose antlers, attach them to the body.

Optional Scarf Addition

Adding accessories to your reindeer and moose can add to the festive feeling of the holiday season. Santa's hats didn't look cohesive as they clashed with the antlers on each animal.

Scarves are a fantastic choice that is customizable with color choices that match your holiday decorations. If you just squinted at me, look at your decorations. The chances are that one color or type of décor is prevalent. We like what we like and tend to purchase similar themes.

I chose a primarily red scarf with a white fringe. You will need red and white felt or construction paper. Create an arch 2 3/8" (6 cm) wide and 2" (5 cm) long.

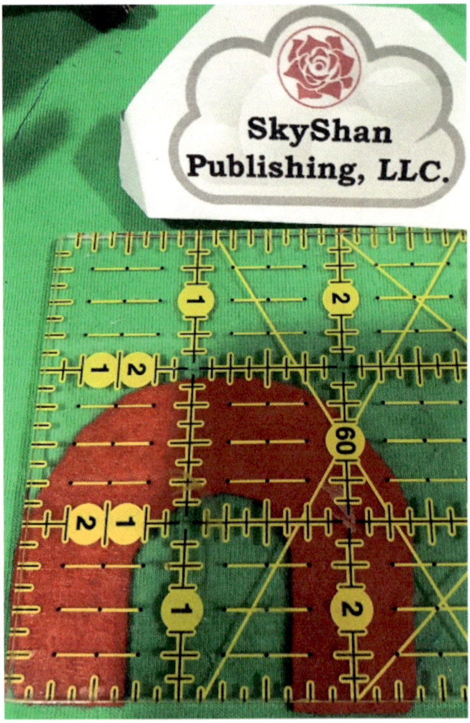

Cut out two pieces of white felt, matching the width of the ends of the scarf, and about ½" (1.25 cm) long. Cut small incisions into the white felt, about half the length, creating a fringe.

Attach the white felt to the bottom of the scarf with glue.

Slide the scarf under the head to abut the center of the arc of the scarf. Glue down the arch to the body. If you don't want to attach the scarf permanently, you can cut a notch into the back of the scarf to let it lay flat.

Napkin Ring Holders

.

Napkin Rings…Really? Now stay with me before you shake your head and dismiss this section out of hand. Adding a customized napkin ring holder that compliments your napkin choice can turn a humdrum place setting into a creative masterpiece.

The options to decorate these holders are endless, but I will show you three unique design choices. I hope you use this as a guide to discovering your inspiration.

A fun and shiny napkin ring can turn a plain white napkin into a festive accompaniment to your holiday meal. This napkin ring uses just yarn and a 1” (2.5 cm) wide piece of toilet paper roll.

This yarn is skinny, and you will need 4 yards (3.65 meters) of thread per napkin ring. Tie the start of the rope around the circle. Begin wrapping the yarn around the ring.

If you are wrapping the yarn, some yarn separates from the rest of the pack; use your finger to align it.

When you reach the end of the yarn, tie the end to one of the interior strings to hide the knot. Make one for each guest attending.

Utilizing fabric for the next napkin ring option is a great choice. You could even use a stained napkin from the set to make multiple rings for your friends and family to enjoy the meal you have planned.

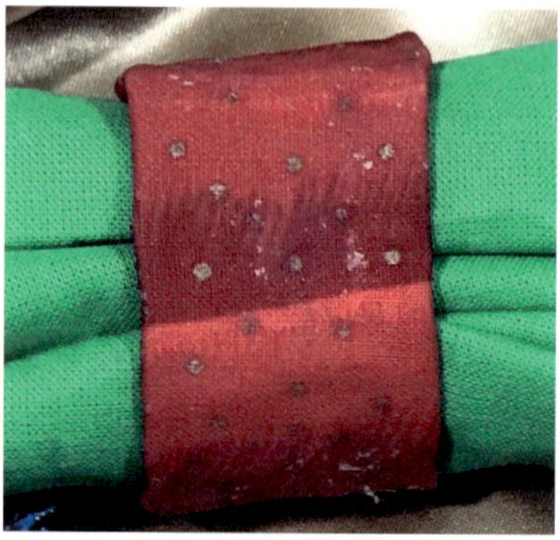

Measure the fabric 2.5" (6.35 cm) wide. (I know the fabric appears to be 3" (7.5 cm) wide, but I did cut that down. The length should be 5 3/8" (13.5 cm).

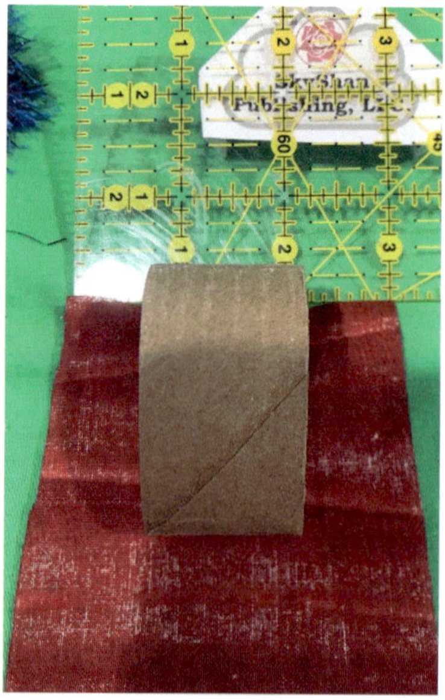

Glue the outside of the ring. Cut notches into the side of the fabric every ¼" (0.635 cm). This process helps the fabric fold into the ring. Glue the inside of the ring, and fold the material into the ring.

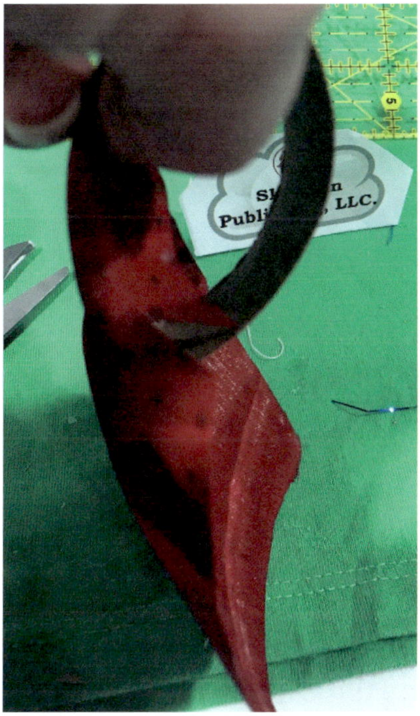

Allow the glue to dry.

Felt might seem like a pedestrian choice, but it works as an excellent background for a fantastic accent piece. Glue a piece of festive colored felt to a napkin ring. Ensure you cut notches into your felt for ease of folding into the napkin ring.

I found a great bow with a bell that I tied to the inside of the ring. A boring piece turns into a formal statement.

Sitting Bear with Vest

If you're scratching your head about why this little animal needs only one TP roll and not two like the other, the answer is the orientation. This great little bear, made with only one toilet paper roll, sits vertically. The bear sitting up can support the head, but a reindeer with a rack of antlers will tip over every time.

Cover and glue a piece of felt or construction paper to a tube of TP. Let dry completely—Mark 1" (2.5 cm) around the circumference. Use scissors to cut it out.

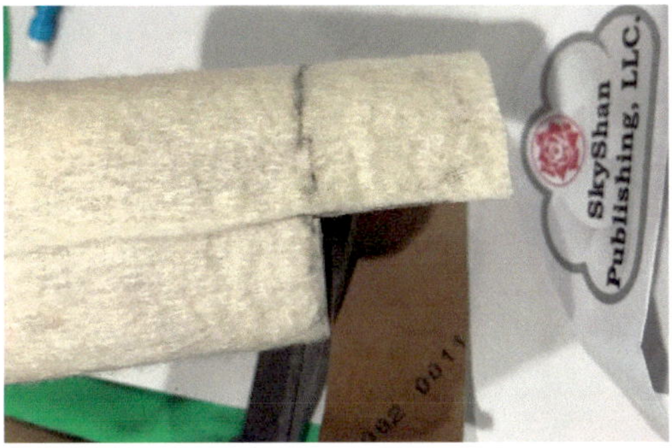

Fold the 1” (2.5 cm) piece to create the face. Glue down if necessary and use a binder clip to hold down the center seam.

Cut out the body. The arms and legs should be ½” (1.25 cm) long and ¼” (0.635 cm) wide. Ensure the tail is cut as well to help it stand.

Cut out the bear's face—round out the bottom of the face and the ears. Add eyes, nose, and ears, and glue.

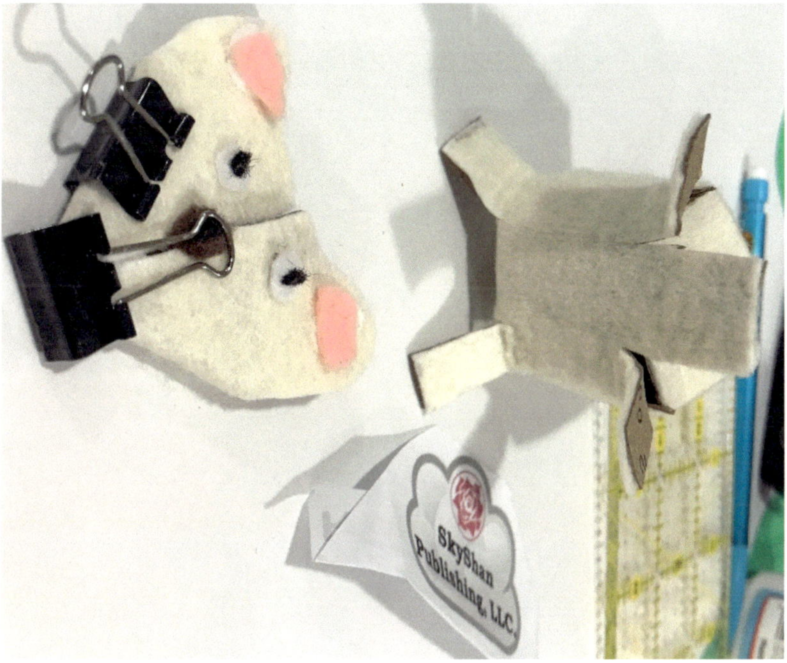

Glue the head onto the neck.

Adding embellishments to this project is a great idea. There is a lot of negative space on the body, so a tie and vest added a festive touch.

Conclusion

We've made it to the end of another crafting journey. I can't thank you enough for reading this book.

Sometimes you must be who you were born to be. I love turning something that usually gets tossed in the trash into something that will occupy you and your kids for hours. If you can look at something from a different perspective, I encourage you to embrace your creativity.

I have many other craft books planned in the coming months and years. What the next book in this series will be is anyone's guess. I can't promise when the next book will be out, as things do not always work out as planned. I hope that you will join me on other journeys.

My only request is to let others know that this book is for them by leaving a review. Unfortunately, the crafting genre is filled with "books" from people that copy and paste blogs instead of putting in the effort themselves. By using your reviewing abilities, small independent authors like myself have a chance of getting noticed.

Happy Crafting!

Also Available from SkyShan Publishing

About the Author

Andrea Reynolds has always had an incredible imagination. She is the internationally bestselling author of over thirty books. Her love of crafting has brought her more joy and peace than any other activity. She just wishes she could draw a stick figure. She is the mother of two beautiful children and one rescue dog. When not writing, her head will be found stuck in her crafting closet.

PROCEEDINGS OF THE 2ND CZECH-CHINA SCIENTIFIC CONFERENCE 2016

Edited by **Jaromir Gottvald** and **Petr Praus**

Proceedings of the 2nd Czech-China Scientific Conference 2016
http://dx.doi.org/10.5772/66773
Edited by Jaromir Gottvald and Petr Praus

Published by InTech

Janeza Trdine 9, 51000 Rijeka, Croatia

Publishing Process Manager Natalia Reinic
Technical Editor SPi Global
Cover InTech Design team

First published December, 2017
Printed in Croatia

Additional hard copies can be obtained from orders@intechopen.com

Proceedings of the 2nd Czech-China Scientific Conference 2016, Edited by Jaromir Gottvald and Petr Praus
p. cm.
Print ISBN 978-953-51-2857-1
Online ISBN 978-953-51-2858-8

Contents

Preface

This book comprises the proceedings of the 2nd Czech-China Scientific Conference 2016 which was held on 7th June 2016 in Ostrava, Czech Republic. The objective of the conference was to present the latest achievements in the fields of advanced science and technology that stem from research activities of VŠB – Technical University of Ostrava and its Chinese partners.

The conference covered multiple topics and allowed young researchers from different scientific areas to present their findings and experience the international conference atmosphere. The conference attracted specialists from the areas of economy, safety in civil engineering and industry, material technologies, environment and computational science. The conference structure corresponds with the structure of the chapters.

Section 1

Computational Sciences

Parallel Iteration Method for Frequency Estimation Using Trigonometric Decomposition

He Wen, Junhao Zhang, Radek Martinek,
Petr Bilik and Zidek Jan

Additional information is available at the end of the chapter

http://dx.doi.org/10.5772/66778

Abstract

The parallel iteration method for frequency estimation based on trigonometric decomposition is presented. First, the multi-frequency signal can be expressed in a matrix form based on the trigonometric decomposition, which implies a possibility to solve the nonlinear mapping functions of frequency estimation by a parallel iteration procedure. Then, frequency estimation with the minimized square errors is achieved by using the gradient-descent method in the parallel iteration procedure, which can effectively restrain the interferences from harmonics and noise. Finally, the workflow is shown, and the efficiency of the proposed method was demonstrated through computer simulations and experiments.

Keywords: frequency estimation, trigonometric decomposition, parallel iteration, signal processing

1. Introduction

With the widespread application of nonlinear loads such as uninterruptible power supply, electric arc furnaces, and other power-electronic devices in power electronic systems, resulting power quality problems have drawn much attention. Frequency estimation is a very important issue in the power system because of the need for an assessment of the power quality [1, 2]. In addition, the frequency of a distribution network can extremely vary during transient events, and it can be very difficult to track the frequency with enough accuracy [3].

In the past years, various methods have been presented to estimate the power system frequency [4, 5]. The DFT-based methods provide a nonparametric approach and require a low computational effort [6–8]. Unfortunately, they have inherent limitations due to the picket fence and spectral leakage effects caused by the noncoherent sampling. Since the system fundamental frequency may deviate from its nominal value because of the power mismatch between the generation and load demands, it is almost impossible to achieve the strict coherent sampling and thus some degree of spectral leakage cannot be avoided. A commonly used frequency estimation methods to compensate both spectral leakage and picket fence effects are the so-called windowed interpolated FFT (WIFFT) method [9]. However, the WIFFT can only reduce, but not completely remove, the estimation errors at the cost of an increment in the computation burden of the frequency estimation.

Other popular frequency estimation methods include the zero-crossing technique, Kalman filter, least-squares method, and artificial neural networks, which are characterized by high resolution. However, they may either suffer from low accuracy or less computational efficiency [10]. For example, the Prony method can provide high-accuracy frequency estimates when the model order is known; however, the determination of the model order is often a difficult task, which requires intensive computational algorithms [5].

Trigonometric decomposition is a well-known technique for determine different frequency components of periodic signals. According to the Fourier analysis, general functions or signals may be represented or approximated by sums of simpler trigonometric functions. So, it is relatively easy to estimate the frequency of a pure sine-wave by trigonometric decomposition. However, estimating frequency of a real signal distorted by harmonics and noise is much more difficult. This is because the frequency estimation errors due to the interference from the harmonics and noise remain significant in trigonometric decomposition, especially when the number of acquired signal cycles is very small. To reduce the frequency estimation errors, an iteration procedure can be applied in trigonometric decomposition; however, the computation would much more complex than the fast Fourier transform. For real-time processing requirements, most of the aforementioned methods offer a tradeoff between accuracy and speed [3]. Therefore, seeking efficient methods to estimate power system frequency for power-quality assessment and solutions has been a significant challenge.

This chapter proposes a parallel iteration method for frequency estimation based on trigonometric decomposition. According to the theory of trigonometric decomposition, a multifrequency signal can be expressed using matrix algebra, which provides a base for the parallel iteration to solve the nonlinear mapping functions of frequency estimation. By using the gradient-descent method in the iteration procedure, the frequency is estimated with the minimized square errors and thus the interferences from harmonics and noise can effectively be restrained. The organization of this chapter is as follows: in Section 2, the proposed method is presented. Simulation and experiment results are provided in Section 3, and the conclusion is made in Section 4.

2. Proposed frequency estimation method

2.1. Trigonometric decomposition of periodic signal

Let us consider a multifrequency electrical waveform $x(n)$ sampled at a known frequency f_s, i.e.,

$$x(n) = U_0 + \sum_{h=1}^{H} U_h \sin(2\pi hfn/f_s + \varphi_h) \tag{1}$$

where $n = 0,1,\ldots, N - 1$; f is the fundamental frequency, H is the number of frequency components, U_h and φ_h are, respectively, the amplitude and phase of the hth component, U_0 is the offset, and N is the acquisition length. To satisfy the Nyquist criterion, the frequency of the Hth harmonic is assumed to be smaller than $f_s/2$.

By applying trigonometric decomposition, it can be rewritten Eq. (1) as follow:

$$x(n) = U_0 + \sum_{h=1}^{H} [U_h \sin\varphi_h \cos(2\pi hfn/f_s) + U_h \cos\varphi_h \sin(2\pi hfn/f_s)] \tag{2}$$

For the sake of simplicity, the a_h and b_h can be used to represent U_h sin φ_h and U_h sin φ_h, respectively, i.e.,

$$a_h = U_h \sin\varphi_h \text{ and } b_h = U_h \cos\varphi_h \tag{3}$$

So, Eq. (1) can be expressed as

$$x(n) = U_0 + \sum_{h=1}^{H} [a_h \cos(2\pi hfn/f_s) + b_h \sin(2\pi hfn/f_s)] \tag{4}$$

By using the matrix operation, Eq. (4) can also be written as

$$X = ZD + AC + BS \tag{5}$$

where $X = [x(1), x(2),\ldots, x(N)]$ denotes the vector of the sampled signal, $Z = U_0$, $D = [1_1,1_2,\ldots, 1_N]$, $A = [a_1,a_2,\ldots, a_H]$, $B = [b_1, b_2, \ldots, b_H]$,

$$C = \begin{bmatrix} \cos(\eta) & \cos(2\eta) & \cdots & \cos(N\eta) \\ \cos(2\eta) & \cos(4\eta) & \cdots & \cos(2N\eta) \\ \vdots & \vdots & \ddots & \vdots \\ \cos(H\eta) & \cos(H2\eta) & \cdots & \cos(HN\eta) \end{bmatrix} \tag{6}$$

and

$$S = \begin{bmatrix} \sin(\eta) & \sin(2\eta) & \cdots & \sin(N\eta) \\ \sin(2\eta) & \sin(4\eta) & \cdots & \sin(2N\eta) \\ \vdots & \vdots & \ddots & \vdots \\ \sin(H\eta) & \sin(H2\eta) & \cdots & \sin(HN\eta) \end{bmatrix} \tag{7}$$

with $\eta = 2\pi f/f_s$.

2.2. Parallel iteration algorithm

From Eq. (5), the frequency estimation for each component can be regarded as a nonlinear mapping problem through N samples, which can be defined as

$$f = F_m(X, Z, D, A, C, B, S) \tag{8}$$

where F_m is the nonlinear mapping functions for the fundamental frequency. The solution procedure for detecting the fundamental frequency of Eq. (8) is not easy. This is because it is difficult to determine the nonlinear mapping functions F_m, which should be implicit function. It has been noticed that the iteration is one of the approaches developed to solve the nonlinear mapping problem. In the following, the procedure with parallel iteration for solving the nonlinear mapping functions F_m of Eq. (8) with a fast convergence is proposed.

Suppose $Y = [y(1), y(2),..., y(N)]$ denote the vector of the estimated signal. According to the least-squares method, the estimated frequency f_e should be the one that minimizes the square error, which can be expressed as

$$f_e = \arg\min[E^2] = \text{argmin}[(X{-}Y)(X{-}Y)^{\mathrm{T}}] \tag{9}$$

where $(\bullet)^{\mathrm{T}}$ denotes the transpose operation.

By using the gradient-descent method, a group of equations can be obtained by setting each partial derivative of E^2 equal to zero. Then, the parallel iteration equations can be obtained as following,

$$\begin{cases} f^{(k+1)} = f^{(k)} - \lambda \dfrac{\partial E^{(k)}}{\partial f^{(k)}} \\ \quad = f^{(k)} + \gamma \dfrac{E^{(k)}}{f_s}\left(\psi . * C^{(k)^{\mathrm{T}}} B^{(k)^{\mathrm{T}}} - \psi . * S^{(k)^{\mathrm{T}}} A^{(k)^{\mathrm{T}}}\right) \\ Z^{(k+1)} = Z^{(k)} - \lambda \dfrac{\partial E^{(k)}}{\partial Z_k} = Z^{(k)} + \lambda E^{(k)} D^{\mathrm{T}} \\ A^{(k+1)} = A^{(k)} - \lambda \dfrac{\partial E^{(k)}}{\partial A_k} = A^{(k)} + \lambda E^{(k)} C^{\mathrm{T}} \\ B^{(k+1)} = B^{(k)} - \lambda \dfrac{\partial E}{\partial B^{(k)}} = B^{(k)} + \lambda E^{(k)} S^{\mathrm{T}} \end{cases} \tag{10}$$

where * denotes the element-by-element multiplication; $\gamma = 2/U_1^{\,2}$ and λ are the descent coefficients, ψ is a matrix of constants

$$\psi = \begin{bmatrix} 1 & 2 & \cdots & N \\ 2 & 4 & \cdots & 2N \\ \vdots & \vdots & \ddots & \vdots \\ H & 2H & \cdots & HN \end{bmatrix} \tag{11}$$

2.3. Convergence of parallel iteration procedure

From Eq. (9), the Lyapunov function can be defined as

$$J_k = \frac{1}{2}\left\| E^{(k)} \right\|^2 \tag{12}$$

and the gradient can be calculated by

$$\Delta J_k = \frac{1}{2} \| E^{(k+1)} \|^2 - \frac{1}{2} \| E^{(k)} \|^2 \tag{13}$$

where $||\bullet||^2$ denotes the square of the **F**-norm.

By using the Taylor expansion, the $E^{(k+1)}$ can be rewritten as

$$E^{(k+1)} = E^{(k)} + \Delta Z_k \frac{\partial E^{(k)}}{\partial Z_k} + \Delta A_k \frac{\partial E^{(k)}}{\partial A_k} + \Delta B_k \frac{\partial E^{(k)}}{\partial B_k} \tag{14}$$

where the partial derivatives can be calculated as

$$\begin{cases} \frac{\partial \Lambda_k}{\partial Z_k} = -D \\ \frac{\partial \Lambda_k}{\partial A_k} = -C \\ \frac{\partial \Lambda_k}{\partial B_k} = -S \end{cases} \tag{15}$$

By using Eq. (9), Eq. (14) can be expressed as

$$E^{(k+1)} = E^{(k)} \left[I - \lambda (D^T D + C^T C + S^T S) \right] \tag{16}$$

where I is an identity matrix. By substituting Eq. (16) into Eq. (13), the gradient can be expressed as

$$\Delta J_k = \frac{1}{2} \| E^{(k)} \|^2 [\| I - \lambda (D^T D + C^T C + S^T S) \|^2 - 1] \tag{17}$$

According to the triangle inequality of the matrix norm, the following inequality can be obtained

$$\begin{aligned} \| I - \lambda (D^T D + C^T C + S^T S) \|^2 &\le \left(\| I \| - \| \lambda (D^T D + C^T C + S^T S) \| \right)^2 \\ &= 1 - 2\lambda \| D^T D + C^T C + S^T S \| \ + \lambda^2 \| D^T D + C^T C + S^T S \|^2 \end{aligned} \tag{18}$$

Substitute Eq. (17) into Eq. (18), one can obtain

$$\Delta J_k \le \frac{1}{2} \| E^{(k)} \|^2 \| D^T D + C^T C + S^T S \| (-2 + \lambda \| D^T D + C^T C + S^T S \|) \tag{19}$$

In Eq. (19), considering the $||E^{(k)}||^2 ||D^T D + C^T C + S^T S|| \ge 0$, to guarantee the convergence of the parallel iteration procedure, i.e., $\Delta J_k < 0$, it is required to hold the following condition

$$-2 + \lambda \| D^T D + C^T C + S^T S \| < 0 \tag{20}$$

That is to say the convergence of parallel iteration procedure can be guaranteed if λ satisfy the following condition

$$0 < \lambda < 2/\|D^TD + C^TC + S^TS\|_F \tag{21}$$

where $||\bullet||_F$ denotes the **F**-norm.

2.4. Workflow of the proposed method

Figure 1 shows the workflow of the parallel iteration method for frequency estimation using trigonometric decomposition, where ε denotes the tolerance. The major steps are as follows:

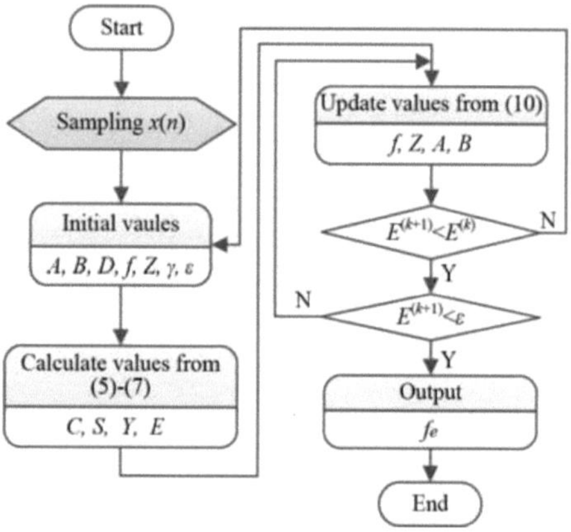

Figure 1. Workflow of the parallel iteration method for frequency estimation using trigonometric decomposition.

1. Initialize the parallel iteration parameters. Set the initial frequency f to be 50 Hz, the initial value of both A and B to be random values within the range (0–1), the initial value of Z to be zero.
2. Calculate the iteration values of C, S, Y, E based on Eqs. (5)–(7).
3. Update iteration values by using Eq. (10).
4. Check if the early-stopping constraint, i.e., $E^{(k+1)}<E^{(k)}$, is met. If not, adjust the λ and return to step (1).
5. Check if the stopping constraint, i.e. $E < \varepsilon$, is met. If not, return to step (3). If yes, keep the value of f_e and output.

It is worth mentioning that the early stopping constraint is to guarantee that the error differences between two iterations are decreasing.

3. Simulation and experimental results

To evaluate the performances of the proposed algorithm, we perform a series of simulations on an electrical power signal with and without white noise in this section. To make comparisons, the WIFFT algorithm and discrete phase difference correction (DPDC) algorithm [7] based on the Hanning window (HNW), 3-term Max decay window (MDW), and the proposed method are adopted. At last, the proposed algorithm is evaluated by practical measurements.

3.1. Comparison with other algorithms without noise

An electrical power signal with 11 orders of harmonics, whose amplitudes are actually measured in an electric power network, is analyzed. This signal model is also used in Ref. [11] and can be expressed as

$$x(n) = \sum_{h=1}^{11} A_h \sin(2\pi hfn + \varphi_h) \tag{22}$$

where the fundamental frequency f is set as 50.2 Hz, the sampling frequency is 3200 Hz, the amplitude A_h and the phase φ_h of each harmonic component are given in **Table 1**.

The WIFFT and DPDC algorithms and the proposed method are adopted to make a comparison. The absolute errors of fundamental and harmonic frequencies by using different algorithms are listed in **Table 2**, where aE-b represents $a \times 10^{-b}$.

Table 2 shows that the accuracy of frequency estimation obtained by the proposed method is higher than those obtained by the WIFFT and DPDC algorithms.

h	1	2	3	4	5	6	7	8	9	10	11
A_h (V)	240	0.1	12	0.1	2.7	0.05	2.1	0	0.3	0	0.6
φ_h (°)	0	10	20	30	40	50	60	–	80	–	100

Table 1. Parameters of each harmonic component of signal (22).

	h	1	2	3	4	5	6	7	9	11
WIFFT	HNW	7E-5	1E1	2E-3	1E-1	1E-3	5E-2	4E-4	9E-4	9E-5
	MDW	1E-6	1E-1	3E-5	8E-3	1E-5	1E-2	2E-5	4E-5	7E-6
DPDC	HNW	2E-4	6E1	2E-3	8E-1	7E-3	9E-1	5E-4	8E-3	4E-3
	MDW	3E-6	4E-1	2E-5	7E-2	1E-5	7E-2	2E-5	1E-6	8E-6
	Proposed	3E-11	3E-11	3E-11	3E-11	3E-11	3E-11	3E-11	3E-11	3E-11

Table 2. Absolute errors of fundamental and harmonic frequencies by using different algorithms.

3.2. With white Gaussian noise

To analyze the influence of white noise, the signal is superposed with zero-mean Gaussian noise. For each SNR value 3000 runs are performed by using $N = 512$ samples.

The estimation variances of fundamental frequency by using different algorithms are listed in **Table 3**. From **Table 3**, it can be seen that the proposed method can achieve the lowest variances of frequency estimation among all the four adopted algorithms making it to be a good choice for high accurate frequency estimation.

SNR	WIFFTHNW	WIFFTMDW	DPDCHNW	DPDCMDW	Proposed
20 dB	3E-8	6E-8	1E-8	3E-8	9E-9
30 dB	3E-9	6E-9	1E-9	3E-9	9E-10
40 dB	3E-10	6E-10	1E-10	3E-10	9E-11
50 dB	3E-11	4E-11	1E-11	3E-11	9E-12
60 dB	4E-12	6E-12	1E-12	3E-12	9E-13
70 dB	4E-13	6E-13	1E-13	3E-13	8E-14
80 dB	4E-14	6E-14	2E-14	3E-14	9E-15

Table 3. Estimation variances of fundamental frequency by using different algorithms with white noise.

3.3. With frequency varying

In addition, the signal is simulated with SNR = 40 dB and frequency varying from 48.5 and 51.5 Hz to investigate the influences of both the white noise and frequency variations on frequency estimation.

The biases of frequency estimation by using different algorithms with frequency variations and SNR = 40 dB are listed in **Table 4**. As shown in **Table 4**, the biases of the frequency estimation obtained by the proposed method are 1–2 orders of magnitude lower than those obtained by the adopted WIFFT and DPDC algorithm.

f/Hz	WIFFTHNW	WIFFTMDW	DPDCHNW	DPDCMDW	Proposed
48.5	−8E-5	2E-6	7E-3	1E-4	−1E-6
49.0	−4E-4	1E-5	6E-3	9E-5	−5E-7
49.5	−7E-4	1E-5	3E-3	4E-5	1E-6
50.0	−7E-4	2E-5	7E-6	7E-5	2E-6
50.5	−5E-4	1E-5	2E-4	3E-6	3E-7
51.0	−2E-4	1E-6	1E-3	2E-5	1E-6
51.5	8E-5	2E-6	4E-3	5E-5	2E-6

Table 4. Biases of fundamental frequency estimation by using different algorithms with frequency variations and SNR = 40 dB.

3.4. Experimental results

The experiments are carried out by using the electrical power standard HBS1030, the data acquisition system ELVIS II of National Instrument. The measurement scheme is depicted in **Figure 2**.

As shown in **Figure 2**, the electrical power standard HBS1030 is used to generate multisine waves with the accuracy 0.05%. The signal is sampled by the data acquisition system ELVIS II with 16-bit analog-to-digital (A/D) converter. The sampling frequency is set as 3.2 kHz. The measurement results of the fundamental frequency are shown in **Table 5**, where the "true" values for calculating the measurement absolute errors are provided by the electrical power standard HBS1030. The experimental results demonstrate that the presented algorithms have high accuracy in practice.

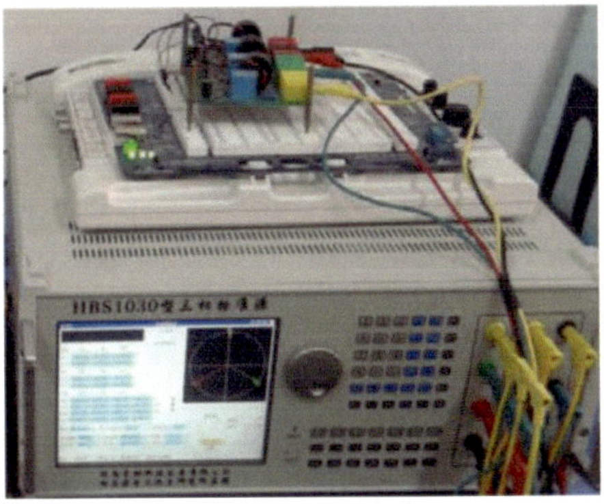

Figure 2. Measurement scheme for the laboratory experiment of the proposed method.

f	49.1	49.2	49.3	49.4	49.5	49.6	49.7	49.8	49.9
Error	4E-4	3E-4	2E-4	3E-4	2E-4	1E-4	2E-4	2E-4	1E-4
f	50.1	50.2	50.3	50.4	50.5	50.6	50.7	50.8	50.9
Error	1E-4	1E-4	2E-4	3E-4	3E-4	3E-4	3E-4	4E-4	4E-4

Table 5. Absolute errors of frequency estimation by experiments.

4. Conclusion

Since frequency estimation errors caused by harmonic interference and noise remain significant when the number of acquired signal cycles is very small, this chapter presents a parallel iteration method for frequency estimation using trigonometric decomposition. Due to the nature of trigonometric decomposition of periodic signal, the parallel iteration can be effectively executed to solve the nonlinear mapping functions of frequency estimation with a fast convergence. Additionally, the minimized square errors can be achieved by using the gradient-descent method in the iteration procedure. By observing the simulation and experimental results, it is seen that the proposed method is more accurate than the WIFFT and DPDC in comparisons.

Acknowledgements

This work has partially been supported by the National Natural Science Foundation of China under grant 61370014.

Author details

He Wen[1]*, Junhao Zhang[1], Radek Martinek[2], Petr Bilik[2] and Zidek Jan[2]

*Address all correspondence to: he_wen82@126.com

1 College of Electrical and Information Engineering, Hunan University, Changsha, Hunan Province, China

2 Faculty of Electrical Engineering and Computer Science, VSB-Technical University of Ostrava, Ostrava, Czech Republic

References

[1] J. Borkowski, D. Kania, and J. Mroczka, "Interpolated-DFT-based fast and accurate frequency estimation for the control of power," *IEEE Transactions on Industrial Electronics*, vol. 61, pp. 7026–7034, 2014.

[2] MISAK, Stanislav; PROKOP, Lukas; BILIK, Petr. Power quality analysis in off-grid power system. In Conference Proceedings: The 10th International Conference, ELEKTRO 2014, Rajecke Teplice, Slovakia,19 May 2014 through 20 May 2014, Category numberCFP1448S-ART, Code 106459, pp. 337-342, ISBN: 978-147993721-9, DOI: 10.1109/ELEKTRO.2014.6848914.

[3] M. D. Kusljevic, J. J. Tomic, and L. D. Jovanovic, "Frequency estimation of three-phase power system using weighted-least-square algorithm and adaptive FIR filtering," *IEEE Transactions on Instrumentation and Measurement*, vol. 59, pp. 322–329, 2010.

[4] H. Wen, Z. S. Teng, and S. Y. Guo, "Triangular self-convolution window with desirable sidelobe behaviors for harmonic analysis of power system," *IEEE Transactions on Instrumentation and Measurement*, vol. 59, pp. 543–552, 2010.

[5] D. Belega, D. Dallet, and D. Petri, "Accuracy of sine wave frequency estimation by multipoint interpolated DFT approach," *IEEE Transactions on Instrumentation and Measurement*, vol. 59, pp. 2808–2815, 2010.

[6] I. Kamwa, S. R. Samantaray, and G. Joos, "Wide frequency range adaptive phasor and frequency PMU algorithms," *IEEE Transactions on Smart Grid*, vol. 5, pp. 569–579, 2014.

[7] H. Wen, Z. S. Teng, Y. Wang, and X. G. Hu, "Spectral correction approach based on desirable sidelobe window for harmonic analysis of industrial power system," *IEEE Transactions on Industrial Electronics*, vol. 60, pp. 1001–1010, 2013.

[8] E. Lavopa, P. Zanchetta, M. Sumner, and F. Cupertino, "Real-time estimation of fundamental frequency and harmonics for active shunt power filters in aircraft electrical systems," *IEEE Transactions on Industrial Electronics*, vol. 56, pp. 2875–2884, 2009.

[9] C. Offelli and D. Petri, "Interpolation techniques for real-time multifrequency waveform analysis," *IEEE Transactions on Instrumentation and Measurement*, vol. 39, pp. 106–111, 1990.

[10] L. Hsiung Cheng, "Power harmonics and interharmonics measurement using recursive group-harmonic power minimizing algorithm," *IEEE Transactions on Industrial Electronics*, vol. 59, pp. 1184–1193, 2012.

[11] Z. Fusheng, G. Zhongxing, and Y. Wei, "The algorithm of interpolating windowed FFT for harmonic analysis of electric power system," *IEEE Transactions on Power Delivery*, vol. 16, pp. 160–164, 2001.

Chapter 2

Use of Regression Analysis to Determine the Model of Lighting Control in Smart Home with Implementation of KNX Technology

Jan Vanus, Radek Martinek, Petr Bilik, Jan Zidek and He Wen

Additional information is available at the end of the chapter

http://dx.doi.org/10.5772/116602

Abstract

To optimize the management of operational and technical functions in the smart home (SH) and for use of effective methods of energy management in SH, it is generally necessary to provide statistics and process relevant data from operational measurement devices. This chapter describes the use of modern methods for statistical data processing using regression analysis techniques. The aim of the analysis is to describe the dependence of single measured values using an appropriate mathematical model that can be efficiently implemented in the control system of SH. This model can be used for the functions of supervision and diagnostics of optimum comfort setting inside the indoor environment of SH. Real experimental measurements of objective parameters of the indoor environment were realized in the selected rooms of unique wooden building in the passive standard. The researched methods were experimentally verified by classifying the behavior of lighting in the SH-selected rooms under specified conditions. The achieved experimental results will be used for the operating and technical functions control in SH for reducing the building operating costs.

Keywords: smart home, energy management, processing, measurement, regression model

1. Introduction

For monitoring, detection, and recognition of operating and technical conditions in SH, it is possible to use information from the measured data using operational sensors. Regression analysis can be used for the measured data processing and mathematical description of dependence between the measured quantities. Results from regression analysis can be used in SH concept [1], SH care concept [2] (for quantify the activity for a complex set of SH activities and predict cognitive health of participants [3]), smart metering concepts [4–6], for example in a new method for estimating the demand response potential of residential air conditioning (A/C), using hourly electricity consumption data ("smart meter" data) from customer accounts in Northern California [7], and smart grid concepts [8, 9]. To visualize [10–13] and monitoring [14] of operational and technical functions in SH [15, 16], it is necessary to use a robust visualization tool with reliable storage of measured data with respect to the needs of the SH inhabitants [17]. In our case, the measured data are read from the individual KNX technology and BACnet technology sensors by means of the Desigo Insight visualization tool. The individual KNX and BACnet technology components are used for the blinds, lighting, cooling, heating, and ventilation control in SH (**Figures 1** and **2**).

Figure 1. Aerial view of the location of SH in terms of the cardinal points on the premises of the Technical University of Ostrava.

Figure 2. Side view of the SH wooden house describing the placement of rooms 202, 203, and 204.

http://dx.doi.org/10.5772/116602

For analysis of the measured data to determine the model of lighting control in SH (**Figures 1** and **2**) built-in within the Moravian-Silesian Wood Cluster (MSDK), we used the regression analysis methods. The measured values of nonelectrical quantities (e.g., illuminance, CO_2, temperature, humidity, etc.) in different rooms (e.g., rooms 202, 203, 204) of the building offer more information about the behavior of the operational and technical system in SH. **Figure 3** shows the measured values of illuminance E_{202} (lx), E_{203} (lx), and E_{204} (lx) in rooms 202, 203, and 204, respectively, and outdoor illuminance E_{out} (lx).

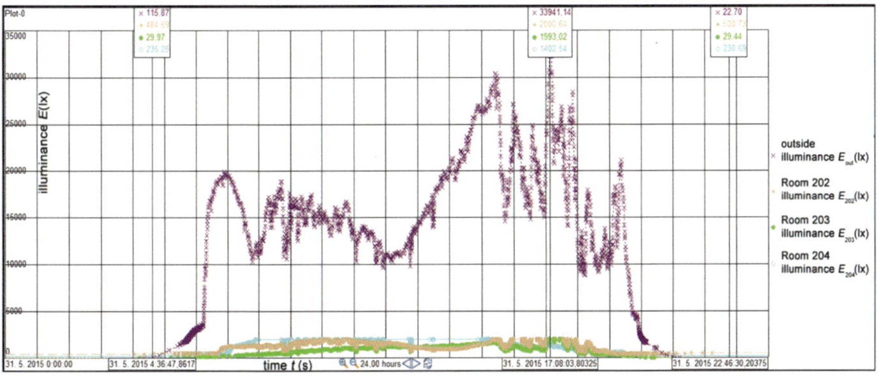

Figure 3. Measured real values of illuminance *E* (lx) in SH on May 31, 2015 in rooms 202, 203, and 204 and outdoor illuminance *E* (lx).

2. Lighting control in SH using KNX technology

The actual lighting system for every room in SH is examined by using the DALI technology for dimming control and the KNX system for control by a closed-loop technology. The testing was performed on small lighting systems, the closed loop approach, which is adequate for illumination control to a constant illuminance *E* level, was used for the control.

2.1. Description of the KNX/DALI gateway system

The control options of the KNX/DALI gateway system (**Figure 4**) are not so extensive.

The sensor is connected directly to the KNX bus, whereby programming is facilitated. It should be noted that a bus-bar sensor must be selected in settings where information from the sensor needs to be transmitted to a large distance and hence, the hazard of data loss due to voltage decrease or interferences exists. The DALI ballasts are controlled by the KNX actuator (KNX/DALI Gateway N 141) and by one bus-bar sensor. The sensor at the ceiling sends information to the actuator, which evaluates light flux requirements in the master line.

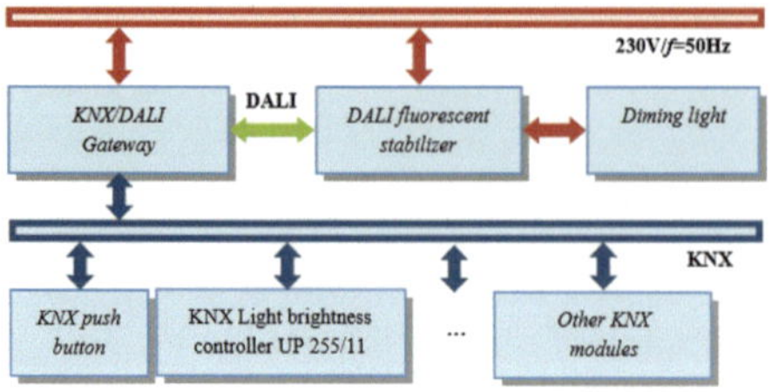

Figure 4. Block diagram of lights control with using of the KNX components: system KNX/DALI gateway.

3. Proposed experiment

The aim of the experiment was to explore the dependence of the measured waveforms of illuminance *E* (lx) in the rooms 202, 203, and 204 on the outdoor illuminance *E* (lx) under the following conditions:

a. The light in each room is turned off, the blinds are pulled up. The effect of the location of rooms 202, 203, and 204 (SH rotation) is compared from the perspective of the cardinal points at the outdoor illuminance E_{out} **> 10,000 lx**. The measurement took place on **September 07, 2014.**
b. Light in rooms 202 and 204 is turned on, the blinds are pulled up. The light in room 203 is turned off. The automatic lighting control is set on the constant value of illuminance $E = 500$ lx in room 202. The automatic lighting control is set on the constant value of illuminance $E = 230$ lx in room 204. The effect of the location of rooms 202, 203, and 204 (SH rotation) is compared from the perspective of the cardinal points at the outdoor illuminance E_{out} **> 10,000 lx**. The measurement took place on **May 31, 2015.**

On the basis of the above-described conditions, a method for determining the optimal regression model to find the suitable mathematical description of states of operating and technical functions in SH was designed, in this case for lighting monitoring and control:

Step 1: calculating the correlation coefficients to determine the strength of dependence between the measured values of illuminance *E* (lx) in each room and outdoors.

Step 2: using linear regression to determine the regression line.

Step 3: selecting the best regression model describing the mathematical relationships between the measured quantities based on the calculated coefficient of determination R^2 (*R*-squared) value.

3.1. Correlation analysis: description

The measured data processing and obtaining the information about the strength of statistical dependence between the measured values of illuminance *E* (lx) in SH (202, 203, and 204 rooms) can be carried out by means of the method of correlation analysis. The force of (linear)

http://dx.doi.org/10.5772/116602

dependence between two measured nonelectrical quantities has been evaluated by means of the value of Pearson's correlation coefficient

$$R_{x,y} = \frac{\Sigma(x_i - \bar{x})(y_i - \bar{y})}{\sqrt{\Sigma(x_i - \bar{x})^2 \Sigma(y_i - \bar{y})^2}} \tag{1}$$

The correlation coefficient $\mathbf{R}_{x,y}$ can take values from the closed interval <−1, +1>. The more the absolute value of the correlation coefficient approaches 1, the stronger the dependence of the random quantities.

3.2. Regression analysis: description

In terms of a mathematical description, this is a research of the relationship between two quantities, in which one of them, the so-called independent variable *X* (outdoor illuminance *E* (lx)), is to influence the other, the so-called dependent variable *Y* (measured illuminance *E* (lx) in each of the rooms 202, 203, and 204). Windows in rooms 203 and 204 are directed to the south and windows in room 202 are directed to the west (**Figure 1**). The results of selected individual measurements were processed into a plot of the fitted model with SW tool Statgraphics, which was used for a statistical analysis (**Figures 5–8**). The expected dependence between the studied variables, the so-called regression, was verified using this comprehensive statistical tool. The so-called linear regression was used which assumes a linear relationship between two quantities. Regression line equation can be written as:

$$Y_i = \beta_0 + \beta_1 x_i + e_i. \tag{2}$$

The estimate of the regression line is written in one of the following ways:

$$\tilde{Y}_i = b_0 + b_1 x_i \tag{3}$$

$$\tilde{Y}_i = b_0^* + b_1(x_i - \tilde{x}) \text{ (called deviation form of record).} \tag{4}$$

$$\tilde{Y}_i = \beta_0 + \beta_1 x_i + e_i, \tag{5}$$

where e_i denotes a residue, error.

Using the SW tool Statgraphics, linear regression model (ANOVA) conditions were verified. Each figure (**Figures 5–8**) shows prediction models (green lines) and single 95% confidence intervals in each plot of fitted models. It is possible to define the interval estimation in regression for the expected value (green lines—confidence limits). The confidence interval for each measurement is identified as the prediction interval. Each of the charts show the equation of the regression line and coefficient of determination R^2 indicating the quality of regression model.

$$R^2 = \frac{SS_R}{SS_R + SS_E}, \tag{6}$$

where SS_R is the model sum of squares and SS_E is the error sum of squares.

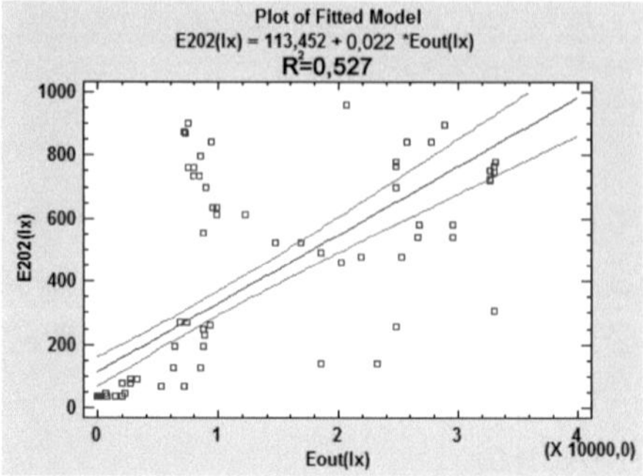

Figure 5. Dependence of E_{202} (lx) in the room 202 on E_{out} (lx)—linear model, $R^2 = 0.527$ (September 07, 2014).

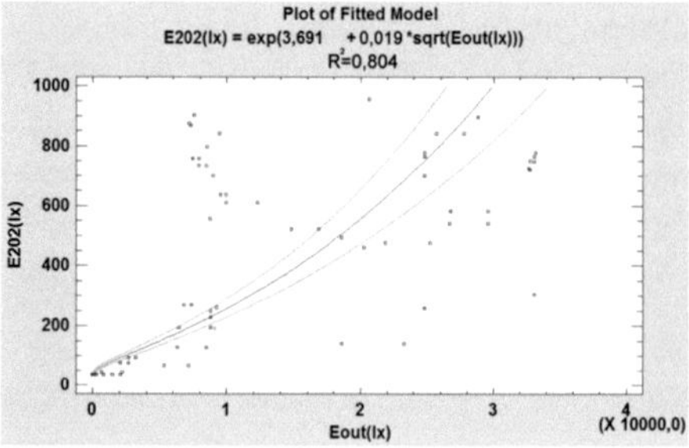

Figure 6. Dependence of E_{202} (lx) in the room 202 on E_{out} (lx)—logarithmic-Y square root-X model, $R^2 = 0.804$ (September 07, 2014).

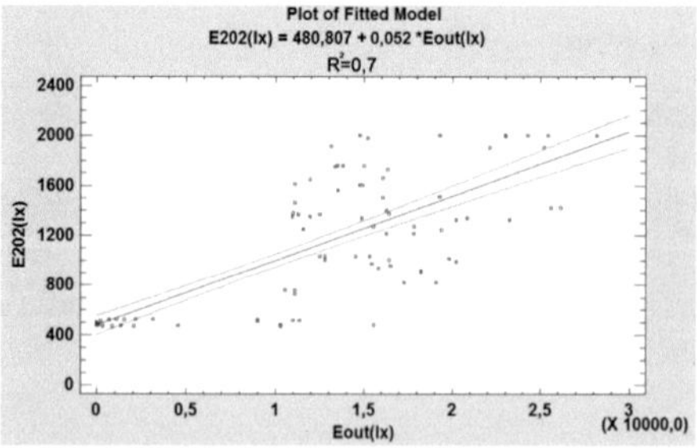

Figure 7. Dependence of E_{202} (lx) in the room 202 on E_{out} (lx)—linear model, $R^2 = 0.7$ (May 31, 2015).

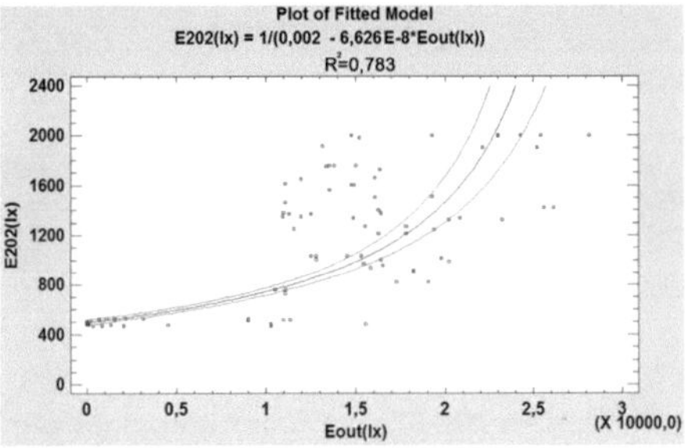

Figure 8. Dependence of E_{202} (lx) in the room 202 on E_{out} (lx)—reciprocal-Y model, $R^2 = 0.783$ (May 31, 2015).

4. Measured values

Based on the above-described conditions, the following values were measured and calculated.

Conditions a: The lighting is off: manual mode (blinds up); measured values: illuminance E (lx) in rooms 202, 203, and 204 and outdoor illuminance E (lx). Measurement day: **September 07, 2014**. The outdoor illuminance $\boldsymbol{E_{out}}$ **> 10,000** (lx) (weather is clear).

In terms of evaluating the correlation analysis, it is to confirm the assumption that the measured waveforms of illuminance E (lx) in rooms 203 and 204 have the greatest degree of similarity because the rooms are oriented to the south (**Table 1**). Comparing the correlation coefficients (**Table 1**) calculated from the measured illuminance E (lx) values in 202 to the remaining rooms 203 and 204, the rotation of the room 202 to the other side of cardinal points (west) is apparent (**Figure 1**). **Figure 5** shows the plot of the fitted model for room 202, representing the linear regression model with $R^2 = 0.527$, it means that the model as fitted explains 52.7% of the variability of the illuminance in room 202.

Figure 6 shows the plot of the fitted model for room 202, representing the logarithmic-Y square root-X regression model with $R^2 = 0.804$.

Based on the calculated value of the coefficient of determination R^2 (**Table 2**), the logarithmic-Y square root-X model (Eq. (7)) (**Figure 6**) was design as the optimal model for individual rooms.

$$Y_i = e^{\beta_0 + \beta_1 \sqrt{x_i}} \tag{7}$$

	E_{out} (lx)	E_{202} (lx)	E_{203} (lx)	E_{204} (lx)
E_{out} (lx)		0.726	0.616	0.611
E_{202} (lx)	0.726		0.843	0.776
E_{203} (lx)	0.616	0.843		0.979
E_{204} (lx)	0611	0.776	0.979	

Table 1. Coefficients of correlation analysis to compare the measured waveforms of illuminance E (lx) in rooms 202, 203, and 204 in SH on September 07, 2014.

Regression Models	R-Squared
Room **202** Logarithmic-Y square root-X	80.4%
Room **203** Logarithmic-Y square root-X	68.0%
Room **204** Logarithmic-Y square root-X	67.1%

Table 2. Comparison of room 202, 203, 204 regression models (7.9. 2014).

Conditions b: The lighting in rooms 202 and 204 is controlled on a constant illuminance (for room 202 E_{const} = 500 lx and room for 204 E_{const} = 230 lx): automatic mode (blinds up); the lighting was turned off in room 203 (**Figure 2**). Measured variables: illuminance *E* (lx) in rooms 202, 203, and 204 and outdoor illuminance E_{out} (lx) (**Table 3**). Measurement day: **May 31, 2015**. The outdoor illuminance $\boldsymbol{E_{out}}$ **> 10,000** (lx) (weather is clear).

	E_{out} **(lx)**	E_{202} **(lx)**	E_{203} **(lx)**	E_{204} **(lx)**
E_{out} (lx)		0.837	0.862	0.763
E_{202} (lx)	0.837		0.845	0.792
E_{203} (lx)	0.862	0.845		0.802
E_{204} (lx)	0.763	0.792	0.802	

Table 3. Coefficients of correlation analysis to compare the measured waveforms of illuminance *E* (lx) in rooms 202, 203, 204 in SH (May 31, 2015).

Based on the calculated value of the coefficient of determination R^2 (**Table 4**), the logarithmic-Y square root-X model was design as the optimal model for rooms 203 and 204. However, the reciprocal-Y model was calculated as the optimal model for room 202 (**Table 4**).

Regression model	R-squared
Room **202** Reciprocal-Y	78.3%
Room **203** Logarithmic-Y square root-X	91.4%
Room **204** Logarithmic-Y square root-X	80.3%

Table 4. Comparison of room 202, 203, 204 regression models (31/05/2015).

The plot of the fitted model for room 202 with linear regression model is shown in **Figure 7**.

The plot of the fitted model for room 202 with the reciprocal-Y regression model is shown in **Figure 8**.

5. Conclusion

This chapter described use of the regression analysis method to determine the regression model (linear regression model, reciprocal-Y regression model and logarithmic-Y square root-X regression model) of lighting control in SH with a potential use for the function of diagnosing the optimal settings for the corresponding comfort of interior lighting control in SH rooms 202, 203, and 204 via the KNX technology. Based on the measured values, it was demonstrated that it is possible to determine a corresponding regression model of mathematical description of operational and technical function behaviors in SH under the specific conditions for the lighting control in this case. Since the *P*-value in the ANOVA table is less than 0.01, there is a statistically significant relationship between illuminance of the room and illuminance of outdoors at the 99% confidence level in each of cases. Since the *P*-value in the *t*-tests for both of parameters intercepts and slope is less than 0.01, there is not a reason to extract these parameters from models. Based on the experiments, the described regression models can be used to further classify and describe the behavior of operational and technical conditions in SH. A database for each of the measured quantities with a precise description of baseline conditions of the SH behavior will be presented with the next chapter. More modern methods for classification and identification [18–23] will be used for compare and optimization of the operational and technical functions control in SH.

Acknowledgements

This chapter has been elaborated within the framework of the project SP2016/146 of the Student Grant System, VSB-TU Ostrava, Czech Republic.

Author details

Jan Vanus[1]*, Radek Martinek[1], Petr Bilik[1], Jan Zidek[1] and He Wen[2]

*Address all correspondence to: jan.vanus@vsb.cz

1 Department of Cybernetics and Biomedical Engineering, Faculty of Electrical Engineering and Computer Science, VSB-Technical University of Ostrava, Ostrava, Czech Republic

2 Hunan University, Changsha, China

References

[1] D. N. Monekosso and P. Remagnino, "Data reconciliation in a SH sensor network," *Expert Systems with Applications*, vol. 40, pp. 3248–3255, Jun 15 2013.

[2] O. Kwon, J. M. Shim, and G. Lim, "Single activity sensor-based ensemble analysis for health monitoring of solitary elderly people," *Expert Systems with Applications*, vol. 39, pp. 5774–5783, Apr 2012.

[3] P. N. Dawadi, D. J. Cook, M. Schmitter-Edgecombe, and C. Parsey, "Automated assessment of cognitive health using SH technologies," *Technology and Health Care*, vol. 21, pp. 323–343, 2013.

[4] A. Kavousian, R. Rajagopal, and M. Fischer, "Determinants of residential electricity consumption: Using smart meter data to examine the effect of climate, building characteristics, appliance stock, and occupants' behavior," *Energy*, vol. 55, pp. 184–194, Jun 15 2013.

[5] A. Kavousian, R. Rajagopal, and M. Fischer, "Ranking appliance energy efficiency in households: Utilizing smart meter data and energy efficiency frontiers to estimate and identify the determinants of appliance energy efficiency in residential buildings," *Energy and Buildings*, vol. 99, pp. 220–230, Jul 15 2015.

[6] F. McLoughlin, A. Duffy, and M. Conlon, "A clustering approach to domestic electricity load profile characterisation using smart metering data," *Applied Energy*, vol. 141, pp. 190–199, Mar 1 2015.

[7] M. E. H. Dyson, S. D. Borgeson, M. D. Tabone, and D. S. Callaway, "Using smart meter data to estimate demand response potential, with application to solar energy integration," *Energy Policy*, vol. 73, pp. 607–619, Oct 2014.

[8] D. Huang, M. Thottan, and F. Feather. Designing customized energy services based on disaggregation of heating usage, *IEEE Transactions on Smart Grid*, vo. 5, pp. 569–579, 2013.

[9] G. Kalogridis, C. Efthymiou, S. Z. Denic, T. A. Lewis, and R. Cepeda, "Privacy for smart meters: towards undetectable appliance load signatures," *2010 IEEE 1st International Conference on Smart Grid Communications (SmartGridComm)*, pp. 232–237, 2010.

[10] J. Vanus, P. Kucera, and J. Koziorek, "The software analysis used for visualization of technical functions control in smart home care," in *3rd Computer Science On-line Conference, CSOC 2014*, vol. 285, R. Silhavy, R. Senkerik, Z. K. Oplatkova, P. Silhavy, and Z. Prokopova, Eds., Berlin: Springer Verlag, pp. 549–558, 2014.

[11] J. Vanus, P. Kucera, and J. Koziorek, "Visualization software designed to control operational and technical functions in smart homes," in *3rd Computer Science On-line Conference, CSOC 2014*, vol. 285, R. Silhavy, R. Senkerik, Z. K. Oplatkova, P. Silhavy, and Z. Prokopova, Eds., Berlin: Springer Verlag, pp. 559–569, 2014.

[12] J. Vanus, P. Kucera, J. Koziorek, Z. Machacek, and R. Martinek, "Development of a visualisation software, Implemented with comfort smart home wireless control system," in *6th FTRA International Conference on Computer Science and its Applications, CSA 2014*, vol. 330, H. Y. Jeong, I. Stojmenovic, J. J. Park, and G. Yi, Eds., Berlin: Springer Verlag, pp. 580–589, 2015.

[13] J. Vanus, P. Kucera, R. Martinek, and J. Koziorek, "Development and testing of a visualization application software, implemented with wireless control system in smart home care," *Human-centric Computing and Information Sciences*, vol. 4, p. 18, 2014.

[14] J. Vanus, R. Martinek, P. Bilik, J. Koziorek, and A. Dracka, "Smart home remote monitoring using PI System Management Tools," in *8th International Scientific Symposium on Electrical Power Engineering, ELEKTROENERGETIKA 2015*, pp. 372–375, 2015.

[15] T. Novak, J. Vanus, J. Sumpich, J. Koziorek, K. Sokansky, and R. Hrbac, "Possibility to achieve the energy savings by the light control in smart home," in *7th International*

Scientific Symposium on Electrical Power Engineering, ELEKTROENERGETIKA 2013, Stara Lesna, pp. 260–263, 2013.

[16] J. Vanus, T. Novak, J. Koziorek, J. Konecny, and R. Hrbac, "The proposal model of energy savings of lighting systems in the smart home care," in *12th IFAC Conference on Programmable Devices and Embedded Systems, PDeS 2013*, Velke Karlovice, pp. 411–415, 2013.

[17] J. Vanus, J. Koziorek, and R. Hercik, "Design of a smart building control with view to the senior citizens' needs," in *12th IFAC Conference on Programmable Devices and Embedded Systems, PDeS 2013*, Velke Karlovice, pp. 422–427, 2013.

[18] J. Vanus and J. Koziorek, "Employment of dynamic time warping method for monitoring states in smart home," in *4th International Workshop on Computer Science and Engineering-Winter, WCSE 2014*, Hong Kong, China, 26–28 December, 2014.

[19] C. J. C. Burges, "A tutorial on support vector machines for pattern recognition," *Data Mining and Knowledge Discovery*, vol. 2, pp. 121–167, Jun 1998.

[20] A. K. Jain, R. P. W. Duin, and J. C. Mao, "Statistical pattern recognition: a review," *IEEE Transactions on Pattern Analysis and Machine Intelligence*, vol. 22, pp. 4–37, Jan 2000.

[21] V. N. Vapnik, "An overview of statistical learning theory," *IEEE Transactions on Neural Networks*, vol. 10, pp. 988–999, Sep 1999.

[22] J. P. Weir, "Quantifying test-retest reliability using the intraclass correlation coefficient and the SEM," *Journal of Strength and Conditioning Research*, vol. 19, pp. 231–240, Feb 2005.

[23] S. Yan, D. Xu, B. Zhang, H.-J. Zhang, Q. Yang, and S. Lin, "Graph embedding and extensions: A general framework for dimensionality reduction," *IEEE Transactions on Pattern Analysis and Machine Intelligence*, vol. 29, pp. 40–51, Jan 2007.

Chapter 3

PFPM: Discovering Periodic Frequent Patterns with Novel Periodicity Measures

Philippe Fournier-Viger, Chun-Wei Lin,
Quang-Huy Duong, Thu-Lan Dam, Lukáš Ševčík,
Dominik Uhrin and Miroslav Voznak

Additional information is available at the end of the chapter

http://dx.doi.org/10.5772/66780

Abstract

Periodic pattern mining is the task of discovering patterns that periodically appear in transactions. Typically, periodic pattern mining algorithms will discard a pattern as being nonperiodic if it has a single period greater than a maximal periodicity threshold, defined by the user. A major drawback of this approach is that it is not flexible, as a pattern can be discarded based on only one of its periods. In this chapter, we present a solution to this issue by proposing to discover periodic patterns using three measures: the minimum periodicity, the maximum periodicity, and the average periodicity. The combination of these measures has the advantage of being more flexible. Properties of these measures are studied. Moreover, an efficient algorithm named PFPM (Periodic Frequent Pattern Miner) is proposed to discover all frequent periodic patterns using these measures. An experimental evaluation on real data sets shows that the proposed PFPM algorithm is efficient and can filter a huge number of nonperiodic patterns to reveal only the desired periodic patterns.

Keywords: frequent pattern mining, periodic patterns, periodicity measures

1. Introduction

In the field of data mining, frequent itemset mining (FIM) [1–3] is widely viewed as a fundamental task for discovering knowledge in databases. Given a transaction database, it consists of discovering sets of items frequently purchased by customers. Besides market basket analysis, FIM has many applications in other fields. Although numerous algorithms have been proposed for FIM [1–3], an inherent limitation of traditional FIM algorithms is that they are not designed to discover patterns that periodically appear in a database. Discovering periodic patterns has many applications such as to discover recurring customer purchase behavior.

Several algorithms have been proposed to discover periodic frequent patterns (PFP) [4–9] in a transaction database (a sequence of transactions). Typically, periodic pattern mining algorithms will discard a pattern as being nonperiodic if it has a single period greater than a maximal periodicity threshold, defined by the user. A major drawback of this approach is that it is not flexible, as a pattern can be discarded based on only one of its periods. In this chapter, we propose a solution to this problem by discovering periodic patterns using three measures: the minimum periodicity, the maximum periodicity, and the average periodicity. This chapter has three main contributions. First, novel measures named *average periodicity* and *minimum periodicity* are proposed to assess the periodicity of patterns. Second, a fast algorithm named PFPM (Periodic Frequent Pattern Miner) is proposed to efficiently discover frequent periodic patterns using these measures. Third, we have conducted several experiments on real-life data sets to evaluate the efficiency of PFPM, and the usage of the novel periodicity measures. Experimental results show that the PFPM algorithm is efficient, and can filter a huge number of nonperiodic patterns to reveal only the desired periodic itemsets. The rest of this chapter is organized as follows. Sections 2, 3, 4, 5, and 6, respectively, present preliminaries related to FIM related work, the novel periodicity measures, the PFPM algorithm, the experimental evaluation, and the conclusion.

2. Related work

The problem of frequent itemset mining is defined as follows. Let I be a set of items (symbols). A *transaction database* is a set of transactions $D = \{T_1, T_2, ..., T_n\}$ such that for each transaction T_c, $T_c \in I$, and T_c has a unique identifier c called its Tid. For example, consider the database of **Table 1**, which will be used as running example. This database contains seven transactions $(T_1, T_2, \ldots, T_7)$. Transaction T_3 indicates that items a, b, c, d and e appear in this transaction. The support of an itemset X in a database D is denoted as $s(X)$ and defined as $|\{t|\ t \in D \wedge X \subseteq t\}|$. In other words, $s(X) = |g(X)|$, where $g(X)$ is defined as the set of transactions containing X. Let there be any total order $\succ$ on items in I. The *extensions* of an itemset X are the itemsets that can be obtained by appending an item y to X such that $y \succ i$, $\forall i \in X$. The *problem of frequent itemset mining* consists of discovering the frequent itemsets [1]. An itemset X is a *frequent itemset* if its support $s(X)$ is no less than a user-specified minimum support threshold *minsup* given by the user. For example, consider that *minsup* = 4 transactions. The set of frequent itemsets is $\{a\}$, $\{a,c\}$, $\{e\}$, $\{c,e\}$, and $\{c\}$, having respectively a support of 4, 4, 5, 4, and 6.

TID	Transaction	TID	Transaction	TID	Transaction	TID	Transaction
T_1	$\{a,c\}$	T_3	$\{a,b,c,d,e\}$	T_5	$\{a,c,d\}$	T_7	$\{b,c,e\}$
T_2	$\{e\}$	T_4	$\{b,c,d,e\}$	T_6	$\{a,c,e\}$		

Table 1. A transaction database.

To discover frequent itemsets, various algorithms have been proposed such as Apriori [1], FP-Growth [10], LCM [2], and Eclat [3]. However, these algorithms are not designed to discover periodic patterns. Inspired by the work on FIM, researchers have designed several algorithms to discover periodic frequent patterns (PFP) in transaction databases [4–9]. Several applications of mining periodic frequent patterns have been reported in previous work [9].

A periodic frequent pattern is defined as follows [9]. Let there be a database $D = \{T_1, T_2, ..., T_n\}$ containing n transactions, and an itemset X. The set of transactions containing X is denoted as $g(X) = \{T_{g_1}, T_{g_2}..., T_{g_k}\}$, where $1 \leq g_1 < g_2 < ... < g_k \leq n$. Two transactions $T_x \supset X$ and $T_y \supset X$ are said to be *consecutive with respect to* X if there does not exist a transaction $T_w \in g(X)$ such that $x < w < y$. The *period* of two consecutive transactions T_x and T_y in $g(X)$ is defined as $pe(T_x, T_y) = (y-x)$, that is, the number of transactions between T_x and T_y (including T_x). The *periods of an itemset* X is a list of periods defined as $ps(X) = \{g_1 - g_0, g_2 - g_1, g_3 - g_2, \cdots g_k - g_{k-1}, g_{k+1} - g_k\}$, where g_0 and g_{k+1} are constants defined as $g_0 = 0$ and $g_{k+1} = n$. Thus, $ps(X) = \bigcup_{1 \leq z \leq k+1}(g_z - g_{z-1})$. For example, consider the itemset $\{a,c\}$. This itemset appears in transactions T_1, T_3, T_5, and T_6, and thus $g(\{a,c\}) = \{T_1, T_3, T_5, T_6\}$. The periods of this itemset are $ps(\{a,c\}) = \{1, 2, 2, 1, 1\}$. The maximum periodicity of an itemset X is defined as $maxper(X) = max(ps(X))$ [9]. An itemset X is a periodic frequent pattern (PFP) if $|g(X)| \geq minsup$ and $maxper(X) \leq maxPer$, where $minsup$ and $maxPer$ are user-defined thresholds [9].

The PFP-tree algorithm is the first algorithm that has been proposed for mining PFPs [9]. It utilizes a tree-based and pattern-growth approach for discovering PFPs, inspired by the FP-Growth algorithm [10]. Thereafter, an algorithm called MTKPP [4] was designed. It relies on a depth-first search and a vertical database representation. To use this algorithm, a user needs to set a parameter k. The algorithm then outputs the k most frequent PFPs in a database. Approximate algorithms for mining periodic patterns have also been developed. Ref. [5] mines PFPs by considering an approximation of the periodicities of patterns. Another approximate algorithm for PFP mining was recently proposed [8]. Other extensions of the PF-Tree algorithm named MIS-PF-tree [6] and MaxCPF [7] were respectively proposed to mine PFPs using multiple *minsup* thresholds, and multiple *minsup* and *minper* thresholds. A drawback of the maximum periodicity measure used by most PFP algorithms is that an itemset is automatically discarded if it has a single period of length greater than the *maxPer* threshold. Thus, this measure may be viewed as too strict.

3. Novel periodicity measures

To address the above limitation of traditional PFP mining algorithms, and provide a more flexible way of evaluating the periodicity of patterns, this chapter proposes the concept of *average periodicity*.

Definition 1 (Average periodicity of an itemset): *The average periodicity of an itemset X is defined as* $avgper(X) = \sum_{g \in ps(X)} / |ps(X)|$.

For instance, the periods of the itemsets $\{a, c\}$ and $\{e\}$ are $ps(\{a,c\}) = \{1, 2, 2, 1, 1\}$ and $ps(\{e\}) = \{2, 1, 1, 2, 1, 0\}$. Thus, their average periodicities are respectively $avgper(\{a,c\}) = 1.4$ and $avgper(\{e\}) = 1.17$.

Lemma 1 (An alternative definition of the average periodicity): *Consider an itemset X. The average periodicity can also be calculated as* $avgper(X) = |D|/(|g(X)| + 1)$.

The proof for that proposed lemma is omitted due to space limitations. This lemma is interesting because it shows that there is a relationship between the average periodicity and the support measure. Moreover, this lemma gives a second method for calculating the average periodicities of itemsets. An interesting observation is that if the term $|D|$ is precalculated, calculating the average periodicity of any itemset only requires obtaining $|g(X)| + 1$, to then divide it by $|D|$. Using this method, average periodicities are calculated more efficiently than using Definition 1.

The average periodicity is a very interesting measure since it measures the average period length of an itemset. However, this measure should always be considered with other measure. The reason is that the average periodicity does not consider if the period lengths vary widely. We illustrate this with an example. In the running example, the average periodicity of $\{b, d\}$ is 2.33. This itemset may seem like a periodic pattern. However, this is not the case. $\{b, d\}$ actually only appears in T_3 and T_4, and its periods $ps(\{T_3, T_4\}) = \{3, 1, 4\}$ vary widely. To ensure that such patterns having periods that vary too much are discovered, we propose to use the average periodicity in combination with other periodicity measure(s), and in particular the following two measures.

The first measure is the *minimum periodicity* and it is used to filter out itemsets having short periods. Let there be an itemset X. The minimum periodicity is defined as $minper(X) = min(ps(X))$. However, a problem with this measure is that the first and last periods of an itemset are respectively equal to 1 or 0 if the itemset respectively appears in the first or the last transactions of the database. For example, because itemset {e} occurs in T_7, its last period is 0, and thus $minper(\{e\}) = 0$. To avoid this problem, we have decided to change the definition of the minimum periodicity of an itemset so that it is calculated by excluding the first and last periods of the itemset. By ignoring the first and last periods, the set of periods may however become empty. In this case, we define the minimum periodicity as ∞.

The second measure that we consider is the *maximum periodicity*, to avoid finding period patterns that do not occur for very long periods of time. This measure is defined as in Section 2 and is denoted as $maxper(X)$ for an itemset X.

In the following, we consider the minimum periodicity, maximum periodicity, and average periodicity measures to discover frequent periodic patterns as they let the user finely specify the type of periodic patterns to be found. But another reason for choosing these measures is that an algorithm can calculate them very efficiently. Consider an itemset X. The three measures can be calculated by browsing the list of transactions $g(X)$ only once, without storing the periods $ps(X)$ in memory. This chapter defines the problem of periodic frequent itemsets using the three measures as follows.

Definition 2 (Periodic frequent itemsets with novel measures): *An itemset X is said to be a periodic frequent itemset if and only if* $minAvg \leq avgper(X) \leq maxAvg$, $minper(X) \geq minPer$, and

http://dx.doi.org/10.5772/66780

maxper(*X*) ≤*maxPer*, where *minAvg*, *maxAvg*, *minPer*, and *maxPer* are thresholds (positive numbers), set by the user.

As an example, assume that *minPer* = 1, *maxPer* = 3, *minAvg* = 1, and *maxAvg* = 2. For these parameters, 11 PFPs are discovered (illustrated in **Table 2**).

Itemset	*support s(X)*	*minper(X)*	*maxper(X)*	*avgper(X)*
{*b*}	3	1	3	1.75
{*b*,*e*}	3	1	3	1.75
{*b*,*c*,*e*}	3	1	3	1.75
{*b*,*c*}	3	1	3	1.75
{*d*}	3	1	3	1.75
{*c*,*d*}	3	1	3	1.75
{*a*}	4	1	2	1.4
{*a*,*c*}	4	1	2	1.4
{*e*}	5	1	2	1.17
{*c*,*e*}	4	1	3	1.4
{*c*}	6	1	2	1.0

Table 2. The set of PFPs for the running example.

To develop an efficient algorithm for mining PFPs, it is important to design efficient pruning strategies. To use the periodicity measures for pruning the search space, the following theorems are presented. Proofs are omitted due to space limitations.

Lemma 2 (Monotonicity of the average periodicity): *Let X and Y be itemsets such that* $X \subset Y$. It follows that *avgper*(*Y*)≥ *avgper*(*X*).

Lemma 3 (Monotonicity of the minimum periodicity): *Let X and Y be itemsets such that* $X \subset Y$. It follows that *minper*(*Y*)≥ *minper*(*X*).

Lemma 4 (Monotonicity of the maximum periodicity): *Let X and Y be itemsets such that* $X \subset Y$. It follows that *maxper*(*Y*)≥ *maxper*(*X*) [9].

Theorem 1 (Maximum periodicity pruning): Let *X* be an itemset appearing in a database *D*. *X* and its supersets are not PFPs if *maxper*(*X*) > *maxPer*. Thus, if this condition is met, the search space consisting of *X* and all its supersets can be discarded. (This follows from Lemma 4.)

Theorem 2 (Average periodicity pruning): Let *X* be an itemset appearing in a database *D*. *X* is not a PFP as well as all of its supersets if *avgper*(*X*) > *maxAvg*, or equivalently if $|g(X)| < (|D|/maxAvg)-1$. Thus, if this condition is met, the search space consisting of *X* and all its supersets can be discarded.

4. The PFPM algorithm

Based on the novel periodicity measures introduced in the previous sections, an efficient algorithm named PFPM (Periodic Frequent Pattern Miner) is proposed to efficiently discover periodic patterns using these measures. The proposed PFPM algorithm is a tid-list-based algorithm, inspired by the Eclat algorithm [3]. The *tid-list* of an itemset X in a database D is defined as the set of transactions $g(X)$ that contains the itemset X. In the proposed algorithm, the tid-list of an itemset X is further annotated with two values: $minper(X)$ and $maxper(X)$. The PFPM (Algorithm 1) takes as input a transaction database, and the *minAvg, maxAvg, minPer,* and *maxPer* thresholds. The algorithm first scans the database to calculate $minper(\{i\})$, $maxper(\{i\})$, and $s(\{i\})$ for each item $i \in I$. Then, the algorithm calculates the value $\gamma = (|D|/maxAvg)-1$ to be later used for pruning itemsets using Theorem 2. Then, the algorithm identifies the set I^* of all items having a periodicity no greater than *maxPer,* and appearing in no less than γ transactions (other items are ignored since they cannot be part of a PFP by Theorems 1 and 2. Items in I^* are then sorted according to the order $\succ$ of ascending support values, as suggested in [3]. A database scan is then performed. During this database scan, items in transactions are reordered according to the total order $\succ$, and the tid-list of each item $i \in I^*$ is built. Then, the depth-first search exploration of itemsets starts by calling the recursive procedure *Search* with the set of single items I^*, γ, *minutil, minAvg, minPer, maxPer,* and $|D|$.

Algorithm 1: The PFPM algorithm

input : a transaction database D, $minAvg$, $maxAvg$, $minPer$ and $maxPer$
output: the set of periodic itemsets

Scan D once to calculate $minper(\{i\})$, $maxper(\{i\})$, and $s(\{i\})$ for each item $i \in I$;
$\gamma \leftarrow (|D|/maxAvg) - 1$;
$I^* \leftarrow$ each item i such that $s(\{i\}) \geq \gamma$ and $maxper(\{i\}) \leq maxPer$;
Let $\succ$ be the total order of support ascending values on I^*;
Scan D to build the tid-list of each item $i \in I^*$;
`PFPMSearch` (I^*, γ, $minAvg$, $minPer$, $maxPer$, $|D|$);

The *PFPMSearch* procedure (Algorithm 2) takes as input extensions of an itemset P (initially assumed that $P = \emptyset$) having the form Pz meaning that Pz was previously obtained by appending an item z to P, γ, *minAvg, minPer, maxPer,* and $|D|$. The search procedure performs a loop on each extension Px of P. In this loop, the average periodicity of Px is calculated by dividing $|D|$ by the number of elements in the tid-list of Px plus one (by Lemma 1). Then, if the average periodicity of Px is in the $[minAvg, maxAvg]$ interval, the minimum/maximum periodicity of Px is no less/not greater than $minPer/maxPer$ according to the values stored in its tid-list, then Px is a PFP and it is output. Then, if the number of elements in the tid-list of Px is no less than γ, and $maxper(Px)$ is no greater than *maxPer,* it means that extensions of Px should be explored (by Theorems 1 and 2). This is performed by merging Px with all extensions Py of P

http://dx.doi.org/10.5772/66780

such that y>x to form extensions of the form *Pxy* containing $|Px| + 1$ items. The tid-list of *Pxy* is then constructed by calling the *Intersect* procedure (cf. Algorithm 3), to join the tid-lists of *P*, *Px*, and *Py*. This latter procedure is similar to the join of tid-list described in the Eclat algorithm [3], with the exception that periods are calculated during tid-list intersection to obtain *maxPer*(*Pxy*) and *minPer*(*Pxy*) (not shown). Then, a recursive call to the *PFPMSearch* procedure with *Pxy* to explore its extension(s). The *PFPMSearch* procedure starts from single items, recursively explores the search space of itemsets by appending single items, and only prunes the search space using Theorems 1 and 2. Thus, it can be easily seen that this procedure is correct and complete to discover all PFPs.

Algorithm 2: The *PFPMSearch* procedure

```
input : ExtensionsOfP: a set of extensions of an itemset P, γ, minAvg, minPer, maxPer, |D|
output: the set of periodic frequent itemsets
1  foreach itemset Px ∈ ExtensionsOfP do
2      avgperPx ← |D|/(|Px.tidlist| + 1);
3      if minAvg ≤ avgperPx ≤ maxAvg ∧ Px.tidlist.minp ≥ minPer ∧ Px.tidlist.maxp ≤ maxPer
        then output Px;
4      if avgperPx ≥ γ and Px.tidlist.maxp ≤ maxPer then
5          ExtensionsOfPx ← ∅;
6          foreach itemset Py ∈ ExtensionsOfP such that y ≻ x do
7              Pxy ← Px ∪ Py;
8              Pxy.tidlist ← Intersect (Px, Py);
9              ExtensionsOfPx ← ExtensionsOfPx ∪ {Pxy};
10         end
11         PFPMSearch (ExtensionsOfPx, γ, minAvg, minPer, maxPer, |D|);
12     end
13 end
```

Algorithm 3: The Intersect procedure

```
input : Px: the extension of P with an item x, Py: the extension of P with an item y
output: the tid-list of Pxy
TidListOfPxy ← ∅;
foreach Tid v ∈ Px.tidlist such that v ∈ Py.tidlist do
    period_v ← calculatePeriod(v,TidListOfPxy);
    UpdateMinPerMaxPer(TidListOfPxy, period_v);
    TidListOfPxy ← TidListOfPxy ∪ {v};
end
return TidListOfPxy;
```

In the implementation of PFPM, two additional optimizations are included. The first optimization is to calculate the support of all pairs of items $\{x, y\}$ occurring in the database during the first database scan, and store it in memory. Then, Line 7 of the search procedure is modified to prune itemset *Pxy* if $s(\{x, y\})$ is less than γ by Theorem 2. The second optimization is in the *Intersection* procedure. During the loop, if it is found that *Pxy* cannot have no more than $\gamma = (|D|/maxAvg)-1$ transactions, the loop can be stopped.

5. Experimental study

To evaluate the performance of the proposed PFPM algorithm, we compared its performance with Eclat, a state-of-the-art algorithm for frequent itemset mining. The Eclat algorithm was chosen for comparison as the PFPM algorithm is based on Eclat. The PFPM and Eclat algorithms are implemented in Java. The experiment was carried out on a sixth generation 64 bit Core i5 processor running Windows 10, and equipped with 12 GB of free RAM. Four benchmark data sets were utilized in the experiment, which are commonly used in the FIM literature. The *retail, chainstore,* and *foodmart* data sets contain anonymized customer transactions from retail stores, while the *mushroom* data set is a dense benchmark data set. The data sets have been chosen because they represent the main types of data encountered in real-life scenarios (dense, sparse, and long transactions). Let $|I|$, $|D|$, and A represent the number of transactions, distinct items, and average transaction length of a data set. Characteristics of the four data sets are *retail* ($|I| = 88,162, |D| = 16,470, A = 10.30$), *mushroom* ($|I| = 8,124, |D| = 119, A = 23.0$), *chainstore* ($|I| = 1,112,949, |D| = 46,086, A = 7.26$), and *foodmart* ($|I| = 4,141$, $|D| = 1,559, A = 4.4$). The experiment consisted of running the PFPM algorithm on each data set with fixed $minPer$ and $minAvg$ values, while varying the $maxAvg$ and $maxPer$ parameters. To be able to compare PFPM with Eclat, Eclat was run with the γ value calculated by PPFM. Execution times, memory consumption, and number of patterns found were measured for each algorithm. All memory measurements were done using the Java API. For each data set, values for the periodicity thresholds have been found empirically for each data set (as they are data set specific), and were chosen to show the trade-off between the number of periodic patterns found and the execution time. Note that results for varying the $minPer$ and $minAvg$ values are not shown because these parameters have less influence on the number of patterns found than the other parameters. Thereafter, the notation *PFPM V-W-X* represents the PFPM algorithm with $minPer = V$, $maxPer = W$, and $minAvg = X$. **Figure 1** compares the execution times of PFPM for various parameter values and Eclat. **Figure 2** compares the number of PFPs found by PFPM for various parameter values, and the number of frequent itemsets found by Eclat.

It can first be observed that mining PFPs using PFPM is generally much faster than mining frequent itemsets. On the *retail* data set, PFPM is up to four times faster than Eclat. On the *mushroom* and *chainstore* data sets, no results are shown for Eclat because it cannot terminate within 1000 seconds or ran out of memory, while PFPM terminates in less than 10 seconds. The reason is that the search space is huge for these data sets when the minimum support is set to γ. The PFPM algorithm still terminates on these data sets because it only searches for periodic patterns, and thus prunes a large part of the search space containing nonperiodic patterns. On the *foodmart* data set, PFPM can be up to five times faster than Eclat depending on the parameters. But it can also be slightly slower in some cases. The reason is that *foodmart* is a sparse data set and thus the gain in terms of pruning the search space does not always offset the cost of calculating the periodicity measures. In general, the more the periodicity thresholds are restrictive, the more the gap between the runtime of Eclat and PFPM increases. A second observation is that the number of PFPs can be much less than the number of frequent itemsets (see **Figure 2**). This demonstrates that a huge amount of nonperiodic patterns are found in real-life data sets. Some of the patterns found are quite interesting as they contain several items. For example, it is found that items with product ids 32, 48, and 39 are periodically bought with an average periodicity of 16.32, a minimum periodicity of 1, and a maximum periodicity of 170. Memory consumption was also compared, although detailed results are not shown. It was observed that

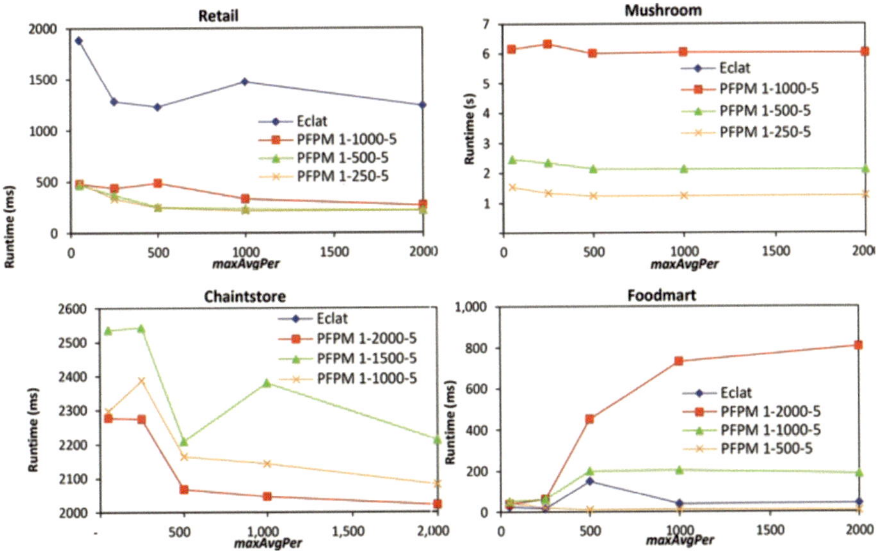

Figure 1. Execution times.

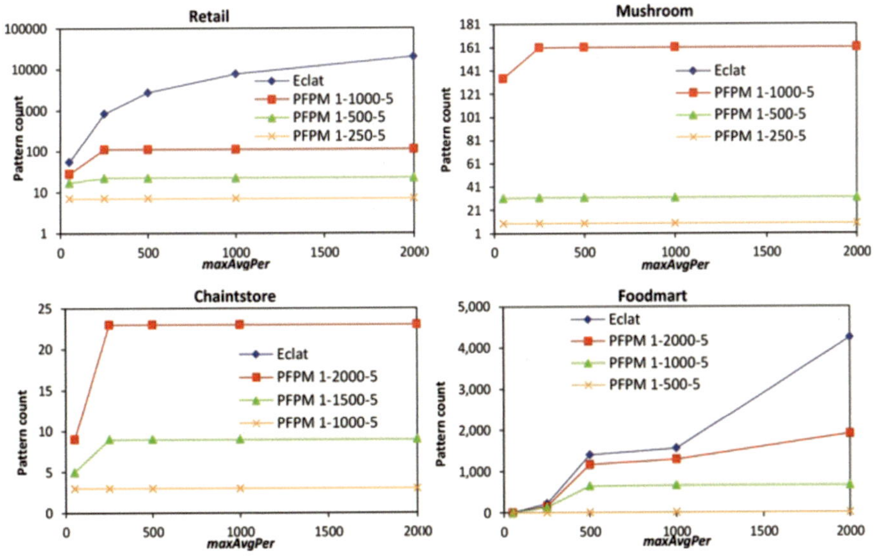

Figure 2. Number of patterns found.

PFPM use up to four and five times less memory than Eclat on the *retail* and *foodmart* data sets, depending on parameter values. For example, on *retail* and $maxAvg = 2,000$, Eclat and PFPM 1-5000-5-500 respectively consumes 900 and 189 MB of memory.

6. Conclusion

In this chapter, an efficient algorithm named PFPM (Periodic Frequent Pattern Miner) was proposed to efficiently discover all frequent periodic patterns using three periodicity measures (the minimum, average, and maximum periodicity). An experimental evaluation on real data sets shows that the proposed PFPM algorithm is efficient and can filter a huge number of nonperiodic patterns to reveal only the desired periodic patterns. Source code and data sets will be made available as part of the SPMF open source data mining library [11] http://www.philippe-fournier-viger.com/spmf/. For future work, we are working on mining correlated periodic patterns [12], and periodic high-utility itemsets [13].

Acknowledgements

This publication has been created within the project support of VŠB-TUO activities with China with financial support from the Moravian-Silesian Region and partially was supported by the grant SGS reg. no. No. SP2016/170 conducted at VSB-Technical University of Ostrava, Czech Republic.

Author details

Philippe Fournier-Viger[1]*, Chun-Wei Lin[2], Quang-Huy Duong[3], Thu-Lan Dam[3,4], Lukáš Ševčík[5], Dominik Uhrin[5] and Miroslav Voznak[5]

*Address all correspondence to: philfv8@yahoo.com

1 School of Natural Sciences and Humanities, Harbin Institute of Technology, Shenzhen Graduate School, Shenzhen, China

2 School of Computer Science and Technology, Harbin Institute of Technology, Shenzhen Graduate School, Shenzhen, China

3 College of Computer Science and Electronic Engineering, Hunan University, China

4 Faculty of Information Technology, Hanoi University of Industry, Hanoi, Vietnam

5 Department of Telecommunications, Faculty of Electrical Engineering and Computer Science, VSB-Technical University of Ostrava, Ostrava-Poruba, Czech Republic

References

[1] R. Agrawal, T. Imielinski, A. Swami, Mining association rules between sets of items in large databases, Sigmod Record 22 (2) (1993) 207–216.

[2] S.-i. Minato, T. Uno, H. Arimura, LCM over ZBDDS: Fast generation of very large-scale frequent item-sets using a compact graph-based representation, in: Proc. Twelfth Pacific-Asia Conf. on Knowledge Discovery and Data Mining, Osaka, Japan, May 20-23, Springer, 2008, pp. 234–246.

[3] M. J. Zaki, K. Gouda, Fast vertical mining using diffsets, in: Proc. ninth ACM SIGKDD Intern. Conf. Knowledge Discovery and Data Mining, Washington, DC, USA, August 24-27, ACM, 2003, pp. 326–335.

[4] K. Amphawan, P. Lenca, A. Surarerks, Mining top-k periodic-frequent pattern from transactional databases without support threshold, in: Proc. Third Intern. Conf. Advances in Information Technology, Bangkok, Thailand, December 1-5, Springer, 2009, pp. 18–29.

[5] K. Amphawan, A. Surarerks, P. Lenca, Mining periodic-frequent itemsets with approximate periodicity using interval transaction-ids list tree, in: Knowledge Discovery and Data Mining, 2010. WKDD'10. Third International Conference on, IEEE, (2010), pp. 245–248.

[6] R. U. Kiran, P. K. Reddy, Mining rare periodic-frequent patterns using multiple minimum supports, Work 5 (6) (2009) 7–8.

[7] A. Surana, R. U. Kiran, P. K. Reddy, An efficient approach to mine periodic-frequent patterns in transactional databases, in: Proc. 15th Pacific-Asia Conf. on Knowledge Discovery and Data Mining, Shenzhen, China, May 24-27, Springer, 2011, pp. 254–266.

[8] R. U. Kiran, M. Kitsuregawa, P. K. Reddy, Efficient discovery of periodic-frequent patterns in very large databases, Journal of Systems and Software 112 (2016) 110–121.

[9] S. K. Tanbeer, C. F. Ahmed, B.-S. Jeong, Y.-K. Lee, Discovering periodic-frequent patterns in transactional databases, in: Proc. Eleventh Pacific-Asia Conf. on Knowledge Discovery and Data Mining, Bangkok, Thailand, April 27-30, Springer, 2009, pp. 242–253.

[10] J. W. Han, J. Pei, Y. W. Yin, Mining frequent patterns without candidate generation, Sigmod Record 29 (2) (2000) 1–12.

[11] P. Fournier-Viger, A. G. Penalver, T. Gueniche, A. Soltani, C.-W. Wu, V. S. Tseng, Spmf: A java open-source pattern mining library, Journal of Machine Learning Research (JMLR) 15 (2014) 3389–3393.

[12] P. Fournier-Viger, C.-W. Lin, T. Dinh, H. B. Le, Mining correlated high-utility itemsets using the bond measure, in: Proc. 11th Intern. Conf. Hybrid Artificial Intelligent Systems, Seville, Spain, April 18-20, 2016, pp. 53–65.

[13] P. Fournier-Viger, C.-W. Lin, Q.-H. Duong, T.-L. Dam, Phm: Mining periodic high-utility itemsets, in: Proc. 16th Industrial Conf. on Data Mining, Newark, USA, July 13–17, 2016.

Chapter 4

Effective Planning and Analysis of Huawei and Cisco Routers for MPLS Network Design Using Fast Reroute Protection

Martin Hlozak, Dominik Uhrin,
Jerry Chun-Wei Lin and Miroslav Voznak

Additional information is available at the end of the chapter

http://dx.doi.org/10.5772/66785

Abstract

This chapter deals with a description of the MPLS traffic engineering technology behavior on two heterogeneous, but nowadays the most commonly used network vendors are Cisco and Huawei. Compatibility and functionality between network devices Huawei and Cisco were verified by testing the appropriate network topology. In this topology, we mainly focused on the useful feature of MPLS TE called Fast Reroute (FRR) protection. It provides link protection, node protection and also bandwidth protection during the failure of the primary link, especially on backbone networks. After successful validation, compatibility and functionality of the network topology between the heterogeneous routers using the Fast Reroute protection will be possible to use this MPLS TE application in the real networks.

Keywords: Cisco, Fast Reroute, Huawei, MPLS

1. Introduction

In the 1990s, asynchronous transfer mode (ATM) was considered an ideal solution in transmission networks to operate with different demands [1]. In earlier times, this technology provided traffic engineering by a virtual channel as well as Frame-Relay. But subsequently IP began to replace the ATM technology, which became the most popular network protocol for transmission. On the other hand, the ATM was still widely used by telecommunication providers at that time. Since 1999, the draft of multiprotocol label switching (MPLS) has become the IETF [2] standard and internet service providers started to use this concept for IP/MPLS transmission over older ATM technology. In this chapter, we focus on the application of MPLS called MPLS traffic engineering. MPLS TE can be understood as "effective planning utilization" [3]. Instead of the normal routing of IP packets, MPLS TE routes traffic according to the source IP addresses. This application can choose the most appropriate links according to the speed of individual lines, delay, delay variability and can also react automatically to the change of these parameters [3, 4]. In addition, the applications of MPLS are also used for an effective creation of separate virtual private networks among the company branches, or for addressing QoS issues in communication networks, such as satellite and mobile cellular networks. This chapter is focused on the most used function of MPLS TE called Fast Reroute. Fast Reroute can be used in the case of a link or node failure in the MPLS network. Both vendors Huawei and Cisco support MPLS TE, but each vendor can use a different function model. The main motivation of this chapter is to bring complex view on usage and cooperation between routers of two different vendors using Fast Reroute protection.

2. State of the art

Multiprotocol label switching (MPLS) is a backbone technology, which uses labels attached to the packets for their transmission. Packets are not transmitted based on the destination IP addresses but according to the MPLS labels. The protocol allows most packets to be forwarded at Layer 2 (switching) rather than at Layer 3 (routing). The term "multiprotocol" means that it can transport various protocols on Layer 3 such as IPv4, IPv6, IPX, and protocols of Layer 2, e.g., Ethernet, HDLC, Frame-Relay, or ATM [5].

As shown in **Figure 1**, source A sends a packet to the router CE1. CE1 handles the packet according to its routing table in a standard way. According to the destination IP address of each packet, the ingress router (PE1) inserts a label in front of the IP header at the edge of the backbone network. All the subsequent routers ignore the IP headers and perform the packet forwarding based on the labels in front of them. This MPLS label determines a path that is used for the routing of a particular packet. Paths through MPLS network are called LSPs [5, 7].

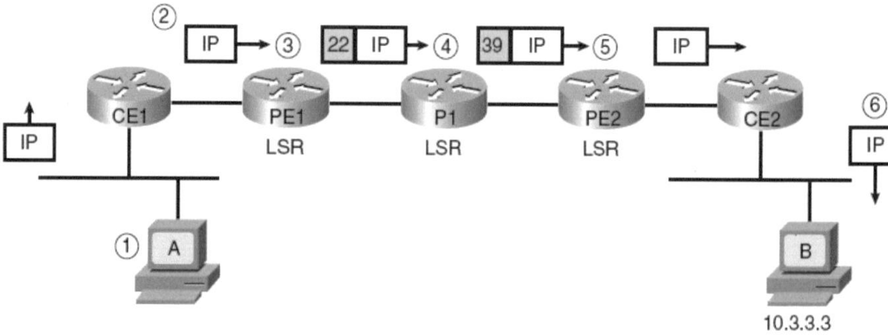

Figure 1. Packet forwarding via MPLS network [6].

http://dx.doi.org/10.5772/66785

Each label has its local importance and every MPLS backbone router processes the packet based on the MPLS label. Finally, the egress router (PE2) removes the label and forwards the original IP packet toward its final destination.

3. Methodology

Nowadays, practically, computer networks are not built only on a homogeneous infrastructure, but they use heterogeneous devices.

As depicted in **Figure 2**, the basic MPLS topology consists of two Huawei routers—the first one AR3200 and the second one AR2200 (marked in the red frame) and two Cisco 2800 series routers. The first goal was to verify MPLS functionality and interoperability among these above-mentioned routers.

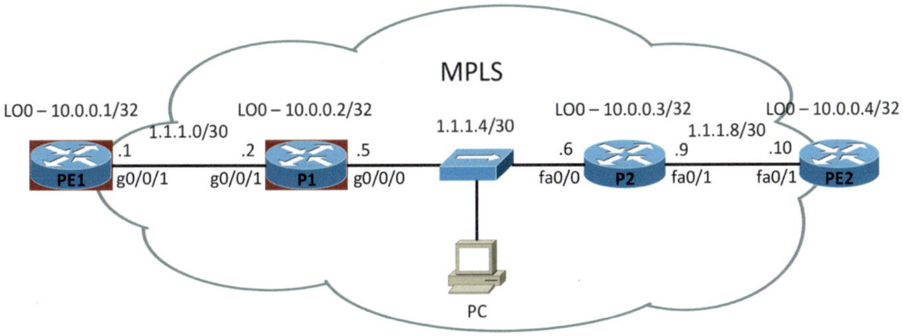

Figure 2. MPLS network.topology.

Huawei routers have only two CLI modes (basic view and the system view). The basic configuration of Huawei routers is as follows:

[Huawei]sysname PE1

[PE1]ospf 1

[PE1-ospf-1]area 0

[PE1-ospf-1-area-0.0.0.0]network 1.1.1.0 0.0.0.3

[PE1-ospf-1-area-0.0.0.0]network 10.0.0.1 0.0.0.0

[P1]mpls lsr-id 10.0.0.2

[PE1]mpls

[PE1-mpls]lsp-trigger all

[PE1]mpls ldp

[PE1]int lo0

[PE1-LoopBack0]ip address 10.0.0.1 255.255.255.255

[PE1]int g0/0/1

[PE1-GigabitEthernet0/0/1]ip address 1.1.1.1 255.255.255.252

[PE1-GigabitEthernet0/0/1]mpls

[PE1-GigabitEthernet0/0/1]mpls ldp

[PE1]int g0/0/0

[PE1-GigabitEthernet0/0/0]ip address 192.168.10.1 255.255.255.0

[PE1-GigabitEthernet0/0/1]mpls

[PE1-GigabitEthernet0/0/1]mpls ldp

All routers use OSPF as a routing protocol. Unlike Cisco routers, LSR identification must be configured on every Huawei router. For identification of Huawei routers, the loopback IP addresses were applied. The command *lsp-trigger all* allocates label for each IP prefix in the routing table. Then the LDP protocol for exchange of MPLS labels had to be activated for each MPLS physical interface.

3.1. Configuration of MPLS TE on Huawei routers

The network topology of the MPLS TE network is depicted in **Figure 3**.

First of all, it is necessary to configure MPLS TE technology and then turn on signalling protocol RSVP-TE. In the case of a link or node failure, we configure mpls rsvp-te hello as well. It is also necessary to enable modified SPF algorithm called CSPF which excludes. Using CSPF algorithm, the ingress MPLS router do not use these lines, which not satisfying the requirements of the data flow.

It is also necessary to explicitly turn on RSVP-TE for each MPLS physical interface. The part of the configuration is setting of the maximum bit rate of a line, which can be reserved. This bit rate cannot exceed the bit rate of a physical interface. The command *mpls te bandwidth bc0 10000* defines maximum total bandwidth for class type 0. In this case, the maximum bit rate of a physical interface is used.

In order to LSR routers could exchange information about set parameters such as maximum bit rate of the line, it is necessary to configure support for a special type of message OSPF LSA 10 for the OSPF area. Then this type of message is used for CSPF algorithm. By the command *opaque-capability enable*, we allow propagation of LSA 10 messages. Next command *enable traffic-adjustment advertise* includes static LSP tunnels into SPF calculation and to the routing table.

[PE1-mpls]mpls te

[PE1-mpls]mpls rsvp-te

http://dx.doi.org/10.5772/66785

[PE1-mpls]mpls rsvp-te hello

[PE1-mpls]mpls te cspf

[PE1-GigabitEthernet0/0/1]mpls rsvp-te

[PE1-GigabitEthernet0/0/1]mpls rsvp-te hello

[PE1-GigabitEthernet0/0/1]mpls te bandwidth max-reservable bandwidth 10000

[PE1-GigabitEthernet0/0/1]mpls te bandwidth bc0 10000

[PE1]ospf 1

[PE1-ospf-1]area 0

[PE1 ospf-1-area-0.0.0.0]mpls-te enable

[PE1-ospf-1]opaque-capability enable

[PE1-ospf-1]enable traffic-adjustment advertise

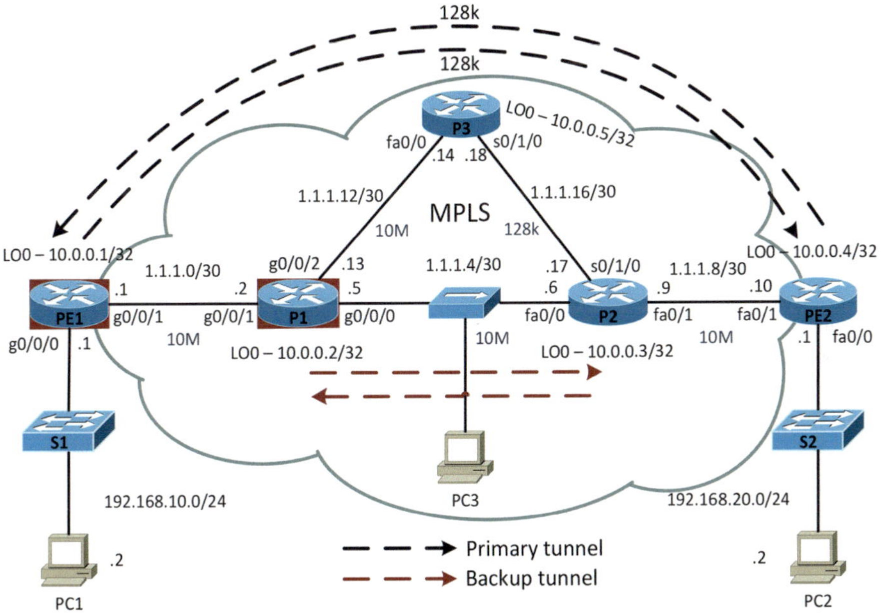

Figure 3. MPLS TE topology.

3.2. Configuration of MPLS TE on Cisco routers

To enable the MPLS TE technology on Cisco routers, it is necessary to configure *mpls traffic-eng tunnels and ip rsvp signaling hello* commands. To achieve establishment of the LDP signaling protocol from the loopback interface, *mpls ldp router-id Loopback0 force* is configured.

A part of the next configuration is to explicitly turn on RSVP-TE for each MPLS physical interface and set the maximum bit rate of a line which can be reserved. Using *mpls traffic-eng area 0* command is configured a special type of message OSPF LSA 10 for the OSPF area. Each Cisco router must be uniquely identified using OSPF router-ID. If the router did not have this identification, OSPF LSA 10 will not be transmitted.

PE2(config)#mpls ldp router-id Loopback0 force

PE2(config)#mpls traffic-eng tunnels

PE2(config)#ip rsvp signalling hello

PE2(config)#interface FastEthernet0/1

PE2(config-if)#mpls traffic-eng tunnels

PE2(config-if)#ip rsvp bandwidth 10000

PE2(config-if)#ip rsvp signalling hello

PE2(config)#router ospf 1

PE2(config-router)#mpls traffic-eng router-id Loopback0

PE2(config-router)#mpls traffic-eng area 0

3.3. Configuration of primary explicit path on Cisco router PE1

To define an explicit path for the primary line through the MPLS network via routers PE1-P1-P3-P2-PE2, each next hop is defined by the IP address of the LSR router.

MPLS TE technology includes configuration of MPLS tunnel connections. As a tunnel source, a loopback interface is defined by *IP address unnumbered interface LoopBack0* command. Last next hop IP address of an explicit path must match the destination of the tunnel. In our case, 10.0.0.4 is used. Identification of the MPLS tunnel is made by *mpls te tunnel-id 1* command. Priority is set by the command *mpls te priority 0,* where zero indicates the highest priority. A part of the configuration must be *mpls te record-route label* which records the links during the initiation of the tunnel.

[PE1]explicit-path PE1-P1-P3-P2-PE2

[PE1-explicit-path-PE1-P1-P3-P2-PE2]next hop 1.1.1.2

[PE1-explicit-path-PE1-P1-P3-P2-PE2]next hop 1.1.1.14

[PE1-explicit-path-PE1-P1-P3-P2-PE2]next hop 1.1.1.17

[PE1-explicit-path-PE1-P1-P3-P2-PE2]next hop 1.1.1.10

[PE1-explicit-path-PE1-P1-P3-P2-PE2]next hop 10.0.0.4

[PE1]interface Tunnel0/0/0

http://dx.doi.org/10.5772/66785

[PE1-Tunnel0/0/0]tunnel-protocol mpls te

[PE1-Tunnel0/0/0]ip address unnumbered interface LoopBack0

[PE1-Tunnel0/0/0]destination 10.0.0.4

[PE1-Tunnel0/0/0]mpls te tunnel-id 1

[PE1-Tunnel0/0/0]mpls te record-route label

[PE1-Tunnel0/0/0]mpls te priority 0

Bit rate 128 kbit/s is assigned to the MPLS tunnel. It is necessary to define an explicit path. During the failover of the primary line, command *mpls te fast-reroute bandwidth* guarantees switching to the backup line with a keeping bit rate of the primary line. By *mplsteigpshortcut* command, tunnel becomes a virtual tunnel line, which will be inserted into the IP routing table. To ensure the tunnel connection precedence over the traditional calculation by OSPF routing protocol, we define an absolute metric for this tunnel using *mpls te igp metric absolute 1*command.

[PE1-Tunnel0/0/0]mpls te bandwidth ct0 128

[PE1-Tunnel0/0/0]mpls te path explicit-path PE1-P1-P3-P2-PE2

[PE1-Tunnel0/0/0]mpls te fast-reroute bandwidth

[PE1-Tunnel0/0/0]mpls te igp shortcut

[PE1-Tunnel0/0/0]mpls te igp metric absolute 1

[PE1-Tunnel0/0/0]mpls te commit

3.4. Configuration of primary explicit path on Huawei router PE2

Because every explicit path is unidirectional, we need to configure MPLS tunnel in the opposite direction via routers PE2-P2-P3-P1-PE1.

Similarly, primary MPLS tunnel is configured on the Cisco router. As a tunnel source, a loopback interface is used. Because the last next hop of the explicit path is IP address 10.0.0.1, this address is defined as a destination address. By *tunnel mpls traffic-eng autoroute announce* command, Cisco router announces the presence of the MPLS tunnel to the IP routing table. The highest priority is set by *mpls te priority 0* command.

PE2(config)#ip explicit-path name PE2-P2-P3-P1-PE1

PE2(cfg-ip-expl-path)#next-address 1.1.1.9

PE2(cfg-ip-expl-path)#next-address 1.1.1.18

PE2(cfg-ip-expl-path)#next-address 1.1.1.13

PE2(cfg-ip-expl-path)#next-address 1.1.1.1

PE2(cfg-ip-expl-path)#next-address 10.0.0.1

PE2(config)#interface Tunnel0

PE2(config-if)#ip unnumbered Loopback0

PE2(config-if)#tunnel destination 10.0.0.1

PE2(config-if)#tunnel mode mpls traffic-eng

PE2(config-if)#tunnel mpls traffic-eng autoroute announce

PE2(config-if)#tunnel mpls traffic-eng priority 0 0

Same bit rate (128 kbit/s) is assigned to MPLS tunnel. The explicit path is then included in the MPLS tunnel interface. During failover of the primary line on the Cisco site, command tunnel *mpls traffic-eng fast-reroute bw-protect* guarantees switching to the backup line with a keeping bit rate of the primary line.

PE2(config-if)#tunnel mpls traffic-eng bandwidth 128

PE2(config-if)#tunnel mpls traffic-eng path-option 1 explicit PE2-P2-P3-P1-PE1

PE2(config-if)#tunnel mpls traffic-eng record-route

PE2(config-if)#tunnel mpls traffic-eng fast-reroute bw-protect

3.5. Configuration of backup path and Fast Reroute on Huawei router P1

In next step, the router P1 is configured. A backup explicit path is defined between routers P1 and P2. This backup tunnel will be used when the primary path PE1-P1-P3-P2-PE2 fails. The function of the backup exit node has a destination router P2 which uses 10.0.0.3 address. This tunnel interface becomes a backup link using *mpls te bypass-tunnel* command. Last command protects the interface GigabitEthernet0/0/2, in the case of failure of the router P3 or link between P1 and P3.

[P1]explicit-path P1-P2

[P1-explicit-path-P1-P2]next hop 1.1.1.6

[P1-explicit-path-P1-P2]next hop 10.0.0.3

[P1]interface Tunnel0/0/0

[P1-Tunnel0/0/0]ip address unnumbered interface LoopBack0

[P1-Tunnel0/0/0]tunnel-protocol mpls te

[P1-Tunnel0/0/0]destination 10.0.0.3

[P1-Tunnel0/0/0]mpls te tunnel-id 1

[P1-Tunnel0/0/0]mpls te record-route

[P1-Tunnel0/0/0]mpls te priority 0

[P1-Tunnel0/0/0]mpls te path explicit-path P1-P2

http://dx.doi.org/10.5772/66785

[P1-Tunnel0/0/0]mpls te bypass-tunnel

[P1-Tunnel0/0/0]mpls te igp shortcut

[P1-Tunnel0/0/0]mpls te igp metric absolute 1

[P1-Tunnel0/0/0]mpls te protected-interface GigabitEthernet0/0/2

[P1-Tunnel0/0/0]mpls te commit

4. Results

4.1. Verification of MPLS TE technology

After the configuration, it is time to verify the correct functionality of the MPLS TE technology. **Figure 4** shows the LFIB table with MPLS labels and also a created primary MPLS TE tunnel. The entry point of the tunnel is the PE1 router with IP address 10.0.0.1, which corresponds to the configured IP address on the loopback interface. The exit point is therouterPE2, which is identified by IPaddress10.0.0.4. Likewise, we can see establishment of the primary-tunnelPE2_t0 to theIPaddress10.0.0.4. Each one-way tunnel route has its own identification (LSPID) and assigned MPLS label.

[PE1]display mpls lsp

LSP Information: RSVP LSP

FEC	In/Out Label	In/Out IF	Vrf Name
10.0.0.1/32	3/NULL	GE0/0/1/-	
10.0.0.4/32	NULL/1077	-/GE0/0/1	

LSP Information: LDP LSP

FEC	In/Out Label	In/Out IF	Vrf Name
10.0.0.2/32	NULL/3	-/GE0/0/1	
10.0.0.2/32	1024/3	-/GE0/0/1	
10.0.0.5/32	NULL/1074	-/GE0/0/1	
10.0.0.5/32	1076/1074	-/GE0/0/1	
1.1.1.4/30	NULL/3	-/GE0/0/1	
1.1.1.4/30	1041/3	-/GE0/0/1	
10.0.0.1/32	3/NULL	-/-	
1.1.1.0/30	3/NULL	-/-	
192.168.10.0/24	3/NULL	-/-	
10.0.0.4/32	1072/NULL	-/-	
192.168.20.0/24	1073/NULL	-/-	
1.1.1.16/30	NULL/1075	-/GE0/0/1	
1.1.1.16/30	1077/1075	-/GE0/0/1	
1.1.1.12/30	NULL/3	-/GE0/0/1	
1.1.1.12/30	1078/3	-/GE0/0/1	

Figure 4. LFIB table of router PE1.

As it can be seen in **Figure 5**, the transmission rate of 128kbit/s is reserved throughout the LSP routersPE1-P1-P3-P2-PE2. The same transmission rate is reserved for the tunnellinePE2-P2-P3-P1-PE1 as well.

[PE1]display mpls te link-administration admission-control

LspID	*In/Out IF*	*S/H*	*Prio*	*CT*	*BW(kbps)*
10.0.0.1:1:13	*--- / GE0/0/1*	*0 /0*		*0*	*128*
10.0.0.4:-:164	*GE0/0/1 / ---*	*0 /0*		*0*	*128*

Figure 5. Reserved transmission rates of MPLS tunnels.

The records marked "T" in the LFIB table of the routerPE2indicated that packets are sent through MPLSTE tunnel. As we can see in **Figure 6**, Huawei router remembers only the IP address of the end of the MPLS tunnel. However, a Cisco router in the LFIB table also records the subnet 192.168.10.0/24.

Figure 7 shows the established primary tunnels, which pass through the router P1, but also established backup tunnels. The entry point of the backup tunnel is the IP address 10.0.0.2 and the exit point is the router P2, which is identified by the IPaddress10.0.0.3. Likewise, we see the backup tunnel(P2_t0), which was defined on the router P2.

PE2#sh mpls forwarding-table

Local Label	*Outgoing Label or VC*	*Prefix or Tunnel Id*	*Bytes Label Switched*	*Outgoing interface*	*Next Hop*
16	*Pop Label*	*10.0.0.3/32*	*0*	*Fa0/1*	*1.1.1.9*
17	*Pop Label*	*1.1.1.4/30*	*0*	*Fa0/1*	*1.1.1.9*
18	*18*	*10.0.0.2/32*	*0*	*Fa0/1*	*1.1.1.9*
19	*Pop Label [T]*	*10.0.0.1/32*	*0*	*Tu0*	*point2point*
20	*No Label [T]*	*192.168.10.0/24*	*0*	*Tu0*	*point2point*
21	*21*	*1.1.1.12/30*	*0*	*Fa0/1*	*1.1.1.9*
22	*22*	*1.1.1.0/30*	*0*	*Fa0/1*	*1.1.1.9*
23	*Pop Label*	*1.1.1.16/30*	*0*	*Fa0/1*	*1.1.1.9*
24	*23*	*10.0.0.5/32*	*0*	*Fa0/1*	*1.1.1.9*

[T] Forwarding through a LSP tunnel.
View additional labelling info with the 'detail' option

Figure 6. LFIB table of router PE2.

[P1]display mpls te tunnel lsr-role all

Ingress LsrId	*Destination*	*LSPID*	*In/Out Label*	*R*	*Tunnel-name*
10.0.0.3	*10.0.0.2*	*31*	*3/--*	*E*	*P2_t0*
10.0.0.4	*10.0.0.1*	*164*	*1076/3*	*T*	*PE2_t0*
10.0.0.1	*10.0.0.4*	*13*	*1077/28*	*T*	*Tunnel0/0/0*
10.0.0.2	*10.0.0.3*	*3*	*--/0*	*I*	*Tunnel0/0/0*

Figure 7. Primary and backup MPLS TE tunnels established on the router P1.

http://dx.doi.org/10.5772/66785

4.1.1. *Verification of MPLS TE Fast Reroute*

Explicitly configured path through MPLS tunnel was verified using trace route command fromPC1toPC2 via PE1-P1-P3-P2-PE2 routers, as depicted in **Figure 8**.

An Ethernet link between routers P1 and P3 was disconnected. Every single second was sent an ICMP message from PC1 to PC2. Because 5ICMP messages were lost, the reconvergence time of Fast Reroute was just 5seconds, which can be seen in **Figure 9**.

As depicted in **Figure 10**, the primary tunnel line used inner MPLS label 28, there is still maintained as the inner label. Value "zero" is used as the outer MPLS label. This explicit NULL label signals to the receiving router P2 to remove the outer MPLS label.

```
root@pcn312h:~# traceroute 192.168.20.2
traceroute to 192.168.20.2 (192.168.20.2), 30 hops max, 60 byte packets
 1  192.168.10.1 (192.168.10.1)  1.446 ms  1.648 ms  1.880 ms
 2  1.1.1.2 (1.1.1.2)  7.365 ms  0.829 ms  7.608 ms
 3  1.1.1.14 (1.1.1.14)  69.859 ms  81.080 ms  92.280 ms
 4  1.1.1.17 (1.1.1.17)  17.246 ms  28.414 ms  39.631 ms
 5  1.1.1.10 (1.1.1.10)  43.780 ms  47.973 ms  52.167 ms
 6  192.168.20.2 (192.168.20.2)  58.398 ms  97.400 ms  11.852 ms
```

Figure 8. Verified explicit path through primary tunnel.

```
root@pcn312h:~# ping 192.168.20.2
PING 192.168.20.2 (192.168.20.2) 56(84) bytes of data.
64 bytes from 192.168.20.2: icmp_req=1 ttl=62 time=13.0 ms
64 bytes from 192.168.20.2: icmp_req=2 ttl=62 time=12.9 ms
64 bytes from 192.168.20.2: icmp_req=3 ttl=62 time=12.9 ms
64 bytes from 192.168.20.2: icmp_req=9 ttl=60 time=0.913 ms
64 bytes from 192.168.20.2: icmp_req=10 ttl=60 time=0.747 ms
64 bytes from 192.168.20.2: icmp_req=11 ttl=60 time=0.738 ms
64 bytes from 192.168.20.2: icmp_req=12 ttl=60 time=0.720 ms
64 bytes from 192.168.20.2: icmp_req=13 ttl=60 time=0.721 ms
^C
--- 192.168.20.2 ping statistics ---
13 packets transmitted, 8 received, 38% packet loss, time 12041ms
rtt min/avg/max/mdev = 0.720/5.355/13.098/5.924 ms
```

Figure 9. Rerouting of ICMP traffic to the backup tunnel from PC1 to PC2.

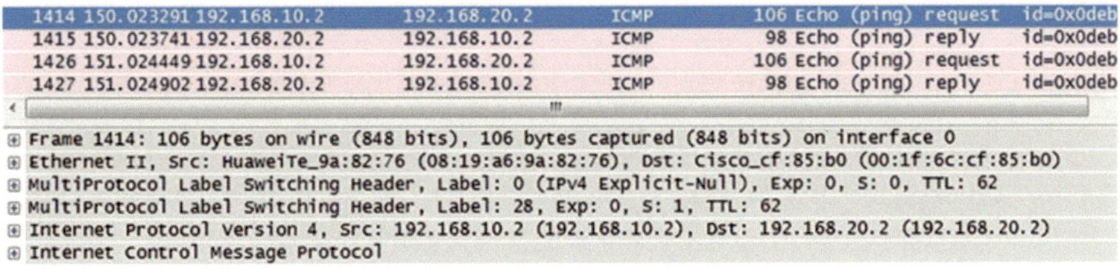

Figure 10. ICMP traffic between routers P1 and P2.

Fast Reroute was also tested on the Cisco site which was subsequently disconnected by means of a serial link between routers P2 and P3. Every single second was sent an ICMP message from PC2 to PC1. Because only two ICMP messages were lost, the convergence time of Fast Reroute was just 2 seconds, which can be seen in **Figure 11**.

```
root@pcn312g:~# ping 192.168.10.2
PING 192.168.10.2 (192.168.10.2) 56(84) bytes of data.
64 bytes from 192.168.10.2: icmp_req=1 ttl=59 time=13.0 ms
64 bytes from 192.168.10.2: icmp_req=2 ttl=59 time=12.9 ms
64 bytes from 192.168.10.2: icmp_req=3 ttl=59 time=15.0 ms
64 bytes from 192.168.10.2: icmp_req=6 ttl=60 time=0.899 ms
64 bytes from 192.168.10.2: icmp_req=7 ttl=60 time=0.906 ms
64 bytes from 192.168.10.2: icmp_req=8 ttl=60 time=0.739 ms
64 bytes from 192.168.10.2: icmp_req=9 ttl=60 time=0.718 ms
^C
--- 192.168.10.2 ping statistics ---
9 packets transmitted, 7 received, 22% packet loss, time 8018ms
rtt min/avg/max/mdev = 0.718/6.338/15.070/6.408 ms
```

Figure 11. Rerouting of ICMP traffic to the backup tunnel from PC2 to PC1.

As depicted in **Figure 12**, the reconvergence time of the OSPF protocol was also measured without the function of Fast Reroute. The measured time was 15 seconds.

```
root@pcn312h:~# ping 192.168.20.2
PING 192.168.20.2 (192.168.20.2) 56(84) bytes of data.
64 bytes from 192.168.20.2: icmp_req=1 ttl=62 time=13.0 ms
64 bytes from 192.168.20.2: icmp_req=2 ttl=62 time=12.9 ms
64 bytes from 192.168.20.2: icmp_req=3 ttl=62 time=12.9 ms
64 bytes from 192.168.20.2: icmp_req=19 ttl=60 time=0.913 ms
64 bytes from 192.168.20.2: icmp_req=20 ttl=60 time=0.728 ms
64 bytes from 192.168.20.2: icmp_req=21 ttl=60 time=0.738 ms
64 bytes from 192.168.20.2: icmp_req=22 ttl=60 time=0.712 ms
^C
--- 192.168.20.2 ping statistics ---
13 packets transmitted, 7 received, 68% packet loss, time 21041ms
rtt min/avg/max/mdev = 0.722/5.924/13.013/5.128 ms
```

Figure 12. Rerouting of ICMP traffic without using Fast Reroute.

5. Conclusion

The goal of this chapter was to test a network scenario of interoperability between different vendor's network devices for MPLS TE technology using the Fast Reroute function. Our goal was to verify the compatibility and functionality between the Cisco and Huawei devices. Although MPLS technology is standardized by RFC, some of our practical experience showed us problems in interoperability between different vendors within various RFC standardized technologies. The basic MPLS configuration was without any problems. The appropriate IP prefixes were successfully exchanged. LIB and LFIB tables were filled up.

The major disadvantage of Huawei routers during the MPLS TE configuration is necessity to have the appropriate license. After the license activation, the MPLS TE technology worked properly and the primary and backup MPLS tunnels were established. Without using the technology MPLS TE, the OSPF reconvergence lasted about 15 seconds, after disconnecting Ethernet cable. Due to function Fast Reroute of MPLS TE, the reconvergence lasted only 5 seconds between routers P1 and P3, which is 1/3of convergence time within the OSPF protocol. When using Fast Reroute, the convergence lasted only 2 seconds after disconnecting serial link between routers P2 and P3. It is 1/8 of convergence time within the OSPF protocol. If more routers were added to the network topology, it would lead to a longer convergence time of OSPF but the reconvergence time within Fast Reroute would remain unchanged.

Because nowadays the fast convergence is very critical, this chapter showed that the ISPs can use these heterogeneous network routers together with Fast Reroute technology, which can greatly reduce the convergence time.

Acknowledgements

This publication was created within the project support of VŠB-TUO activities with China with financial support from the Moravian-Silesian Region and partially was supported by the grant SGS reg. no. SP2016/170 conducted at VSB-Technical University of Ostrava, Czech Republic.

Author details

Martin Hlozak[1*], Dominik Uhrin[1], Jerry Chun-Wei Lin[2] and Miroslav Voznak[1]

Address all correspondence to: martin.hlozak@vsb.cz

1 Department of Telecommunications, Faculty of Electrical Engineering and Computer Science, VSB-Technical University of Ostrava, Ostrava-Poruba, Czech Republic

2 School of Computer Science and Technology, Harbin Institute of Technology, Shenzhen Graduate School, Shenzhen, China

References

[1] K. I. Park, "QoS in packet networks", *The Kluwer International Series in Engineering and Computer Science*, vol. 779, pp. 213–231, 2005.

[2] IETF, *RFC 3031: Multiprotocol Label Switching Architecture* [online], 2001. Available from: https://www.ietf.org/rfc/rfc3031.txt

[3] Ramadža, J. Ožegović, V. Pekić, "Network performance monitoring within MPLS traffic engineering enabled networks", in *Software, Telecommunications and Computer Networks (SoftCOM) 2015*, Croatia, IEEE pp. 315–319, 2015.

[4] B. Dekeris, L. Narbutaite, "Traffic control mechanism within MPLS networks", *IEEE Information Technology Interfaces*, pp. 603–608, 2004.

[5] M. Hlozak, J. Frnda, Z. Chmelikova, M. Voznak, "Analysis of Cisco and Huawei routers cooperation for MPLS network design", *Telecommunications Forum Telfor (TELFOR)*, pp. 115–118, 2014.

[6] Yoo-Hwa Kang, J. Lee, "The implementation of the premium services for MPLS IP VPNs", in *IEEE Advanced Communication Technology, ICACT 2005*, South Korea, IEEE pp. 1107–1110, 2005.

[7] T. Almandhari, F. Shiginah, "A performance study framework for Multi-Protocol Label Switching (MPLS) networks", in *GCC Conference and Exhibition (GCCCE)*, Oman, IEEE pp. 1–6, 2015.

Chapter 5

Design of Emotion Recognition System

Dominik Uhrin, Pavol Partila, Jaroslav Frnda,
Lukas Sevcik, Miroslav Voznak and
Jerry Chun Wei Lin

Additional information is available at the end of the chapter

http://dx.doi.org/10.5772/116607

Abstract

The chapter deals with a speech emotion recognition system as a complex solution including a Czech speech database of emotion samples in a form of short sound records and the tool evaluating database samples by using subjective methods. The chapter also involves individual components of an emotion recognition system and shortly describes their functions. In order to create the database of emotion samples for learning and training of emotional classifier, it was necessary to extract short sound recordings from radio and TV broadcastings. In the second step, all records in emotion database were evaluated using our designed evaluation tool and results were automatically evaluated how they are credible and reliable and how they represent different states of emotions. As a result, three final databases were formed. The chapter also describes the idea of new potential model of a complex emotion recognition system as a whole unit.

Keywords: classifier, database, emotion, neural network, emotion, recognition system, sample

1. Introduction

There are many fields which require the information about the emotional state. Technological development puts more pressure on the greater accuracy and simplicity of communication between man and computer. Current applications use the speech as an input-output interface, and this trend is broadening increasingly. This type of interaction can develop two problems caused by the absence of information on the emotional state. The first one is an incorrect recognition of a sentence or a command from a person who is facing stress situation. The machine recognizes human speech differently than a human with hearing. The accuracy is affected by changes in the voice signal which are caused by stress in the vocal tract. The second problem is an absence of emotional state regarding the machine speech of the loudspeaker. Typical application, such as Text-To-Speech, combines truly correct parts of speech sounds but on the other side, this signal does not contain any emotion. Such a speech influences the human and is synthetically unreliable.

2. State of the art

Psychological research confirmed that emotional state has an impact on human speech and also on the physiological state of body. Noticeable improvement has been made in a field of automatic classification of human emotion as well. These achievements have been attained by recognition techniques mainly in past few decades. In comparison to the past 10 or 20 years, contemporary computation power of processors has reached very different level. Thus, this new hardware allows us to real-time use of methods for emotion recognition. A lot of secondary information obtained from speech could not have been processed previously due to the lack of computation power and method used for the process. But the bad quality of training samples remained the significant issue. Nowadays, there is a lot of emotional recordings databases. However, a significant number of databases are created based on simulated emotions by actors instead of real-life emotions. On the other hand, quality of sound recordings is very high because of the use of studio recording. Therefore, the recordings do not contain any unnecessary noise. Creating such a type of database is much easier in comparison to the real emotion database samples. This kind of samples has to be manually cut out from sound recordings which contain real emotions and recording processing is much more time consuming. Working with simulated emotion recordings is simpler because each of them is labeled. The labels contain information on features like the kind of recorded emotion, gender of an actor, etc. The fact that actors are pronouncing mostly the same sentences also guarantees the same context of recordings.

These recordings are more efficient in terms of training emotional classifier. The following recording databases can be considered as some of the most known and recent ones: Humane [1], Emotional Prosody Speech and Transcripts, Danish Emotional Speech Corpus, Berlin Emotional Speech Database [2], Serbian Emotional Speech Database [3].

3. Methodology

The emotion recognition system can be divided into two parts: the first part is emotional classifier used for classification of emotion from a sound recording and the second part of the system is the emotion database intended for learning and training of emotional classifier. Emotional classifier is very important component of the emotion recognition system, and it is also the core of the system. Three parts of neural classifier are shortly described. The first part describes the process of a sound sample preparation, the second part discusses feature extraction from a sample, and the last part of emotional classifier describes a specific type of neural networks and the way how these networks work. **Figure 1** shows the block diagram

http://dx.doi.org/10.5772/116607

of emotional classifier. Three more subchapters are dedicated to emotion database. First subchapter deals with the creation of a sample database, its extraction and technical parameters. In the second part, we describe a tool for subjective evaluation. It describes certain parts of the tool and also the process of evaluation altogether with the processing of evaluation results. Last subchapter is dedicated to the future vision of a complex automated emotion recognition system and its use.

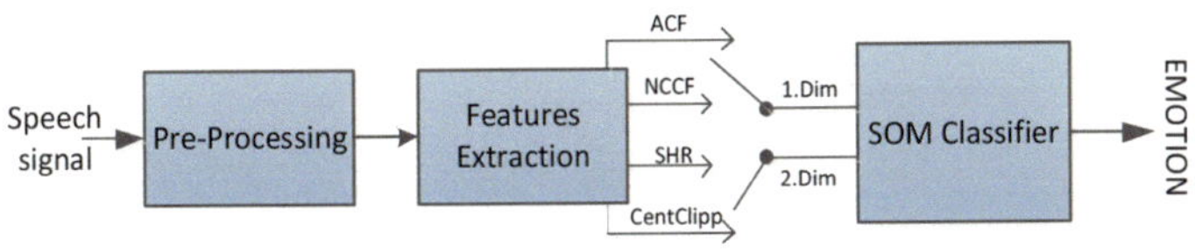

Figure 1. Speech emotion recognition system.

3.1. Preprocessing

Speech signal is stochastic by nature. However, the speech signal has a number of characteristics that may be considered as unwanted during the processing. This part of the chapter deals with a process called preprocessing—an important part of the digital speech signal processing. These few steps prepare the signal for subsequent extraction of signal parameters. The values of these parameters could be wrong without the preprocessing process. **Figure 2** shows the preprocessing. Speech signal digitizing and sound cards processing may also have side effects. Sound cards insert the DC component into the speech signal. It may not be suitable for the calculation of parameters such as signal energy or others. Therefore, removing of the DC component is part of the preprocessing. Unwanted DC offset is removed by subtracting the mean of each sample. In real-time applications that represent many cases, we do not have all the audio, which means that the true mean cannot be estimated. Thus, in real-time processing it is necessary to calculate the mean value for each sample. The mean value for the current sample can be determined based on the mean value of the previous sample. In the end, the DC component is removed by a simple subtraction of the mean value [4, 5]. The energy of the signal decreases with increasing frequency is another characteristic of the speech signal. Most of the speech signal energy is included in the first 300 Hz of spectra, which means that the information of the higher frequencies expires compared to higher energies from the bottom of the spectrum. Saving of the higher end of the spectrum is achieved by increasing the energy in the higher part of spectra artificially, which represents the second part of the preprocessing. Spectrum part energy increase is performed using pre-emphasis. As mentioned above, speech signal has a stochastic character. From a mathematical point of view, it is very difficult to find dependency and frequency in this signal. Because of that it is necessary to divide the speech signal into smaller parts called frames. Frame length is selected between 20 and 30 ms. This length is derived from the lag of the human vocal tract. Division of the signal is the third part of the preprocessing. Values of samples in neighboring frames may vary rapidly, therefore frame overlap is appropriate. The frame overlap is selected in half. Processing speech signals between the frames can have a side effect because the edges of neighboring frames may have sharp transitions. It can have a bad influence mostly on speech processing and frequency analysis. The above-mentioned disadvantage can be removed by applying the window function on each frame. Many window functions are used in speech processing, and the choice depends on the characteristics of the following processing methods. Hamming window function is used in most such cases due to its suitable properties in both the time domain and frequency [5, 6].

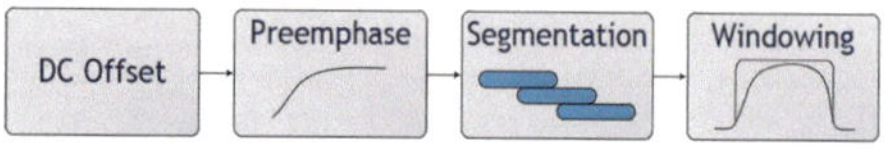

Figure 2. Preprocessing process.

3.2. Speech processing parameters

Volume, intonation, and tempo are speech characteristics that can be recognized by a human ear. In DSP (digital signal processing) and speech processing in particular, we use some other parameters which characterize speech signal and human vocal tract. Other two parameters signal energy and ZCR (zero crossing rate) are also important in speech processing. These parameters were used as a voice activity detector intended to eliminate silence or noise. Signal energy is characterized by intensity, and the human ear can sense it as a volume. Energy is influenced by the way of recording and digitizing speech, speaker distance from the microphone, and other features. Voiced and unvoiced parts can be separated using a sound energy profile. ZCR describes how many times the polarity of the signal changes, in other words how many times it crosses zero. This parameter can also carry information about F0 change. ZCR carries the information on both the voice activity and the energy [8, 9]. The fundamental frequency of vocal cords (F0) is one of the most important parameters within the speech processing because it carries a lot of information about the vocal tract, and thus also the basic features of man. Age, gender, speech errors, and emotional state of a man can be determined using this parameter. There are several methods of signal processing that enable to estimate the fundamental frequency. Human speech consists of voiced and unvoiced speech sounds. The vocal cords are almost completely open in the creation of voiceless phonemes. The basic tone does not arise with opened vocal cords, and therefore F0 can be calculated only from the nonvoiceless parts. Each of the methods used to calculate F0 has its advantages and disadvantages. The following methods can be used to calculate F0: auto-correlation function, normalized cross-correlation function, autocorrelation function with central clipping and sub-harmonic-to-harmonic ratio [9, 10].

3.3. Self-organizing feature map

The emotional state classifier is based on self-organizing maps (SOM). These maps represent a specific type of neural networks with uncontrolled competitive learning. There are generally two-dimensional maps of neurons. The learning process of SOM is uncontrolled, which means that the input data do not need to know the output. In the process of learning, SOMs determine for themselves, how to classify the inputs [7]. At the beginning of learning, the weights of all the inputs of neurons can be set randomly. Randomly selected input vectors are applied to neurons and then analyzed in order to find the one which is the most similar to an input. This neuron is called the winner. The weights of neighboring neurons are adjusted according to the following rule in Eq. (1). The equation describes the weight between neurons i and j for $t + 1$ iteration and input $x\_j\ (t)$ [10].

$$w_{ij}\ (t+1) = w_{ij}\ (t) + \gamma\left(x_j\ (t) + w_{ij}\ (t)\right) \tag{1}$$

3.4. Sample database creation

Next iteration means a new vector for input, finding new winner and changing weights between neurons again. When the learning process is completed, the map has a shape that represents the characters of input parameters. In order for the emotional classifier to be as precise as possible, sample database of real emotion for training and learning has to be created. Of course, we are speaking about Czech emotional sample database. It is very difficult to determine the emotion of sound recordings. So the precision of emotional classifier within determination of the emotion depends on how many real emotion samples it learns. A few hours recording from two Czech radio broadcastings was the first step to create real emotion database not simulated by actors. Some of the recordings have been available on the official web page archive of the third radio station, and some of the recordings of Czech television broadcasting were downloaded from share video portal YouTube. Out of television broadcasting, only sound part was cut for the creation of database samples. Parameters of database samples have to fulfill the following conditions:

- Sample had to have a duration from 6 to 6 seconds.
- Sample should not have contained environment noise.
- Sample had to contain human speech in a form of few words or a full sentence.
- Sample had to be uniquely named.

Name of the database sample consists of three parameters: the first one is state of the emotion. There are seven basic emotion states. Database samples have been made for four emotion states because, as for the rest of the states, it is difficult to find real emotion recordings, or it is hard to recognize it by using subjective methods (boredom, disgust, and fear) [1]. As an output format for database samples, waveform audio-file with 16-bit PCM (pulse code modulation) coding, monochannel and sample frequency at the level of 16 kHz have been used. These parameters are sufficient enough, taking into consideration that the source broadcasting recording has been recorded to MPEG-2 Audio Layer III audio file with a bit rate of 128 kbps. Some of the source recordings from which database samples have been obtained were recorded using VideoLAN Client media player. For editing and cutting of source recordings, software Audacity was used. By default, Audacity was unable to edit mp3 audio file; therefore, LAME Encode library had to be installed [11].

3.5. Emotional database formation

Next step to build the emotional database was the creation of the tool for evaluation of database samples from which emotional database has been created after evaluation. As methods for evaluation of emotion samples, the subjective one has been chosen. Subjective methods represent using people to evaluate a small amount of samples in this case. The web page represents the evaluation tool as a direct tool for evaluation connected to MySQL database in order to save the results of subjective evaluation. The web page consists of four pages: the first one is invitation page, it invites evaluating subject and gives it short instruction. The second page is the evaluating page. It is the core of whole evaluating tool. It consists of html5 audio player for playing database sound samples and rollout menu for selection of the state of emotion. Subject simply plays the recording selected by an algorithm, and consequently selects the emotion in rollout menu. The result is sent to and saved into MySQL database. The next two pages of the tool are the final page announcing the end of the process of evaluation to the subject, and the error page announcing to the subject that something went wrong during the evaluation process. The tool also consists of page used to insert samples to the system. It makes inserting much easier and less time consuming. Besides the web page, the tool also consists of MySQL database. As mentioned above, database was used to save the results of subjective evaluation. Database consists of two connected tables. These tables are shown in **Figure 3**. The first table provides information about individual emotional samples and contains four kinds of information. Number column represents how many times the sample has been loaded to audio player. Emotion column represents the first letter from English expression of a selected sample. Ref_id column represents a unique name for a sample. The second table provides the information about evaluation of sample, and it includes seven kinds of information. Meaning of first one, ref_id, is same as in table one.

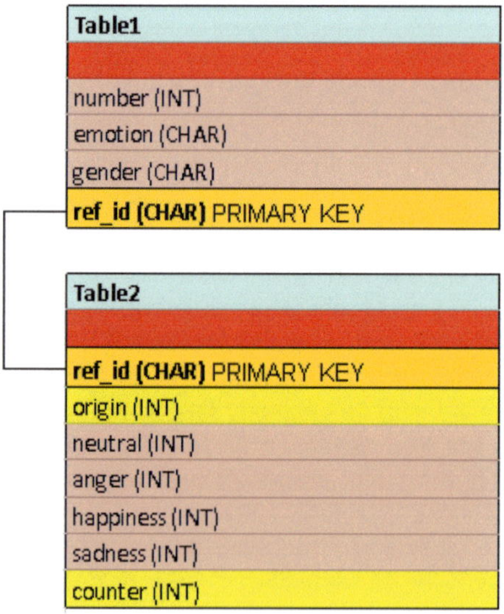

Figure 3. MySQL database tables.

Origin column shows how many times the same emotion has been selected by the subject for sample as was originally selected by the author during the process of creating database samples. Information included in columns three to six represents the emotional state selected by the subject during the process of evaluation. And finally the last column, counter, shows the total number of sample evaluation. Logically, the web page contains also scripts to perform queries by using mysql_query() with SELECT or UPDATE queries. It was also necessary to create a customs function for loading and saving data from or to the database named database_load(). Furthermore, the web page uses POST and GET forms to obtain data from previous page loads. It was necessary to create custom functions generate_sample() and audio(), too. The first function is used to generate the name of the sample that is loaded into html5 audio player using second function audio(). Using htlm5 audio player is much easier. Flash or any other plug-in does not have to be installed. This all is the part of web browser, and it uses much less computation performance. Custom functions were created to insert the obtained results of sample evaluation to database tables. In the first version of this tool, it was necessary to manually process results by exporting tables with results and use external scripts to achieve statistic results. But in this modified version, we created tool function evaluate_results() that uses MySQL scripts to automatically processed evaluation results. The diagram of processing from evaluation till final database is shown in **Figure 4**.

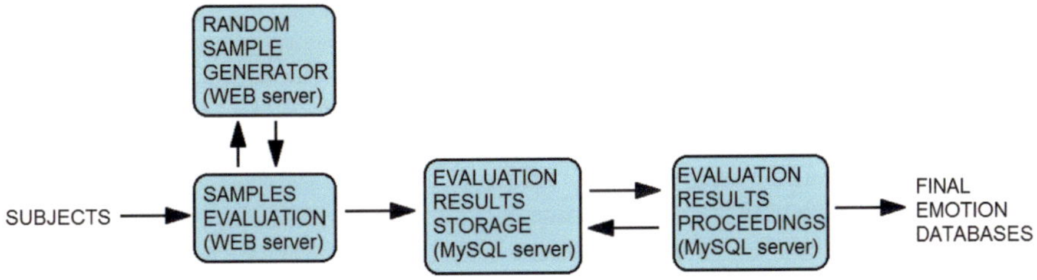

Figure 4. Sample evaluation processing diagram.

3.6. Vision of future

Emotion recognition of Czech spoken language or the attempts to recognize the emotion from Czech spoken words are not at beginnings of their development. Thus, the actual status of an emotion recognition system does not allow the automatized system to be created for emotion recognition which could be used as real application. It is caused by the fact that emotional classifier has abilities to determine emotion variety with precision about of 70–75%. But this is not applicable in real life. The level of precision can be increased by choosing and also modifying the method of learning emotion states, as well as by using real, not acted, emotion samples to learn the emotion. Due to this we are trying, but it remains the close future. For the future when the precision level of emotion determination will be useable in real application, the model of the automated recognition system should appear same as indicated in **Figure 5**. For example, in call centers, employees can be divided into groups. Each group would take care of customers with different emotional state. Before the customer would be forwarded to call center employee, they would be asked to repeat some sentence or to answer a question.

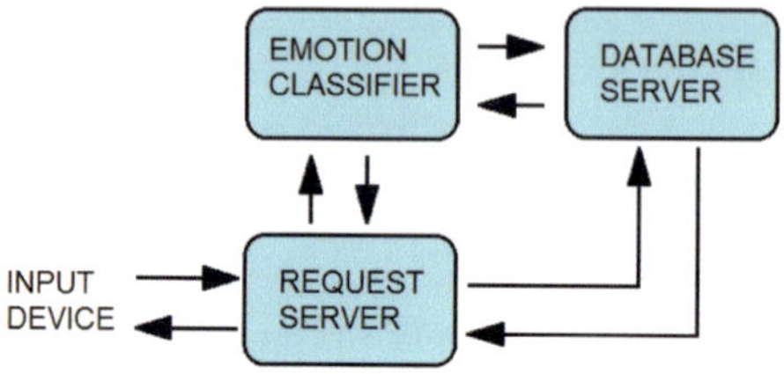

Figure 5. Automatized emotion recognition system.

Based on the results of emotion analysis of their voice made by automatized emotion recognition system, the customer would be then forwarded to the assigned employee. All this would be done real time. Police special force could use this automatized system as well during the negotiation with a kidnapper or terrorist. Their voice could be analyzed for emotion state, which could help the police to make the right steps when solving the problem. The above-mentioned examples are only few from many possible real applications. In general, the automatized system receives the request by a network from some device (such as PC, smart phone, or server) to recognize the emotion from the voice. First, request is registered by the request server and recorded to the database server. Afterward, the file is sent to emotional classifier for analysis. The classifier takes file through all the components described above and sends the result to the both database and request servers. This would result in the customer device from which request was received. It could be a potential scheme for the future automatized emotion recognition system. Emotional classifiers can also be used during the evaluation of samples instead of subjective evaluation methods in the future. It should be easier and less time consuming, but it is a sound of the future.

4. Results

The evaluation of database samples was made by subjects represented by students in age range from 18 to 26 years. The selection rate of each emotion kind for random sample is listed in **Table 1**. As mentioned before, the automatized system was used for sample evaluation. The veracity value was determined for each sample from the database as well. **Table 2** shows first five samples of database with percentage of veracity, state of emotion, and a level of veracity. After determining the veracity value, three levels of veracity were assigned to samples: low, medium, and high. Levels have been set to the following ranges: a low level with a range from 0 to 75, a medium level range from 75 to 90, and a high level percentage range from 90 up to 100. Three final emotion databases have been created based on these three levels. Names of these emotion database samples were formed under

http://dx.doi.org/10.5772/116607

conditions, as shown in **Figure 6**. First database with high veracity of samples is suitable for learning of neural emotional classifier. This classifier has been developed by Mr. Partila at our university [8, 9] and its components are shortly described in the chapter. Second database of samples with medium veracity is suitable for training of neural classifier and verifying of its learning skills. The last database of emotion samples with low level of veracity is formed by samples that contain mixed emotions. As for this emotion database, it was difficult to determine the emotion state of samples. In order to determine the emotion state better, more evaluations of samples have to be performed, or some samples are simply not suitable for it.

Emotion	Probability for one sample (%)
Neutral	34
Anger	28
Happiness	20
Sadness	18

Table 1. Probability with what emotion was selected.

Evaluation veracity (%)	Emotion kind	Level of veracity
71.34	Neutral	Low
38.21	Anger	Low
48.77	Neutral	Low
82.85	Happiness	1. Medium
82.85	Neutral	Medium

Table 2. Veracity table for first five samples with emotion kind determined.

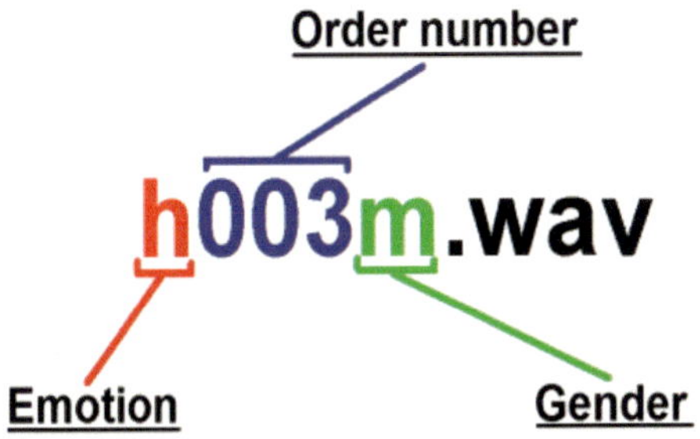

Figure 6. Database sample name.

5. Conclusion

This chapter focused on the emotional classifier as the complex thing including creation of training and learning database. Furthermore, the chapter try to solve the problem of samples preprocessing and also the feature extracting process, which are both important for the next procedure of emotion recognition. Last but not least, it describes functioning of self-organizing feature maps. The SOM classifier has the lowest error rate, and thus the best resolving power between normal and stress emotional states. The tool for subjective sample evaluation has been upgraded for automatized result evaluation that made it easier and less time consuming. All created samples have been evaluated by the specific group of subjects and based on the results, three final databases were formed. Two of them are usable for learning and training the classifier. As for further development, automatic evaluation of samples is an option to be used instead of subjective evaluation in a form of neural classifier.

Acknowledgements

This publication was created within the project Support of VŠB-TUO activities with China with financial support from the Moravian-Silesian Region and partially was supported by the grant SGS reg. no. No. SP2016/170 conducted at VSB-Technical University of Ostrava, Czech Republic.

Author details

Dominik Uhrin[1,*], Pavol Partila[1], Jaroslav Frnda[1], Lukas Sevcik[1], Miroslav Voznak[1] and Jerry Chun Wei Lin[2]

*Address all correspondence to: dominik.uhrin@vsb.cz

1 Department of Telecommunications, Faculty of Electrical Engineering and Computer Science, VSB-Technical University of Ostrava, Ostrava, Poruba, Czech Republic

2 School of Computer Science and Technology, Harbin Institute of Technology Shenzhen Graduate School, Shenzhen, China

References

[1] Ververidis, D., Kotropoulos, C., 2010. A review of emotional speech databases. *In Proc. Panhellenic Conference on Informatics (PCI)*, pp. 560–574.

[2] Burkhardt, F., Paeschke, A., Rolfes, M., Sendlmeier, W., Weiss, B., 2003. A Database of German Emotional Speech. *In Interspeech 2005 – Eurospeech, 9th European Conference on Speech Communication and Technology*, pp. 1517–1520.

[3] Jovičić, S.T., Kašić, Z., Đorđević, M., Rajković, M., 2004. Serbian emotional speech database: design, processing and evaluation. *In 9th Conference Speech and Computer (SPECOM 2004)*, pp. 77–81.

[4] Nicholson, J., Takahashi, K., Nakatsu, R., 2006. Emotion Recognition in Speech Using Neural Networks. *In Neural Computing & Applications*, Volume 9, Issue 4, Springer Verlag, pp. 290–296.

[5] Partila, P., Voznak, M., Mikulec, M., Zdralek, J., 2012. Fundamental Frequency Extraction Method using Central Clipping and its Importance for the Classification of Emotional State. *In Advances in Electrical and Electronic Engineering*, Volume 10, Issue 4, pp. 270–275.

[6] Kasi, K., Zahorian, S. A., 2002. Yet Another Algorithm for Pitch Tracking. *In IEEE International Conference on Acoustics, Speech, and Signal Processing (ICASSP)*, vol.1, pp. 361–364, IEEE.

[7] Roussinov, D., Chen, H. A., 1998. Scalable Self organizing Map Algorithm for Textual Classification. *A Neural Network Approach to Thesaurus Generation. In Communication Cognition and Artificial Intelligence*, Volume 15, pp. 81–111.

[8] Lee, C.M., Narayanan, S.S., 2005. Toward detecting emotions in spoken dialogs. *IEEE Transactions on Speech and Audio Processing*, Volume 13, Issue 2, pp. 293–303.

[9] Voznak, M., Partila, P., Mehic, M., Jakovlev, S., 2013. Recognizing Emotions from Human Speech Using 2-D Neural Classifier and Influence the Selection of Input Parameters on its Accuracy. *In 21st Telecommunications Forum Telfor*, art no. 6716272, pp. 482–485.

[10] Partila, P., Voznak, M., 2013. Speech Emotions Recognition Using 2-D Neural Classifier, *Nostradamus 2013: Prediction, Modeling and Analysis of Complex Systems, Advances in Intelligent Systems and Computing* Volume 210, Springer International Publishing, pp. 221–231.

[11] Uhrin, D., Partila, P., Voznak, M., Chmelikova, Z., Hlozak, M., Orcik, L. 2014. Design and implementation of Czech database of speech emotions. *22nd Telecommunications Forum, TELFOR 2014 - Proceedings of Papers*, art. no. 7034463, pp. 529–532.

Chapter 6

Controllable Subspace as a New Characterization of Influence of Nodes in Complex Networks

Jiuhua Zhao

Additional information is available at the end of the chapter

http://dx.doi.org/10.5772/66783

Abstract

In this chapter, we investigate the influence of a node on a network. By virtue of the classical control theory, the influence of a node is represented by the controllable subspace, which is further transformed into a specified graph named general-cacti. An algorithm is developed to calculate the influence, which contains the searching of different kinds of circles and the longest path in the directed graph. Moreover, eight real networks are studied and simulations show that (1) a node in dense and homogeneous networks could have more influence comparing with nodes in sparse and heterogeneous networks, (2) any single studied classical centrality measure including output degree, betweenness and PageRank could not rank the influence and (3) a node with high rank in all these three measures could have more influence.

1. Introduction

With the development of society and technology, nowadays lots of industry and manmade systems have become larger and larger. Different from traditional systems, these systems have a large amount of state variables and complex connection between states. Usually the variables, which capture the connection between states, are unable to measure and at the same time these variables can be easily changed for reasons of temperature, humidity and so on. Hence, these complex systems are difficult to analyze by using classical control theory. In the 1970s, a new control conception, structural controllability, was first introduced by Ching-Tai Lin in 1974 [1]. In Ref. [1] the complete structural controllability of systems with single input was studied. Shields and Pearson [2] extended the result of Lin to multiple input structural systems. Glover and Silverman [3] simplified the work of Shields and Pearson. The conception of structural controllability allows part of systems of the same structure, though with measure of zero, be uncontrollable. Mayeda and Yamada [4] introduced the conception of strong structural controllability to eliminate this kind of uncontrollability. Reinschke et al. [5] studied the conception in multiinput and multioutput systems. Jarczyk et al. [6] revisited the previous results and gave out a new graph-theoretic characterization of strong structural controllability. The solid foundation of structural controllability had been laid down from then on. For those structural uncontrollable systems, the structural controllable subspace is studied by Hosoe [7] and Poliak [8]. All these studies shed light on the study on the systems where the variables, which capture the connections between states, are unmeasured or mutable. When the system becomes larger, the methods and algorithms mentioned by the above literature studies become inapplicable for the reason of complexity of computation. Recently, Liu et al. [9] combined the structural controllability theory with the complex network, by using the algorithm of maximum match to calculate the minimum input nodes for complete control and methods of complex network science to analyze the properties of the input nodes. Since then more and more studies have been dedicated to controllability of complex network. Nepusz and Vicsek [10] used edge dynamics to describe the controllability of complex networks. Banerjee and Roy [11] exploited the property of driver nodes and made some supplement for the result of Liu et al. All the existing literature studies about the controllability of complex networks are focused on complete controllability. How about those complex networks, who are not controllable with the given diver nodes? Among hundreds even thousands or billions nodes, what role does a single node actually play? How many nodes will it affect, if it is chosen as a driver node? What properties do those nodes with powerful control ability have? All these questions motivated us to study the problem when networks are not fully controlled.

The chapter is organized as follows. In Section 2, some useful preliminaries and the model descriptions are introduced. In Section 3, an algorithm to search for the influence of a single node is proposed and we apply it to several real complex networks in Section 4. Finally, Section 5 contains concluding remarks.

2. Preliminaries and model description

Consider a linear time-invariant control system

$$\dot{x} = Ax(t) + Bu(t) \tag{1}$$

where $x(t) \in R^n$, $u(t) \in R^m$ and we assume that A and B are structural matrices, which means all elements in the matrix are either fixed (zero) or unfixed (indeterminate) and the indeterminate parameters are unrelated. Hereafter Eq. (1) is called a structural system and denoted by (A, B). Let $[AB]$ denote the matrix of the system and $G(V, E)$ denote the digraph of the system. We generate a digraph $G(V, E)$ from (A, B), with $n + m$ nodes, $V = v_1, \ldots, v_n, v_{n+1}, \ldots, v_{n+m}$, where $v_1, \ldots, v_n$ represent the state nodes of (A, B) and $v_{n+1}, \ldots, v_{n+m}$ represent the control nodes. And all those edges are generated as follows: for every fixed entry m_{ij} of $[AB]$, the graph contains an oriented edge (v_i, v_j). For a structural system (A, B), his corresponding fixed system is denoted by $(\hat{A}, \hat{B})$, where unfixed parameters in (A, B) are filled by one class of unrelated real numbers.

http://dx.doi.org/10.5772/66783

Definition 1: A system (A, B) is structural controllable if all its possible fixed systems $(\hat{A}, \hat{B})$ are controllable.

The above definition was first introduced by C-T Lin. He reduced the property of structural property to the property of the graph of the system.

Definition 2 (stem) [1]: A stem is an elementary path originating from an input vertex (see **Figure 1**).

Definition 3 (bud) [1]: A bud is an elementary cycle with an additional edge that ends, but not begins, in a vertex of the cycle. Vertex $v1$ is called the origin of the bud (see **Figure 2**).

Definition 4 (cactus) [1]: A cactus is defined recursively as follows. A stem is cacti. Given a stem S_0 and buds $B_1, B_2, \cdots, B_p$, then $S_0 \cup B_1 \cup \cdots \cup B_p$ is a cacti if for every $i(1 \leq i \leq p)$ the origin vertex of B_i is also the origin of an edge in $S_0 \cup B_1 \cup \cdots \cup B_i$ and is the only vertex belonging the same time to B_i and $S_0 \cup B_1 \cup \cdots \cup B_i$. A set of disjoint cacti is called cactus (see **Figure 3**).

We call the red node in **Figure 3**, i.e., the input vertex, driver node and the nodes connected to driver node are defined as driven nodes(e.g., node $d1$ and $d2$ in **Figure 3**.

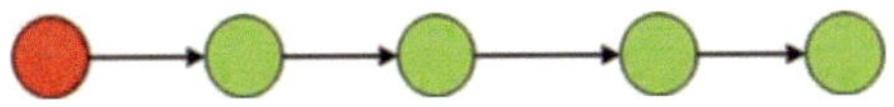

Figure 1. Stem.

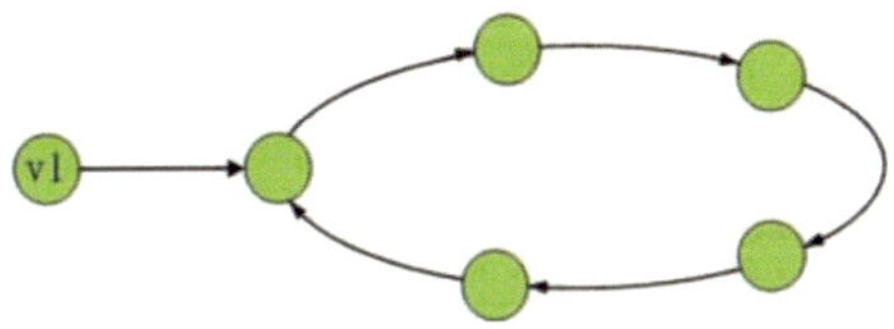

Figure 2. Bud.

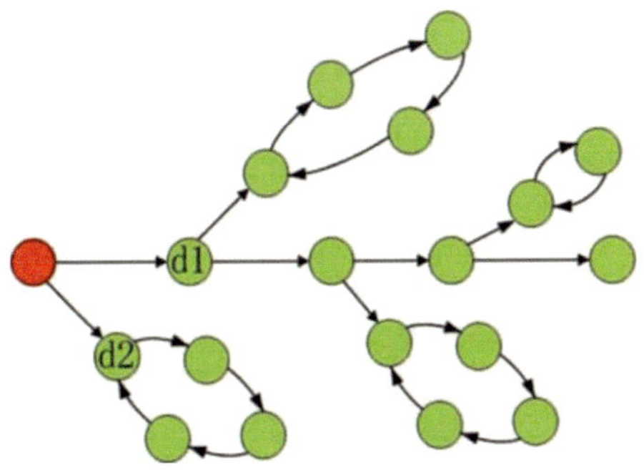

Figure 3. Cacti.

Theorem 1 (Lin) [1]: The single input system pair (A, b) is structurally controllable iff the graph of (A, b) is spanned by cacti.

Many works have been done to study the complete controllability of a system. Hosoe considered the problem for the first time, where the system is not fully controllable. He defined the maximal dimension of controllability matrix of the structural system as the generic dimension of controllable subspace, denoted by $d_c = \text{gennericrank}[A \quad AB \quad \cdots \quad A^{n-1}B]$. And give out two methods for determining the generic dimension of controllable subspace of a system. We show the graph-theoretic method here (see Theorem 1).

Theorem 2 [2]: Let d_c be the generic dimension of the controllable subspace of the structural system (A, B) and $G(A, B)$ represents the graph of (A, B). Assume that $[A\ B]$ is irreducible. Then $d_c = \max_{\tilde{G} \in \mathcal{G}}\{|E(\tilde{G})|\}$ where $\mathcal{G}$ denotes the subgraphs of $G(A, B)$ which is defined by and $|E(\tilde{G})|$ represents the number of edges in graph $\tilde{G}$. Comparing with the graphic properties of complete controllable system in Lin [1], the graph of controllable subspace contains more circles than cactus. Those circles are not directly connected to the cactus in the form of buds. They are connected to the cactus through at least one node. We define the circle in $\tilde{G}$ as a general-bud. Then replacing all the buds in cacti by general-buds, we can have the definition of general-cacti (see **Figure 4**). In **Figure 4**, those circles with dash lines are the defined general-bud.

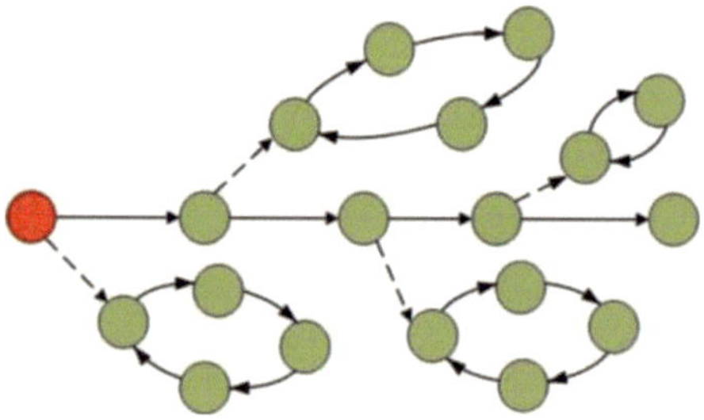

Figure 4. General-cacti.

In our research, we consider the influence of a single node in the complex network. By the influence of node i, we mean the controllable subspace of the system when we only control node i. Then we denote the structural input vector as b_i, where all elements in vector b_i are fixed zero except for the ith element. Therefore, we can just study the controllable subspace of structural system (A, b_i), whose corresponding structural matrix and digraph are denoted as $[A\ b_i]$ and G_i. Let us define the controllability of a node i in a complex network as P_i. Then we can easily derive P_i from Theorem [1], i.e.,

$$P_i = \max_{\tilde{G}_i \in \mathcal{G}_i}\{|V(\tilde{G}_i)|\} \tag{2}$$

where $\mathcal{G}_i$ denotes the set of general-cacti originated from node i in digraph G_i. To ignore the influence of the network size, the influence of node i is normalized as $p_i = P_i/N$.

3. Algorithm

To find the largest general-cacti, we classify the circle in the network into three categories. The first one is circles with no inbound (see **Figure 5a**). This kind of circles cannot be arrived by the input, so they are not in our general-cacti. The second one is circles with no outbound edges (see **Figure 5b**). This kind of circles can only act as general-buds in the general-cacti without affecting the competition of two different general-cactus. The third one is circles with both inbound and outbound edges (see **Figure 5c**). These circles can also be part of a stem, which means these nodes affect the dimension of general-cacti.

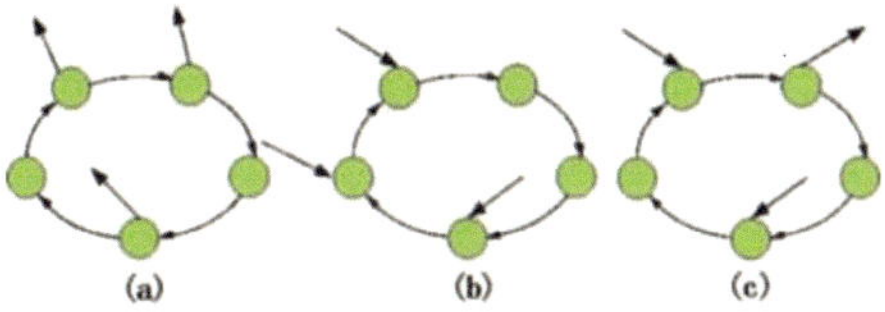

Figure 5. Circles of three categories.

Our instinct is exploring the longest path to act as a stem and then add all the circles as the general-buds. Nevertheless, the third kind of circles makes it difficult to search for the largest general-cacti. As shown in **Figure 6**, the blue stem in (b) is longer than the red one in (a), but the circle can also act as a general-bud to increase the dimension of controllable subspace. Hence, the dimension of general-cacti in (a) is larger than the one in (b).

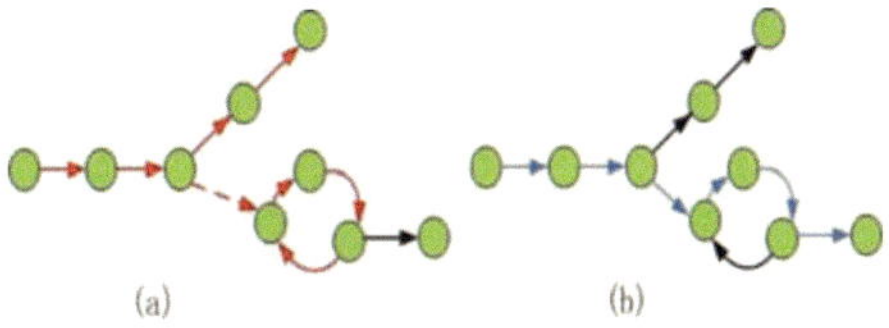

Figure 6. (a, b) Two different cacti in the same network.

However, if the circle of the third kind is not large enough, the choice of the longest path will not affect the influence of the origin node significantly. When the circle is large enough, we set it as a general-bud to prevent it from participating in the stem forming. The definition of "large" affects the accuracy of our method. However, from the real database we find that the circles are either very small or very large. Our method is efficient in real networks. Thus far we can give an approach to discover the influence of a single node in the complex network.

3.1. Procedure to discover the influence of a single node

- Search the reachable nodes from node i in the original graph and delete all those unreachable nodes.
- Search for the second category circles in the current graph, count the number (N_1) of their nodes, then delete all those nodes and the edges connected to them.
- Search for large circles in the current graph, count the number (N_2) of their nodes, then delete all those nodes and the edges connected to them.
- Search for the longest path originate from the input in the current graph, count the number (N_3) of the nodes, then delete all those nodes and the edges connected to them.
- Search for all the circles left in the current graph, count the number (N_4) of the nodes, then delete all those nodes and the edges connected to them.
- $P_i = N_1 + N_2 + N_3 + N_4$

The first, second and fourth steps of this procedure are all about search for the directed circle in the digraph. We recursively delete the nodes without any input or without any output. Then we can get all the nodes in circle. Using the depth-first search method, we can separate these nodes into different circles and count for the number of nodes of each circle. For the circle, which has small circles inside, we just count the largest circle.

The third step of this procedure is the most difficult one. To the best of our knowledge, there exist no optimal algorithms to search for the longest path in the digraph. The existing algorithms of searching for the shortest path cannot be modified to search for the longest path, for the existing of directed circles. We modify the classical depth-first search method to search for the longest path in a digraph with directed circles (see Algorithm 1).

Algorithm 1 Algorithm to search for the longest path in a digraph

Require: $v \neq NULL$

Ensure: $Long(v)$

$deep \Leftarrow deep + 1$, $visited(v) \Leftarrow 1$,

$vexdeep(v) \Leftarrow deep$,

$vex \Leftarrow$ the vertex v directs to

while $vex \neq NULL$ **do**

if $visited(v) = 0$ and ($circle(v) \neq 0$ or $deep \geq vexdeep(v)$) **then**

$Long(vex)$

$visited(v) \Leftarrow 0$

vex ⇐ the vertex *v* directs to

deep ⇐ *deep* − 1

else if $visited(v) = 1$ **then**

for all vertex *i* in the current circle **do**

$circle(i) \Leftarrow 1$

end for

vex ⇐ the vertex which has the same origin as *vex*

else

vex ⇐ the vertex which has the same origin as *vex*

end if

end while

In the algorithm, we use vector *visited* and *circle* to mark the node. Node *i* is visited (in a circle) if $visited(i) = 1$ ($circle(i) = 1$), otherwise unvisited (not in a circle). We use *deep* to record the current depth and vector *vexdeep* to record the depth of every vertex. The most significant difference between our method and the depth-first search method is that we allow one vertex to be detected more than one time and at the same time we record the circles in the network for searching in other directions. The time complexity of our algorithm is $O(NM)$, where N is the number of nodes in the network and M denotes the number of edges.

4. Simulation results

We now apply the proposed algorithm to several kinds of real networks. It is infeasible to search all the nodes to find out the most influential node, for the unbearable complexity. Hence, we rank the node by their ODCR (outcome-degree centrality rank), BCR (betweenness centrality rank) and PR (page rank), then analysis the first three nodes of every categories. Form **Table 1**, we can find that nodes in dense and homogeneous networks, like network Caenorhabditis elegans, network Seagrass, network s838, network s420 and network s208, can control more percentage of nodes than nodes in sparse and heterogeneous networks. This phenomenon implies that nodes are more influential in dense and homogeneous networks. We think it is reason why dense and homogeneous networks need less driver nodes for full structural control which is observed in [9].

Type	Regulatory			Neuronal	Food web	Electronic circuits		
Name	TRN-Yeast-1 [12]	TRN-Yeast-2 [12]	TRN-EC-2 [12]	Caenorhabditis elegans [13]	Seagrass [14]	S838 [12]	S420 [12]	S208 [12]
N	4441	688	418	297	49	512	252	122
L	12873	1079	519	2345	226	819	399	189
n_D	96.5%	82.1%	75.1%	16.5%	26.5%	23.2%	23.4%	23.8%
p_{ODR1}	0.97%	0.29%	0.96%	58.59%	24.49%	31.64%	32.54%	34.43%
p_{ODR2}	1.10%	0.44%	0.48%	58.59%	22.45%	2.73%	31.35%	24.60%
p_{ODR3}	1.15%	0.29%	0.72%	56.9%	16.33%	26.95%	3.17%	31.97%
p_{BCR1}	1.15%	0.73%	0.96%	0.34%	24.49%	31.64%	3.17%	5.74%
p_{BCR2}	1.10%	0.44%	1.20%	58.59%	14.29%	7.42%	3.17%	34.43%
p_{BCR3}	1.04%	0.29%	0.48%	58.59%	10.20%	1.76%	3.57%	18.03%
p_{PR1}	1.15%	0.29%	0.96%	0.34%	10.20%	31.64%	32.54%	34.43%
p_{PR2}	1.10%	0.44%	0.48%	58.59%	14.29%	2.34%	31.35%	5.74%
p_{PR3}	0.97%	0.44%	0.72%	58.59%	12.24%	2.73%	5.95%	12.30%

Table 1. The influence of the studied nodes in eight real networks.

From **Figure 7**, we find influence of node does not rank according to any single studied centrality measure. In fact, in network Yeast-1, nodes ODR3, BCR1 and PR1 are the same node and it has largest influence 1.15% among the considered nodes. In network Yeast-2, nodes ODR2, BCR2 and PR2 are the same node and it has the largest influence 0.44%. In network EC-2, nodes ODR1, BCR1 and PR1 are the same node and it has the largest influence 0.96%. In network elegans, nodes ODR1, BCR2 and PR2 are the same node and it has the largest influence 58.59%. In network Seagrass, nodes ODR1 and BCR1 are the same node and it has the largest influence 24.49%. In network s838, nodes ODR1, BCR1 and PR1 are the same node and it has the largest influence 31.64%. In network s240, nodes ODR1 and PR1 are the same node and it has the largest influence 32.54%. In network s208, nodes ODR1, BCR2 and PR1 are the same node and it has the largest influence 34.43%. Hence, we can conclude that if a node has a high ODCR, BCR and PR at the same time, this node can have lager influence.

5. Conclusions

In this chapter, we have introduced the notion of the influence of a node in complex networks and gave out a useful algorithm to count for that property. Meanwhile we applied this method to exploit the influence of nodes in some real networks. Results have shown that nodes in dense and homogeneous networks were more influential than nodes in sparse and heterogeneous networks. And in the same network node which has high rank in ODR, BCR and PR simultaneously could be more influential. Next we will try to study the characteristic of the most influential node in a complex network and the influence of a set of nodes.

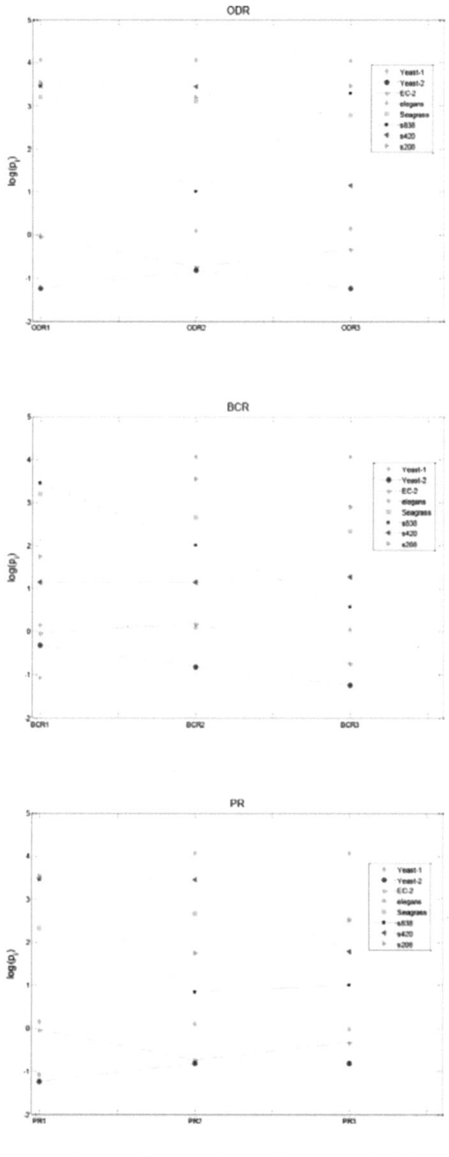

Figure 7. The influence of top three nodes according to each centrality measure in eight real networks.

Author details

Jiuhua Zhao

Address all correspondence to: jiuhuadandan@sjtu.edu.cn

Department of Automation, Shanghai Jiao Tong University and Key Laboratory of System Control and Information Processing, Ministry of Education of China, Shanghai, China

References

[1] C.-T. Lin, "Structural controllability," IEEE Transactions on Automatic Control, vol. 19, no. 3, pp. 201–208, 1974.

[2] R. W. Shields and J. B. Pearson, "Structural controllability of multiinput linear systems," IEEE Transactions on Automatic Control, vol. 21, no. 2, pp. 203–212, 1976.

[3] K. Glover and L. M. Silverman, "Characterization of structural controllability," IEEE Transactions on Automatic Control, vol. 21, no. 4, pp. 534–537, 1976.

[4] H. Mayeda and T. Yamada, "Strong structural controllability," SIAM Journal on Control and Optimization, vol. 17, no. 1, pp. 123–138, 1979.

[5] K.J. Reinschke, F. Svaricek and H.D. Wend, "On strong structural controllability of linear systems," in Proceedings of the 31st Conference on Decision and Control, pp. 203–208, 1992.

[6] J. C. Jarczyk, F. Svaricek and B. Alt, "Strong structural controllability of linear systems revisited," in Proceedings of the 50th Conference on Decision and Control and European Control Conference, pp.1213–1218, 2011.

[7] S. Hosoe, "Determination of generic dimensions of controllable subspace and its application," IEEE Transactions on Automatic Control, vol. 25, no. 6, pp. 1192–1196, 1980.

[8] S. Poliak, "On the generic dimension of controllable subspace," IEEE Transactions on Automatic Control, vol. 35, no. 3, pp. 367–369, 1990.

[9] Y.-Y. Liu, J.-J. Slotine and A.-L. Barabasi, "Controllability of complex networks," Nature, vol. 473, no. 7346, pp. 167–173, 2011.

[10] T. Nepusz and T. Vicsek, "Controlling edge dynamics in complex networks," Nature Physics, vol. 8, pp. 568–573, 2012.

[11] S. J. Banerjee and S. Roy, "Key to network controllability," vol. http://arxiv.org/abs/1209.3737, 2012.

[12] Milo, Ron, et al. "Network motifs: simple building blocks of complex networks." Science, vol. 298, no. 5594, pp. 824-827, 2002.

[13] Modha, Dharmendra S. and Raghavendra Singh. "Network architecture of the long-distance pathways in the macaque brain." Proceedings of the National Academy of Sciences, vol. 107, no. 30, pp. 13485–13490, 2010.

[14] Christian, Robert R. and Joseph J. Luczkovich. "Organizing and understanding a winter's seagrass foodweb network through effective trophic levels." Ecological Modelling, vol. 117, no. 1, pp. 99–124, 1999.

Chapter 7

Distributed Consensus-Based Estimation with Random Delays

Dou Liya

Additional information is available at the end of the chapter

http://dx.doi.org/10.5772/66784

Abstract

In this chapter we investigate the distributed estimation of linear-invariant systems with network-induced delays and packet dropouts. The methodology is based on local Luenberger-like observers combined with consensus strategies. Only neighbors are allowed to communicate, and the random network-induced delays are modeled as Markov chains. Then, the sufficient and necessary conditions for the stochastic stability of the observation error system are established. Furthermore, the design problem is solved via an iterative linear matrix inequality approach. Simulation examples illustrate the effectiveness of the proposed method.

1. Introduction

The convergence of sensing, computing, and communication in low cost, low power devices is enabling a revolution in the way we interact with the physical world. The technological advances in wireless communication make possible the integration of many devices allowing flexible, robust, and easily configurable systems of wireless sensor networks (WSNs). This chapter is devoted to the estimation problem in such networks.

Since sensor networks are usually large-scale systems, centralization is difficult and costly due to large communication costs. Therefore, one must employ distributed or decentralized estimation techniques. Conventional decentralized estimation schemes involve all-to-all communication [1]. Distributed schemes seem to fit better. In this class of schemes, the system is divided into several smaller subsystems, each governed by a different agent, which may or may not share information with the rest. There exists a vast literature that study the distributed estimation for sensor networks in which the dynamics induced by the communication network (time-varying delays and data losses mainly) are taken into account [2–10]. Millan et al. [6]

have studied the distributed state estimation problem for a class of linear time-invariant systems over sensor networks subject to network-induced delays, which are assumed to have taken values in $[0, \tau_M]$.

One of the constraints is the network-induced time delays, which can degrade the performance or even cause instability. Various methodologies have been proposed for modeling and stability analysis for network systems in the presence of network-induced time delays and packet drop-outs. The Markov chain can be effectively used to model the network-induced time delays in sensor networks. In Ref. [11], the time delays of the networked control systems are modeled by using the Markov chains, and further an output feedback controller design method is proposed.

The rest of the chapter is organized as follows. In Section 2, we analyze the available delay information and formulate the observer design problem. In Section 3, the sufficient and necessary conditions to guarantee the stochastic stability are presented first and the equivalent LMI conditions with constraints are derived. Simulation examples are given to illustrate the effectiveness of the proposed method in Section 4.

Notation: Consider a network with p sensors. Let $\upsilon = \{1, 2, \cdots, p\}$ be an index set of p sensor nodes, $\varepsilon \subset \upsilon \times \upsilon$ be the link set of paired sensor nodes. Then the directed graph $G = (\upsilon, \varepsilon)$ represents the sensing topology. The link (i, j) implies that the node i receives information from node j. The cardinality of ε is equal to l. Define $q = g(i, j)$ as the link index. $N_i = \{j \in \upsilon | (i, j) \in \varepsilon\}$ denotes the subset of nodes that communicating to node i.

2. Problem formulation

Assume a sensor network intended to collectively estimate the state of a linear plant in a distributed way. Every observer computes a local estimation of the plant's states based on local measurements and the information received from neighboring nodes. Observers periodically collect some outputs of the plant and broadcast some information of their own estimation. The information is transmitted through the network, so network-induced time delays and drop-outs may occur.

In this work, the system to be observed is assumed to be an autonomous linear time-invariant plant given by the following equations:

$$x(k+1) = Ax(k) \tag{1}$$

$$y_i(k) = C_i x(k) \quad \forall i = 1, 2, \cdots, p, \tag{2}$$

where $x(k) \in R^n$ is the state of the plant, $y_i(k) \in R^{m_i}$ is the system's outputs and p is the number of the observers. Assume (A, C) is observable, where $C = [C_1, \cdots, C_p]$.

Besides the system's output $y_i(k)$, observer i receives some estimated outputs $\hat{y}_{ij}(k) = C_{ij}\hat{x}_j$ from each neighbor $j \in N_i$. The matrix C_{ij} is assumed to be known for both nodes. Define $\overline{C}_i$ as a matrix stacking the matrix C_j and the matrices C_{ij} for all $j \in N_i$. It is assumed that $(A, \overline{C}_i)$ is observable $\forall i$.

2.1. Delays modeled by Markov chains

The communication links between neighbors may be affected by delays and/or packet dropouts. The equivalent delay $\tau_{ij}(k) \in N$ (or $\tau_q(k)$, with $q = g(i,j) \in \{1, \cdots, l\}$) represents the time difference between the current time instant k and the instant when the last packet sent by j was received at node i. The delay includes the effect of sampling, communication delay, and packet dropouts. The number of consecutive packet dropouts and network-induced delays are assumed to be bounded, so $\tau_{ij}(k)$ is also bounded.

The Markov chain is a discrete-time stochastic process with Markov property. One way to model the delays is to use the finite state Markov chain as in Refs. [7–9]. The main advantages of the Markov model are that the dependencies between delays are taken into account since in real networks the current time delays are usually related to the previous delays [8]. In this note, $\tau_{ij}(k)$ $(\forall i, j \in N_i)$ are modeled as l different Markov chains that take values in $W = \{0, 1, \cdots, \tau_M\}$. And their transition probability matrices are $\Lambda_q = [\lambda_{qrs}]$, $q = 1, 2, \cdots, l$. That means $\tau_{ij}(k)$ jump from mode r to s with the probability λ_{qrs}:

$$\lambda_{qrs} = \Pr\left(\tau_q(k+1) = s | \tau_q(k) = r\right), q = 1, 2, \cdots, l, \tag{3}$$

where $\lambda_{qrs} \geq 0$ and $\sum\limits_{s=0}^{\tau_M} \lambda_{qrs} = 1$ for all $r, s \in W$.

Remark 1: In the real network, the network-induced delays are difficult to measure. Using the stochastic process to model the delays is more practical. For sensor networks, the communication link between different pairs of nodes is also different, so the data may experience different time delays. It is more reasonable to model the delays by different Markov chains.

2.2. Observation error system

The structure of the observers described in the following is inspired by that given in Ref. [6]. To estimate the state of the plant, every node is assumed to run an estimator of the plant's state as:

$$\begin{aligned} \hat{x}_i(k+1) &= A\hat{x}_i(k) + M_i\left(\hat{y}_i(k) - y_i(k)\right) \\ &+ \sum_{j \in N_i} N_{ij}\left(C_{ij}\hat{x}_j\left(k - \tau_{ij}(k)\right) - C_{ij}\hat{x}_i\left(k - \tau_{ij}(k)\right)\right) \end{aligned} \tag{4}$$

$$\hat{y}_i(k) = C_i\hat{x}_i(k), \quad \forall i = 1, 2, \cdots, p, \tag{5}$$

The observers' dynamics are based on both local Luenberger-like observers weighted with M_i matrices, and consensus with weighting matrices N_{ij}, which takes into account the information received from the neighboring nodes.

The observation error of observer i is defined as $e_i(k) = \hat{x}_i(k) - x(k)$. From Eqs. (1)–(5), the dynamics of the observation errors can be written as:

$$e_i(k+1) = (A + M_iC_i)e_i(k) - \sum_{j\in N_i} N_{ij}C_{ij} \times e_i\left(k-\tau_{ij}(k)\right) + \sum_{j\in N_i} N_{ij}C_{ij}e_j\left(k-\tau_{ij}(k)\right) \tag{6}$$

Define $e(k) = [e_1^T(k) \quad e_2^T(k) \quad \cdots \quad e_p^T(k)]^T$, $X(k) = [e^T(k) \quad e^T(k-1) \quad \cdots \quad e^T(k-\tau_M)]^T$, then we have the observation error system:

$$X(k+1) = \left(\Psi(M) + \Phi\left(N, \tau_1(k), \cdots, \tau_l(k)\right)\right)X(k) \tag{7}$$

where

$$\Psi(M) = \begin{bmatrix} \varphi(M) & 0 & \cdots & 0 & 0 \\ I & 0 & \cdots & 0 & 0 \\ 0 & I & \cdots & 0 & 0 \\ \vdots & \vdots & \ddots & \vdots & \vdots \\ 0 & 0 & \cdots & I & 0 \end{bmatrix}$$

$$\Phi\left(N, \tau_1(k), \cdots, \tau_l(k)\right) = \begin{bmatrix} \phi\left(N, \tau_1(k), \cdots, \tau_l(k)\right) \\ 0 \\ \vdots \\ 0 \end{bmatrix},$$

$$\phi(M) = \begin{bmatrix} A + M_1C_1 & 0 & 0 & 0 \\ 0 & A + M_2C_2 & 0 & 0 \\ \vdots & \vdots & \ddots & \vdots \\ 0 & 0 & \cdots & A + M_pC_p \end{bmatrix} \tag{8}$$

$$\phi\left(N, \tau_1(k), \cdots, \tau_l(k)\right) = \phi_1\left(N, \tau_1(k)\right) + \cdots + \phi_l\left(N, \tau_l(k)\right),$$

$$\phi_q\left(N, \tau_q(k)\right) = \underbrace{[0 \quad \cdots \quad 0 \quad \Pi_q \quad 0 \quad \cdots \quad 0]}_{\text{the } \left(1+\tau_q(k)\right)\text{-th element is } \Pi_q}, \; q = 1, 2, \cdots, l.$$

Π_q are block matrices in correspondence with each of the links q communicating the observer i with j, in which the only blocks different from zero are $-N_{ij}C_{ij}$ and $N_{ij}C_{ij}$ in the (i,i) and (i,j) positions, respectively. $M = \{M_i, i\in\upsilon\}$, $N = \{N_{ij}, i\in\upsilon, j\in N_i\}$ are observer matrices to be designed.

Remark 2: The observation error system (Eq. (7)) depends on the delays $\tau_1(k), \cdots, \tau_l(k)$. This makes the analysis and design more challenging. The objective of this note is to design the observers to guarantee the stochastic stability of Eq. (7).

Definition 1 [7]: The system in Eq. (7) is stochastically stable if for every finite $X_0 = X(0)$, initial mode $\tau_1(0), \cdots,$

$\tau_l(0)\in W$, there exists a finite $Z > 0$ such that the following holds:

$$E\left\{\sum_{k=0}^{\infty} \|X(k)\|^2 \Big|_{X_0,\tau_1(0),\cdots,\tau_l(0)}\right\} < {X_0}^T Z X_0 \tag{9}$$

3. Observers' design

In this section, we first derive the sufficient and necessary conditions to guarantee the stochastic stability of system (Eq. (7)) with Definition 1. For ease of presentation, when the system's delays are

$$\tau_1(k) = r_1, \ \ldots, \tau_l(k) = r_l(r_1, \ldots r_l \in W), \tag{10}$$

we denote $\Phi\big(N, \tau_1(k), \cdots, \tau_l(k)\big)$ as $\Phi(N, r_1, \cdots, r_l)$.

Theorem 1: Under the observer (Eqs. (4) and (5)), the observation error system (Eq. (7)) is stochastically stable if and only if there exists symmetric $P(r_1, r_2, \cdots, r_l) > 0$ such that the following matrix inequality:

$$\begin{aligned} L(r_1, r_2, \cdots, r_l) = &\sum_{s_1=0}^{\tau_M} \sum_{s_2=0}^{\tau_M} \cdots \sum_{s_l=0}^{\tau_M} \lambda_{1r_1s_1} \lambda_{1r_2s_2} \cdots \lambda_{1r_ls_l} \\ &\times [\Psi(M) + \Phi(N, r_1, r_2, \cdots, r_l)]^T P(s_1, s_2, \cdots, s_l) \\ &\times [\Psi(M) + \Phi(N, r_1, r_2, \cdots, r_l)] - P(r_1, r_2, \cdots, r_l) < 0 \end{aligned} \tag{11}$$

holds for all $r_1, r_2, \cdots, r_l \in W$.

Proof: *Sufficiency:* For the system Eq. (7), construct the Lyapunov function

$$V\big(X(k), k\big) = X(k)^T P\big(\tau_1(k), \tau_2(k), \cdots, \tau_l(k)\big) X(k) \tag{12}$$

Calculating the difference of $V\big(X(k), k\big)$ along system Eq. (7) and taking the mathematical expectation, we have

$$\begin{aligned} &E\Big\{\Delta\Big(V\big(X(k), k\big)\Big)\Big\} \\ &= E\Big\{V\big(X(k+1), k+1\big) - V\big(X(k), k\big)\Big\} \\ &= E\Big\{X(k+1)^T P\big(\tau_1(k+1), \cdots, \tau_l(k+1)\big) X(k+1). |_{X_k, \tau_1(k)=r_1, \cdots, \tau_l(k)=r}\Big\} \\ &\quad - X(k)^T P\big(\tau_1(k), \cdots, \tau_l(k)\big) X(k) \end{aligned} \tag{13}$$

Define $\tau_1(k+1) = s_1, \cdots, \tau_l(k+1) = s_l$. To evaluate the first term in Eq. (13), we need to apply the probability transition matrices for $\tau_1(k) \to \tau_1(k+1), \cdots, \tau_l(k) \to$

$\tau_l(k+1)$, those are $\Lambda_q, q = 1, 2, \cdots, l$.

Then, Eq. (13) can be evaluated as

$$\begin{aligned}
&E\left\{\Delta\left(V\left(X(k),k\right)\right)\right\}\\
&= X(k)^T\left\{\sum_{s_1=0}^{\tau_M}\sum_{s_2=0}^{\tau_M}\cdots\sum_{s_l=0}^{\tau_M}\lambda_{1r_1s_1}\lambda_{1r_2s_2}\cdots\lambda_{1r_ls_l}\right.\\
&\times[\Psi(M)+\Phi(N,r_1,r_2,\cdots,r_l)]^T P(s_1,s_2,\cdots,s_l)\\
&\left.\times[\Psi(M)+\Phi(N,r_1,r_2,\cdots,r_l)]-P(r_1,r_2,\cdots,r_l)\right\}\\
&\times X(k)
\end{aligned} \tag{14}$$

Thus, if $L(r_1,r_2,\cdots,r_l)<0$, then

$$\begin{aligned}
&E\left\{\Delta\left(V\left(X(k),k\right)\right)\right\} = X(k)^T L(r_1,r_2,\cdots,r_l)X(k)\\
&\leq -\lambda_{\min}\left(-L(r_1,r_2,\cdots,r_l)\right)X(k)^T X(k)\\
&\leq -\beta\,\|X(k)\|^2
\end{aligned} \tag{15}$$

where $\beta=\inf\left\{\lambda_{\min}\left(-L(r_1,r_2,\cdots,r_l)\right)\right\}>0$. From Eq. (15), we can see that for any $T\geq 1$

$$\begin{aligned}
&E\left\{V\left(X(T+1),T+1\right)\right\}-E\left\{V(X_0,0)\right\}\\
&\leq -\beta E\left\{\sum_{t=0}^{T}\|X(t)\|^2\right\}
\end{aligned} \tag{16}$$

Then we have

$$\begin{aligned}
&E\left\{\sum_{t=0}^{T}\|X(t)\|^2\right\}\\
&\leq \frac{1}{\beta}\left(E\{V(X_0,0)\}-E\left\{V\left(X(T+1),T+1\right)\right\}\right)\\
&\leq \frac{1}{\beta}E\{V(X_0,0)\}\\
&= \frac{1}{\beta}X(0)^T P\left(\tau_1(0),\cdots,\tau_l(0)\right)X(0)
\end{aligned} \tag{17}$$

According to Definition 1, the observation error system Eq. (7) is stochastically stable.

Necessity: For necessity, we need to show that if the system Eq. (7) is stochastically stable, then there exists symmetric $P(r_1,\cdots,r_l)>0$ such that Eq. (11) holds. It suffices to prove that for any bounded $Q\left(\tau_1(k),\cdots,\tau_l(k)\right)>0$, there exists a set of $P\left(\tau_1(k),\cdots,\tau_l(k)\right)$ such that

$$\begin{aligned}
&\sum_{s_1=0}^{\tau_M}\sum_{s_2=0}^{\tau_M}\cdots\sum_{s_l=0}^{\tau_M}\lambda_{1r_1s_1}\lambda_{1r_2s_2}\cdots\lambda_{1r_ls_l}\\
&\times[\Psi(M)+\Phi(N,r_1,r_2,\cdots,r_l)]^T P(s_1,s_2,\cdots,s_l)\\
&\times[\Psi(M)+\Phi(N,r_1,r_2,\cdots,r_l)]-P(r_1,r_2,\cdots,r_l)\\
&= -Q(r_1,r_2,\cdots,r_l)
\end{aligned} \tag{18}$$

Define

$$X(t)^T\tilde{P}\left(T-t,\tau_1(t),\cdots,\tau_l(t)\right)X(t)$$
$$=E\left\{\sum_{k=t}^{T}X(k)^TQ\left(\tau_1(k),\cdots,\tau_l(k)\right)X(k)|_{X_t,\tau_1(t),\cdots,\tau_l(t)}\right\} \tag{19}$$

Assuming that $X(k)\neq 0$, since $Q\left(\tau_1(k),\cdots,\tau_l(k)\right)>0$, as T increases, $X(t)^T\tilde{P}\left(T-t,\tau_1(t),\cdots,\tau_l(t)\right)$ $X(t)$ is monotonically increasing, or else it increases monotonically until $E\left\{X(k)^TQ\left(\tau_1(k),\cdots,\tau_l(k)\right)X(k)|_{X_t,\tau_1(t),\cdots,\tau_l(t)}\right\}=0$ for all $k\geq k_1\geq t$. From Eq. (9), $X(t)^T\tilde{P}\left(T-t,\tau_1(t),\cdots,\tau_l(t)\right)X(t)$ is bounded. Furthermore, its limit exists

$$X(t)^TP(r_1,\cdots,r_l)X(t)$$
$$=\lim_{T\to\infty}X(t)^T\tilde{P}\left(T-t,\tau_1(t)=r_1,\cdots,\tau_l(t)=r_l\right)X(t)$$
$$=\lim_{T\to\infty}E\left\{\sum_{k=t}^{T}X(k)^TQ\left(\tau_1(k),\cdots,\tau_l(k)\right)X(k)|_{X_t,\tau_1(t),\cdots,\tau_l(t)}\right\} \tag{20}$$

Since it is valid for any $X(t)$, we have

$$P(r_1,\cdots,r_l)=\lim_{T\to\infty}\tilde{P}\left(T-t,\tau_1(t)=r_1,\cdots,\tau_l(t)=r_l\right). \tag{21}$$

From Eq. (20), we obtain $P(r_1,\cdots,r_l)>0$ since $Q\left(\tau_1(k),\cdots,\tau_l(k)\right)>0$. Consider

$$E\left\{X(t)^T\tilde{P}\left(T-t,\tau_1(t),\cdots,\tau_l(t)\right)X(t)-X(t+1)^T\right.$$
$$\times\tilde{P}\left(T-t-1,\tau_1(t+1),\cdots,\tau_l(t+1)\right)X(t+1)$$
$$\left.|_{X_t,\tau_1(t)=r_1,\cdots,\tau_l(t)=r_l}\right\}$$
$$=X(t)^TQ(r_1,\cdots,r_l)X(t). \tag{22}$$

The second term in Eq. (22) equals to

$$E\left\{X(t+1)^T\tilde{P}\left(T-t-1,\tau_1(t+1),\tau_2(t+1),\cdots,\tau_l(t+1)\right)\right.$$
$$\left.\times X(t+1)|_{X_t,\tau_1(t)=r_1,\tau_2(t)=r_2,\cdots,\tau_l(t)=r_l}\right\}$$
$$=X(t)^T\left\{\sum_{s_1=0}^{\tau_M}\sum_{s_2=0}^{\tau_M}\cdots\sum_{s_l=0}^{\tau_M}\lambda_{1r_1s_1}\lambda_{1r_2s_2}\cdots\lambda_{1r_ls_l}\right.$$
$$\times[\Psi(M)+\Phi(N,r_1,r_2,\cdots,r_l)]^T\tilde{P}(T-t-1,s_1,\cdots,s_l)$$
$$\left.\times[\Psi(M)+\Phi(N,r_1,r_2,\cdots,r_l)]\right\}X(t) \tag{23}$$

Substituting Eq. (23) into Eq. (22) gives rise to

$$\begin{aligned}
&X(t)^T\Big\{\tilde{P}\Big(T-t,\tau_1(t),\cdots,\tau_l(t)\Big)\\
&-\sum_{s_1=0}^{\tau_M}\sum_{s_2=0}^{\tau_M}\cdots\sum_{s_l=0}^{\tau_M}\lambda_{1r_1s_1}\lambda_{1r_2s_2}\cdots\lambda_{1r_ls_l}\\
&\times[\Psi(M)+\Phi(N,r_1,r_2,\cdots,r_l)]^T\tilde{P}(T-t-1,s_1,\cdots,s_l)\\
&\times[\Psi(M)+\Phi(N,r_1,r_2,\cdots,r_l)]\Big\}X(t)\\
&=X(t)^TQ(r_1,r_2,\cdots,r_l)X(t)
\end{aligned}\tag{24}$$

Letting $T\to\infty$ and noticing Eq. (21), it is shown that Eq. (11) holds. This completes the proof.

As it is clearly seen from Eq. (11) that the matrix inequality to be solved in order to design the observers is nonlinear. To handle this, Proposition 1 gives the equivalent LMI conditions with nonconvex constraints. It can be solved by several existing iterative LMI algorithms. Product reduction algorithm in Ref. [10] is employed to solve the following conditions.

Proposition 1: There exist observers Eqs. (4) and (5) such that the observation error system Eq. (7) is stochastically stable if and only if there exists matrices $\varphi(M)$, $\phi_1(N,r_1)$, $\phi_2(N,r_2),\cdots,\phi_l(N,r_l)$, and symmetric matrices $\overline{X}(s_1,s_2,\cdots,s_l)>0$, $P(r_1,r_2,\cdots,r_l)>0$, satisfying

$$\begin{bmatrix}-P(r_1,r_2,\cdots,r_l) & V(r_1,r_2,\cdots,r_l)^T\\ V(r_1,r_2,\cdots,r_l) & -X(r_1,r_2,\cdots,r_l)\end{bmatrix}<0\tag{25}$$

$$\overline{X}(s_1,s_2,\cdots,s_l)P(s_1,s_2,\cdots,s_l)=I\tag{26}$$

for all $r_1,\cdots,r_l\in W$, with

$$\begin{aligned}
V(r_1,\cdots,r_l)&=[\,V_0(r_1,\cdots,r_l)^T\ \ \cdots\ \ V_{\tau_M}(r_1,\cdots,r_l)^T\,]^T\\
V_{s_1}(r_1,\cdots,r_l)&=[\,V_{s_1,0}(r_1,\cdots,r_l)^T\ \ \cdots\ \ V_{s_1,\tau_M}(r_1,\cdots,r_l)^T\,]^T\\
V_{s_1s_2}(r_1,\cdots,r_l)&=[\,V_{s_1s_2,0}(r_1,\cdots,r_l)^T\ \ \cdots\ \ V_{s_1s_2,\tau_M}(r_1,\cdots,r_l)^T\,]^T\\
&\ \ \vdots\\
V_{s_1\cdots s_{l-1}}(r_1,\cdots,r_l)&=\begin{bmatrix}(\lambda_{1r_1s_1}\cdots\lambda_{lr_l0})^{\frac{1}{2}}[\Psi(M)+\Phi(N,r_1,\cdots,r_l)]\\(\lambda_{1r_1s_1}\cdots\lambda_{lr_l1})^{\frac{1}{2}}[\Psi(M)+\Phi(N,r_1,\cdots,r_l)]\\ \vdots\\(\lambda_{1r_1s_1}\cdots\lambda_{lr_l\tau_M})^{\frac{1}{2}}[\Psi(M)+\Phi(N,r_1,\cdots,r_l)]\end{bmatrix}\\
X(r_1,\cdots,r_l)&=diag\{X_0(r_1,\cdots,r_l),\cdots,X_{\tau_M}(r_1,\cdots,r_l)\}\\
X_{s_1}(r_1,\cdots,r_l)&=diag\{X_{s_1,0}(r_1,\cdots,r_l),\cdots,X_{s_1,\tau_M}(r_1,\cdots,r_l)\}\\
&\ \ \vdots\\
X_{s_1\cdots s_{l-1}}(r_1,\cdots,r_l)&=diag\{\overline{X}(s_1,\cdots,s_l),\cdots,\overline{X}(s_1,\cdots,s_l)\}.
\end{aligned}\tag{27}$$

Proof: As we know $\overline{X}(s_1,\cdots,s_l)>0$, we have $X(r_1,\cdots,r_l)>0$ by the construction of it. By applying the Schur complement, Eq. (25) is equivalent to

$$-P(r_1,\cdots,r_l)+V(r_1,\cdots,r_l)^TX^{-1}(r_1,\cdots,r_l)V(r_1,\cdots,r_l)<0\tag{28}$$

Since $\overline{X}(s_1,\cdots,s_l)=P(s_1,\cdots,s_l)^{-1}$, we can derive Eq. (11).

4. Numerical example

Consider a plant whose dynamics is given by:

$$x(k+1) = \begin{bmatrix} 0.99 & 0 \\ 0 & 1.01 \end{bmatrix} x(k). \tag{29}$$

Assume the network has two nodes, with two links, one is from node 1 to node 2, and the other is from node 2 to node 1. The matrices are given as follows:

$$\begin{array}{l} C_1 = [1 \quad 0], \; C_2 = [1 \quad 1], \\ C_{12} = C_2, \quad C_{21} = C_1, \end{array} \tag{30}$$

The random delays are assumed to be $\tau_q(k) \in \{0, 1\} (q = 1, 2)$, and their transition probability matrices are given by

$$\Lambda_1 = \begin{bmatrix} 0.4 & 0.6 \\ 0.5 & 0.5 \end{bmatrix}, \quad \Lambda_2 = \begin{bmatrix} 0.3 & 0.7 \\ 0.4 & 0.6 \end{bmatrix}. \tag{31}$$

Figure 1 shows part of the simulation run of the delay $\tau_2(k)$ governed by its transition probability matrix Λ_2.

By using Proposition 1, we design the observers with the following matrices:

$$\begin{array}{ll} M_1 = \begin{bmatrix} -0.9900 \\ 0.0673 \end{bmatrix}, & M_2 = \begin{bmatrix} -0.4620 \\ -0.5387 \end{bmatrix}, \\ N_{12} = \begin{bmatrix} 0.0071 \\ 0.3865 \end{bmatrix}, & N_{21} = \begin{bmatrix} 0.1320 \\ -0.1347 \end{bmatrix} \end{array} \tag{32}$$

The initial values of the plant and the observers are $x(0) = [2 \quad 0.5]^T$, $\hat{x}_1(0) = \hat{x}_2(0) = [0 \quad 0]^T$ and $\hat{x}_1(-1) = \hat{x}_2(-1) = [0 \quad 0]^T$. **Figure 2** represents the evolution of the plant's states (solid lines)

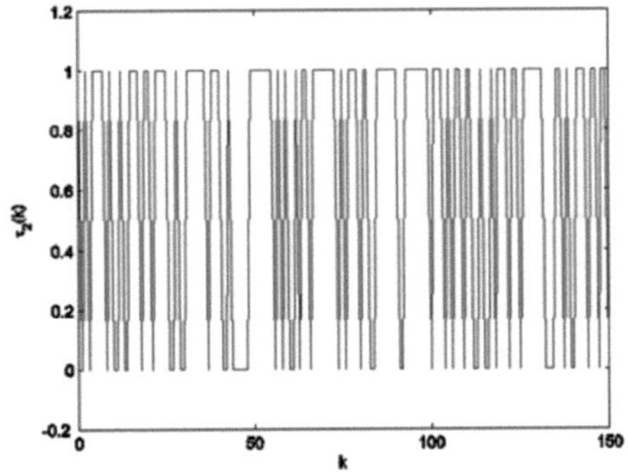

Figure 1. The random delays $\tau_2(k)$.

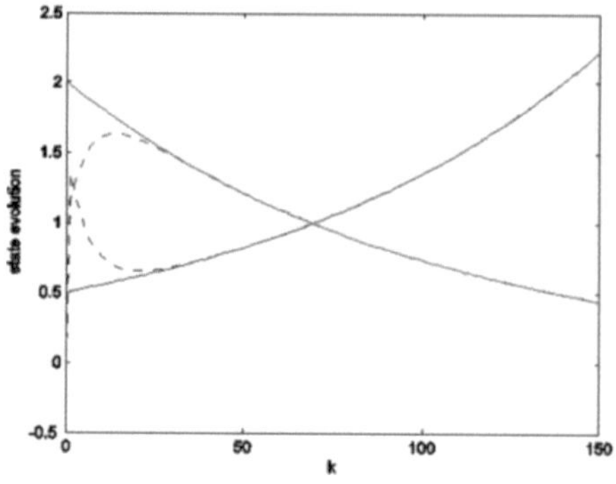

Figure 2. Evolution of the estimates for observer 2.

and the estimated states (dashed lines) for observer 2. It is observed that the estimation of the observers converge to the plant's state.

5. Conclusion

This chapter addresses the problem of distributed estimation considering random network-induced delays and packet dropouts. The delays are modeled by Markov chains. The observers are based on local Luenberger-like observers and consensus terms to weight the information received from neighboring nodes. Then the resulting observation error system is a special discrete-time jump linear system. The sufficient and necessary conditions for the stochastic stability of the observation error system are derived in the form of a set of LMIs with nonconvex constraints. Simulation examples verify its effectiveness.

Author details

Dou Liya

Address all correspondence to: douliya@sjtu.edu.cn

Department of Automation, Shanghai Jiao Tong University, Shanghai, China

References

[1] R. Olfati-Saber, "Distributed Kalman filter with embedded consensus filters," in *44th Conference on Decision and Control and the European Control Conference*, Seville, Spain, December 12–15, 2005, 8179–8184.

[2] R. Olfati-Saber, "Distributed Kalman filtering for sensor networks," in *46th Conference on Decision and Control*, New Orleans, Lousiana, USA, December 2007, 5492–5498.

[3] M. Kamgarpour, C. Tomlin, "Convergence properties of a decentralize Kalman filter," in *47th Conference on Decision and Control*, Cancun, Mexico, December 2008, 3205–3210.

[4] J.M. Maestre, D. Munoz de la Pena, E.F. Camacho, "Wireless sensor network analysis through a coalitional game: application to a distributed Kalman filter," in *8th IEEE International Networking, Sensing and Control*, 2011, 227–232.

[5] M. Farina, G. Ferrari-Trecate, R. Scattolini, Distributed moving horizon estimation for linear constrained systems, *IEEE Transactions on Automatic Control*, 55(11), 2462–2475, 2010.

[6] P. Millan, L. Orihuela, C. Vivas, et al., Distributed consensus-based estimation considering network induced delays and dropouts, *Automatica*, 48(10), 2726–2729, 2012.

[7] J. Nilsson, *Real-Time Control Systems with Delays*, Department of Automatic Control, Lund Institute of Technology, vol. 1049, 138 p, 1998.

[8] L. Zhang, Y. Shi, T. Chen, et al., A new method for stabilization of networked control systems with random delays, *IEEE Transactions on Automatic Control*, 50(8), 1177–1181, 2005.

[9] L. Zhang, B. Huang, J. Lam, H8 model reduction of Markovian jump linear systems, *Systems and Control Letters*, 50(2), 103–118, 2003.

[10] L. Yan, X. Zhang, Z. Zhang, Y. Yang, Distributed state estimation in sensor networks with event-triggered communication, *Nonlinear Dynamics*, 76(1), 169–181, 2014.

[11] S. Yang, Y. Bo, Output feedback stabilization of networked control systems with random delays modeled by Markov chains, *IEEE Transactions on Automatic Control*, 54(7), 1668–1674, 2009.

Section 2

Economic, Financial and Managerial Aspects of Sino-European Relations

Chapter 8

An Influence of Relative Income on the Marginal Propensity to Consume: Evidence from Shanghai

Ondřej Badura, Tomáš Wroblowský and Jin Han

Additional information is available at the end of the chapter

http://dx.doi.org/10.5772/66785

Abstract

This chapter deals with the question whether there is a relationship between the marginal propensity to consume and the status of the household in income distribution represented by a relative income. If so, then the current assumption of mainstream theory of consumption about the constant marginal propensity to consume could no longer be considered realistic and it will be necessary to take the element of relative income as a new key determinant of general consumption function. The aim of this work is to identify, describe, and prove an influence of relative income on the marginal propensity to consume using data for urban residents of Shanghai and to prove the correctness of Duesenberry's relative income hypothesis. To achieve this goal, we use a panel regression, through which the results clearly confirm the validity of the initial hypothesis about the existence of functional dependence of the marginal propensity to consume on the relative income and so it fully supports the idea of interdependent concept of utility and consumption.

Keywords: relative income, marginal propensity to consume, Duesenberry's hypothesis, interdependent utility, consumption function

JEL: D11, D12

1. Introduction

Consumption represents a key determinant of economic thought in many ways, not so much for its immense practical significance, but rather because it *de facto* represents the essence of economics itself, the essence of the issue of infinite needs and finite resources. Both in terms of microeconomics, that consumption hypotheses are always necessarily based on, and within a macroeconomic approach the widely accepted theory of consumption of mainstream economics seems to be very well formulated and developed and as such it has remained virtually unaltered for nearly 60 years. But is this theoretical concept entirely accurate and complete? Could not even here be one of the major determinants of the general consumption function omitted? Now these questions are a starting point of this chapter.

Since the 1950s of the twentieth century, the approach of permanent income theory and lifecycle hypothesis has prevailed in professional circles of economic theory. This mainstream view of the basic economic laws determining household consumption is in professional economic texts established to such an extent that the different approaches are practically not visible. However, this does not mean that there are not any alternative hypotheses of consumer behavior. We can find many critical perspectives on the standard theory of consumption, but often it is only a solution of narrowly focused issues, the pieces of a mosaic of complex alternative theory, that as a whole remains fragmented across countless of professional studies as poited out by Ackerman (1997). And if this comprehensive theory arose after all, still it was ignored for various reasons. And that is exactly also the case of Duesenberry's relative income hypothesis—consumer concept, based on the idea of interdependent utility, which has the potential with theoretical way to challenge a complete validity of the consumption theory of mainstream economics, and ultimately and primarily to significantly enrich the basic pattern of generally accepted consumption function of life cycle-permanent income hypothesis (LC-PIH)[1] (Mason, 2000).

Income and price are the key determinants of consumer choice as for mainstream economics. Relative income hypothesis, however, points out the fact that if the consumer is also affected by consumption habits of his surroundings, then the income itself must be seen in two ways: in absolute and in relative terms. From these two concepts of the basic economic determinant of general consumption function, it stems also two channels of influence on the total amount of consumption. An absolute concept of income implies a direct effect, already well known from the Keynesian consumption function. Higher disposable income will lead to a proportionately greater amount of consumer spending. A variable of disposable income then figures in the functional form of consumer equation simply as the independent variable directly explaining the level of consumption. While the relative concept of income, at least according to the principles of Duesenberry's hypothesis, implies an indirect effect. Higher disposable income will lead to higher position of household across income distribution and according to interdependent concept of utility and consumption also to a lower value of the marginal propensity to consume (MPC). The decline of MPC then, as an element transforming disposable income into consumption, negatively affects the ultimate level of consumption. The position of household in income distribution is then represented in the consumption pattern as an independent variable, which indirectly through MPC affects the level of consumption.

The problem is that while the absolute (direct) income effect is a well-known matter and the virtually undisputed, relative (indirect) income effect remains often completely ignored by professional economic communities, whether in the form of Keynesian consumption function or access of LC-PIH. It is true that every relevant and really applicable model must be extracted of elements that have not a major impact on it. However, is also

[1]A theoretical approach to consumption, based on original works: Modigliani and Brumberg (1954) and Friedman (1957). In the case of adding an element of rational expectations, then it can be primarily referred to the so-called random walk model, as defined by Hall (1978).

http://dx.doi.org/10.5772/66785

the relative income effect the insignificant element, which should be completely removed out of the consumption function without a trace? Is the interdependent concept of utility from the consumption a matter totally irrelevant? If so, then this whole work is a pointless effort.

To explore this matter, we use data from China, which has been undergoing significant structural changes recently. The shift from investment and export-oriented economy into consumption-oriented economy is one of the biggest changes. Although the consumption contribution to the country's GDP is still lower than in all developed countries, the change of government's policy (hence the whole economy) is significant. Such development turns the attention of researchers to the consumption and its determinants.

Unfortunately, only few research studies have been carried out in this field earlier. There are several studies analyzing the factors influencing very low consumption rates in China (see for example Horioka and Wan, 2007; Yang et al., 2011) or studies generally describing the consumption determinants on the macro level as Guo and Papa (2010). Many studies also focus on inequality of income distribution as a factor affecting the consumption (see Lou and Li, 2011). However, none of these studies mention the relative income as one of the possible consumption determinants. The interdependence of consumers seems to be analyzed much more by marketing specialists, see for example the studies of Zhang and Kim (2013); Yu (2014). These studies can provide useful insight into the relations among the consumers, but they cannot provide any evidence of the influence of "keeping up with the Joneses" effect on the final general consumption function.

The aim of this work is to identify, describe, and prove an influence of relative income on the marginal propensity to consume using data of urban residents of Shanghai and thus to prove the correctness of Duesenberry's hypothesis.

2. Relative income hypothesis

"Professor Duesenberry's study of the impact of budgetary and aggregative empirical consumption data on the received theory of consumer behavior is one of the most significant contributions of the postwar period to our understanding of economic behavior" written in his review by Arrow (1950, p. 906), his time respected neoclassical economist and later Nobel prize laureate in economics.

The relative income hypothesis is fundamentally built on criticism of established neoclassical preconditions for the creation of demand and Keynesian theory of consumption based on them. The main and fundamental idea with which Duesenberry (1949) comes to the field of knowledge in order to confront these established relationships of mainstream economics is a complex social concept of consumer and revision of Veblen's demonstration effect (Veblen, 1899), which the author gives a particular dimension through income distribution of households.

We can find two fundamental propositions in the work of James Duesenberry, let's say postulates, on which the theory of relative income stands and which are the basis for its further implications (Palley, 2010, p. 6):

1. "The strength of any individual's desire to increase his consumption expenditure is a function of the ratio of his expenditure to some weighted average of the expenditures of others with whom he comes into contact."

2. "The fundamental psychological postulate underlying our argument is that it is harder for a family to reduce its expenditure from a higher level than for a family to refrain from making high expenditures in the first place."

The real foundation of the new model is, however, the first claim. The author himself called this effect as *keeping up with the Joneses* or the effect of relative income. The principle is mainly simple. The consumer is not isolated from others, he lives in a world where he every day meets his friends, colleagues, family, his neighbors, and so on. And not only he meets them, especially he is confronted with their consumption. He sees what they buy, what they spend for, by what they form their standard of living, and their position in society. He sees what Veblen saw in his theory, the so-called pompous ("pointless") consumption. Unlike Veblen (1899), for the majority of the population, these consumer expenditures are not pointless, because it allows them to reach the intangible social values—a status. And that is what this is about. Our consumer shall see how people around him buy goods for their ceremonial value, before his eyes they increase the value of their status, strengthen their social position and even he does not want to be left behind. Therefore, if the consumer belongs to low-income households (his disposable income (Y_D) is under the society-wide weighted average ($\bar{Y}_D$)), then he spends more of his disposable income just to demonstrate that he can afford it, just to catch up with social status of others. His MPC is then relatively high. Conversely, high-income households[2] (whose Y_D is above the society-wide weighted average) usually already have valuable status, therefore they have not such a motivation to "catch up with someone", they do not have to spend so much of their income and vice versa they save more, simply because they can afford it. So we come to the first simple implication:

$$\mathrm{MPC}_1 > \mathrm{MPC}_2 > \ldots > \mathrm{MPC}_n \qquad (1)$$

where the higher value of the index n stands for a household with a higher value of relative disposable income (Y_{RD}), most simply expressed as:

$$Y_{\mathrm{RD}} = \frac{Y_D}{\bar{Y}_D} \qquad (2)$$

Put simply marginal propensity to consume can be written as a negative functional dependence of relative (disposable) income, as similarly shows (Palley, 2010):

$$\mathrm{MPC} = c\left(Y_{\mathrm{RD}}\right) \; 0 < c < 1; \; \grave{c} < 0 \qquad (3)$$

[2]As you can see, for simplicity, there is described a mechanism of functioning at only two types of households: high income and low income. This is however only a demonstration of the principle, which otherwise could be applied to any number of categories (social classes), as shown in Eq.(1).

http://dx.doi.org/10.5772/66785

The total amount of household consumption C is then given by the product of disposable income and the marginal propensity to consume, which is not constant now (as naively assumes the mainstream theory of consumption), but it depends on the position of the entity in the curve of income distribution:

$$C = c(Y_{\mathrm{RD}}) \cdot Y_D \tag{4}$$

Plain view on the derivation of the final rule of general consumption function, especially on the relationship between MPC and Y_{RD} (Eq.(3)), can logically evoke questions like: Is not such a general notation too trivial? Would it be possible at this point to express the dependence of marginal propensity to consume on relative disposable income in particular functional form? We find in the later part of this work that the real version of this relationship is not such a trivial matter, it depends on the number of other factors and it simply cannot be expressed in the general shape like this. There are only a number of methods by which this relationship can be approximated into particular form. One of these ways is, as we could see, the central theme of this work.

3. Methods and data

The first thing we need to realize at this point is that the marginal propensity to consume of households does not change due to the amount of disposable income, but depending on the relative disposable income, as shown in Eq. (3). This is essentially a central idea of discussed hypothesis, a key contribution to the debate on the form of consumption function. As literally written by Alvarez-Cuadrado and Long (2011, p. 1489): "For any given relative income distribution, the percentage of income saved by a family will tend to be unique, invariant, and increasing function of its percentile position in the income distribution. The percentage saved will be independent of the absolute level of income. It follows that the aggregate saving ratio will be independent of the absolute level of income." An important factor is that although the MPC and therefore also APC of households differ substantially across the Lorenz curve of distribution of disposable income (which we can figure out even with simplest common sense, but no longer with the standard theory of consumption), it is this way only because of the effect of relative income, which does not exist at the aggregate level.[3] Average propensity to consume for the whole economy is then constant in the long term, thus the relative income hypothesis is entirely consistent with the observation presented by Kuznets et al. (1946)[4] 70 years ago.

[3]The indicator may only be relative compared with another value. But an aggregate scale only shows one type of household—the "aggregate" one. Therefore, disposable income has nothing to be compare with, respectively, is equal to the average disposable income. After substituting into Eq. (2), YRD is always equal to one and whether the MPC inferred form it takes any value, it will be constant throughout the progress of consumption function. And because it is linear and based on the origin of coordinates, the average propensity to consume is also constant taking equality MPC = APC.

[4]Widely appreciated and respected study that using macroeconomic data from the US for nearly 70 years proves that even during rapid long-term growth of real income, the average propensity to consume had not virtually changed, just as the autonomous component of consumption would not exist. This discovery thus de facto entirely denies a validity of Keynesian consumer theory in the long run.

Whatever is the strength of the effect of relative income throughout income distribution in the society, the MPC for every household, more precisely a category to which the household belongs, is always given by a functional relationship due to its relative disposable income. And as it is well known that the function generates for any given situation only one result, therefore, each type of household also has only one marginal propensity to consume. Maybe the above sounds trivial and like a commonplace, but it is important to realize that the MPC of different groups of households does not change over time ceteris paribus,[5] it is independent on the absolute amount of income and so it has for each Y_D a constant value. But first and foremost, as the previous lines try to imply and how sadly Palley (2010) himself, whose model we use as a basis, forgot to mention, the above applies to types of households, to the categories to which they belong, not to individual households themselves and their individual consumption functions. This is a fundamental difference!

The biggest shortcoming of the standard model of consumption in the form of LC-PIH can therefore be seen in a constant characteristic of value of marginal propensity to consume for all kinds of income categories. To refute this erroneous assumption is then precisely the goal of the following analysis.

3.1. Methods

Let us recall at this point that the main motive of this work is to prove an influence of relative income on the value of its marginal propensity to consume, particularly by formulation of a specific form of its possible functional form. The term of relative income thus still remains the key concept for us. From the principle point of view it is *de facto* quantification and therefore the possibility of mathematical-economic interpretation of the issue of household's position in the distribution of disposable income. From the definitional point of view, it is a ratio of disposable income to the society-wide weighted average, as shown in Eq. (2). Now we have only left to specify precisely the variable of $\bar{Y}_D$. From the perspective of the principles of relative income hypothesis, it seems to be the best solution to set the weights as the average numbers of household members in the given income category, which would epitomize the best a frequency of individual income cases in society. However, due to limited data source we have to settle for determining variable $\bar{Y}_D$ as the simple arithmetic average of disposable incomes for considered income categories. Therefore, this point can be considered as a necessary simplification given by the availability of empirical data and potentially a weaker place of the following analysis, but not weak enough to make it impossible to achieve the stated objective.

For actual try of expressing a specific form of assumed functional dependence, we use a regression analysis by estimation of regression coefficients using the least squares method. Due to the nature of the input data, in particular the limited number of statistically measured income categories (small number of observations), the classical regression could lead to distorted results, therefore, will use panel regression.

[5]Changing values of MPC in time could in our case characterize only one thing, a change in the distribution of disposable income, thus de facto enlargement or reduction of income inequality.

http://dx.doi.org/10.5772/66785

The general formula of the required univariate linear regression model depends on whether we use a panel regression method for fixed or random effects. Which of these panel regression methods is more suitable for expression of wanted dependency will be shown up by Hausman's test at a later stage of the analysis, so it is necessary now to still consider both the options. In the case of using fixed effects the regression equation is given by:

$$\mathrm{MPC}_{i,t} = \alpha_i + \beta \cdot Y_{\mathrm{RD}_{i,t}} + u_{i,t} \tag{5}$$

where $\mathrm{MPC}_{i,t}$ is the marginal propensity to consume for the category i at time t, α_i is the level constant (an intercept) for the ith income category, the product of $Y_{\mathrm{RD}_{i,t}}$ is relative disposable income for the ith category at the time t and the regression coefficient β expressing the sensitivity of the marginal propensity to consume to the relative disposable income. Variable $u_{i,t}$ symbolizes the random component. In a more detailed breakdown, the level constant α_i for each category is divided into two subfolders, where:

$$\alpha_i = \beta_0 + \gamma_i \tag{6}$$

where β_0 is the basic level constant to which it applies $\beta_0 = \alpha_1$. A constant γ_i is then an added fixed impact for given income category for $i \in \{2; \ldots; I\}$, where I is the number of categories. By simply rewriting α_i according to Eq. (6) we get new more detailed form of the general expression of wanted regression equation using fixed effects:

$$\mathrm{MPC}_{i,t} = \beta_0 + \gamma_i + \beta \cdot Y_{\mathrm{RD}_{i,t}} + u_{i,t} \tag{7}$$

Since in the case of using the fixed effects method (for a given entity) we subsequently need also to verify the appropriateness using time fixed effects, it is necessary to consider other, 1 order of magnitude more detailed breakdown of level constants α_i, which could now be broken down to the given shape:

$$\alpha_i = \beta_0 + \gamma_i + \tau_t \tag{8}$$

where for newly level constant it applies the condition $\beta_0 = \alpha_1$ only for $t = 1$ and where τ_t is an added fixed impact due to the time period for $t \in \{2; \ldots; T\}$, where T stands for the number of such time periods. Moreover, by a new rewriting α_i in Eq.(5) we can write down a general expression of wanted regression equation using fixed effects for given categories and time:

$$\mathrm{MPC}_{i,t} = \beta_0 + \gamma_i + \tau_t + \beta \cdot Y_{\mathrm{RD}_{i,t}} + u_{i,t} \tag{9}$$

For regression estimation based on random effects the wanted relationship is characterized more simply and clearly in the form:

$$\mathrm{MPC}_{i,t} = \alpha + \beta \cdot Y_{\mathrm{RD}_{i,t}} + u_{i,t} + \varepsilon_{i,t} \tag{10}$$

Where newly α represents a level constant for all categories, $u_{i,t}$ is a random component between categories and $\varepsilon_{i,t}$ is a random component within an income category.

Either way, an important prerequisite of any possible resulting variations of panel regression is a negative value of the coefficient β, because according the principles of Duesenberry's hypothesis with increasing relative disposable income the marginal propensity to consume must necessarily decline, as demonstrated by Eq. (3).

3.2. Data

The prerequisite of negative linear dependency of MPC on Y_{RD} is tested here using the example of data for the budgetary situation of urban households in China's Shanghai; therefore, all the input data for the aforementioned analysis were taken from the database of the Shanghai Municipal Statistics Bureau (2016). The original input data are annual statistics between the years 2000 and 2014, which resulted in essentially two time series, which are further divided into five another subfolders. Followed 15 observations are then basically written in two variables:

Y_D = average nominal disposable income of household per capita in CNY,

C = average nominal consumption of household per capita in CNY.

As can be seen, we work with the mean values per person. For better demonstration of the validity of Duesenberry's hypothesis, this procedure is certainly preferable. An important finding is also mentioned in the secondary division of basic variables. Indicators Y_D and C are both equally divided into five other subfolders reflecting income and consumption situation of different types of households arranged ascendingly by quintiles of disposable income. Finally, we register 10 input time series here, divided into five panels by the types of income categories. Indicators directly entering the subsequent panel regressions are Y_{RD} calculated according to Eq. (2) and APC expressed by formula:

$$\text{APC} = \frac{C}{Y_D} \tag{11}$$

It is then necessary at this point to realize that we work with income categories here (not with individual households), for which the value of APC is independent on Y_D and in the absence of an intercept it is at any point equal to MPC. That is why we could use this simple equivalence, where MPC values are substituted by the average propensity to consume. In conclusion, we note that although the original input data in this study are nominal expression of consumption and disposable income, but due to the relative nature of indicators MPC and Y_{RD} the unwanted effect of changes in the price level is to be fully canceled out anyway.

4. Results

Chart 1 is used for preliminary visual assessment of the expected dependence. Although the linear dependence of both followed quantities is quite obvious at this point, only a graphical

http://dx.doi.org/10.5772/66785

analysis is obviously not enough for us. The aim here is to mathematically approximate this relationship by the regression equation.

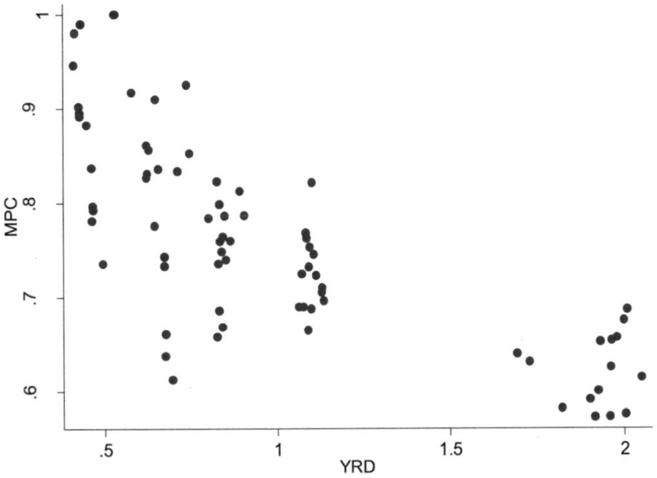

Chart 1. Visual assessment of linear dependence of MPC and Y_{RD}. *Source*: own calculations and processing in Stata 12.

Before it can be proceeded to the actual final estimate of regression parameters of mentioned dependency, it is necessary for the panel nature of the data to decide whether it should be used as a method of fixed or random effects, in other words, whether there are differences significant enough in the wanted functional relationship between the categories that they must be captured in a separate level constant just for each category. This dilemma is unambiguously solved by executed Hausman's test when its results indicate that a suitable panel regression in this case is in the method of random effects, at least at the 5% significance level, which we also use for further analysis.

The results of the final panel regression using random effects are summarized in **Tables 1** and **2**. In this final estimation of the desired functional form, we use a robust method of estimation of standard error using White's estimator, thereby the model was protected against a possible autocorrelation and heteroskedasticity. An important finding is that considering the inclusion of only one explanatory variable, a relatively high value of the coefficient of determination was achieved, which indicates that approximately 57% of the variability of MPC was explained just by Y_{RD}. This fact then clearly confirms the main initial assumption about the influence of relative disposable income on the marginal propensity to consume.

There is no doubt that the model as a whole is statistically significant, as well as the regression coefficient and level constant. The key element—the wanted regression coefficient β

achieves exactly according to our expectations a negative value, which cannot be influenced nor by potential standard error. The resulting model corresponds to an initial economic theory and predicts that a change in the relative disposable income by 0.1 also changes the value of the marginal propensity to consume of any income category in the opposite direction by 0.0155.

Number of observations	**75**
$F(10, 149)$	36.69
Prob > F	0.000
R^2	0.575

Source: own calculations and processing in Stata 12.

Table 1. Estimation of Eq. (10) using panel regression with random effects, part 1.

MPC	**Coefficient**	**Robust Std. Err.**	*z*	*P*>|*z*|	**[95% Conf. Interval]**
β	−0.155	0.026	−6.06	0.000	−.206, −.105
α	0.912	0.037	24.75	0.000	.84, .984

Source: own calculations and processing in Stata 12.

Table 2. Estimation of Eq. (10) using panel regression with random effects, part 2.

In conclusion, let us emphasize that the result of Hausman's test significantly influenced (and positively) the very predictive ability of the resulting model. The final use of the random effects method means that the regression relationship between the MPC and Y_{RD} can be expressed in a fully general way and elegantly by only one equation (which could not be possible using fixed effects) and therefore it does not depend on what income category we are situated. The final functional dependence of the marginal propensity to consume on relative disposable income has then a following form:

$$\mathrm{MPC}_{i,t} = 0.912 - 0.155 \cdot Y_{\mathrm{RD}_{i,t}} + u_{i,t} + \varepsilon_{i,t} \quad (12)$$

5. Conclusion

The primary goal of this work was to find and prove an influence of relative (disposable) income on the value of marginal propensity to consume. To achieve this goal, we have used primarily a panel regression for data from the Chinese province of Shanghai. There is no doubt that relative income affects the marginal propensity to consume, which concurrently means that validity of "keeping up with the Joneses" effect ("keeping up with the Wangs" as we say in the context of China) is finally proved. And as indicated by the relatively high value

http://dx.doi.org/10.5772/66785

of the coefficient of determination (relative to one explanatory variable), this dependence must become a new key factor of the general consumption function.

The mainstream theory of consumption, mainly represented by the concept of LC-PIH, assumes a constant value of MPC for all types of income categories. However, as it is shown by the results of our study, this assumption can no longer be considered realistic. Marginal propensity to consume remains unchanged in relation to disposable income only for a given income category, not for individual households. If the income situation of household changes, it will shift to the new income category and at the same time it will fix the new value of MPC. Household consumption function then does not have a constant slope (opposed to the consumption function of income categories) as mistakenly assumed by the mainstream theory of consumption, but it is under a concave characteristic. This is occurring due to the effect of relative income, it is appropriate at this point to emphasize again that the mainstream microeconomics distinguishes only between the income and substitution effect. Duesenberry's theory, as well as the conclusions of this study, requires to add further subdivision and so to distinguish between the income effect of absolute (direct) and relative (indirect).

Although the impact of relative income on the marginal propensity to consume was unequivocally confirmed, the issue of its precise nature still remains open. Approximation of followed dependency, of course, depends on a functional form, which is used for it, and here utilized linear function is certainly not the only option. Moreover, it may not even be the most appropriate. It is important to realize that, at least in terms of statistics, there is not only one correct and objective functional form, this is only just what we define it. And the definition of a new, elegant and more convenient functional relationship of MPC and Y_{RD} better and more accurately describing consumer behavior of households so remains the motive for further scientific research.

Acknowledgements

This chapter was supported by a grant from Students Grant Project EkF, VŠB—TU Ostrava within the project SP2016/112 and within Operational Programme Education for Competitiveness—Project No. CZ.1.07/2.3.00/20.0296.

Author details

Ondřej Badura[1*], Tomáš Wroblowský[1] and Jin Han[2]

*Address all correspondence to: ondrej.badura@vsb.cz

1 VŠB-Technical University Ostrava, Department of Economics, Sokolská třída Ostrava, Czech Republic

2 Shijiazhuang University of Economics, Shijiazhuang, Hebei, China

References

Ackerman, F. (1997). Consumed in Theory: Alternative Perspectives on the Economics of Consumption. *Journal of Economics Issues*. 31(3): 651–664. ISSN 0021-3624.

Alvarez-Cuadrado, F. and N. V. Long. (2011). The Relative Income Hypothesis. *Journal of Economic Dynamics and Control*. 35: 1489–1501. DOI: 10.1016/j.jedc.2011.03.012.

Arrow, K. J. (1950). Income, Saving and the Theory of Consumption Behavior. By James S. Duesenberry (book review). *American Economic Review*. 40(5): 906–911. ISSN 0002-8282.

Duesenberry, J. S. (1949). *Income, Saving, and the Theory of Consumer Behavior*. Cambridge: Harvard University Press. ISBN 978-0674447509.

Friedman, M. (1957). *A Theory of the Consumption Function*. New York: Princeton University Press. ISBN 0-691-04182-2.

Guo, K. and N. D. Papa (2010). Determinants of China's Private Consumption: An International Perspective. IMF Working Paper, WP/10/93.

Hall, R. E. (1978). Stochastic Implication of the Life Cycle – Permanent Income Hypothesis: Theory and Evidence. *Journal of Political Economy*. 86(6): 971–987. ISSN 0022-3808.

Horioka, C. Y. and J. Wan (2007). The Determinants of Household Saving in China: A Dynamic Panel Analysis of Provincial Data. *Journal of Money, Credit and Banking*. 39(8): 2077–2096.

Kuznets, S., L. Epstein and E. Jenks (1946). *National Product Since 1869*. New York: National Bureau of Economics Research. ISBN 0-87014-045-0.

Mason, R. (2000). The Social Significance of Consumption: James Duesenberry's Contribution to Consumer Theory. *Journal of Economic Issues*. 34(3): 553–572. ISSN 0021-3624.

Modigliani, F. and R. Brumberg (1954). Utility Analysis and the Consumption Function: An Attempt and Integration. In: Kurihara, Kenneth (ed.), *Post-Keynesian Economics*. New Brunswick: Rutgers University Press. ISBN 978-0415607896.

Palley, T. I. (2010). The Relative Income Theory of Consumption: A Synthetic Keynes-Duesenberry-Friedman Model. *Review of Political Economy*. 22(1): 41–56. DOI: 10.1080/09538250903391954.

Shanghai Municipal Statistics Bureau (2016). Shanghai: SMSB. [20. 3. 2016]. Available at: http://tjj.sh.gov.cn/

Veblen, T. (1899). *The Theory of the Leisure Class: An Economic Study of Institutions*. New York: MacMillan.

Lou, F. and X. Li, (2011). An Empirical Analysis of Income Disparity and Consumption in China. *Frontiers of Economics in China*. 6(1): 157–170.

http://dx.doi.org/10.5772/66785

Yang, D. T., J. Zhang, et al. (2011). Why Are Saving Rates So High in China? NBER Working Paper. No. 16771, Cambridge: National Bureau of Economic Research.

Yu, D. (2014). Motivations of Luxury Consumption in America vs. China. Graduate Theses and Dissertations. Iowa State University, Ames, Iowa, USA, Paper 13854.

Zhang, B. and J. Kim (2013). Luxury Fashion Consumption in China: Factors Affecting Attitude and Purchase Intent. *Journal of Retailing and Consumer Services*. 20(1): 68–79.

Chapter 9

The Quality Perceived by the Young Customer Versus Coca Cola Zero Advertisement

Pavel Blecharz and Hana Stverkova

Additional information is available at the end of the chapter

http://dx.doi.org/10.5772/66786

Abstract

Nowadays, the advertisement is one of the key activities of sale. An advertisement is not always true and sometimes it could be even misleading or false. This chapter is focused on the exemplary verification of the truth of the concrete advertisement. The verification has been provided based on the data coming from the research being provided among 300 young customers from the Czech Republic. The data have been processed statistically, and based on the hypothesis testing using one-way analysis of variance (ANOVA), the data are evaluated whether the advertisement is true or misleading. Research findings provide testimony of unfair practice of advertisement.

Keywords: advertisement, customer, unfair practices, quality perceived by the customer

1. Introduction

In these days, the struggle for a customer belongs to the daily company activities. The companies compete among each other in the field of the quality, price, payment methods, sales, and in many other ways. One of these fields, where the companies try to win customers, is the advertisement. The customer is bombed by the producers' advertisement and services providers in newspapers, magazines, TV, and in other media. Due to recent IT development, the effective marketing communication is via social networking site, more has been given by Shen et al. (2016) and Dehghani et al. (2016). The advertisement producers invent more or less perfect "stories" in order to change the customer's mind and to get him to their side. View more by El Ouardighi et al. (2016), Dehghani et al. (2016), Shen et al. (2016) and Chang (2009).

Is the content of advertisements true or the customer is only being misled? It is not a simple answer to this question. Of course, like everywhere, both will be true and it will depend on the concrete case. Following text deals with the analysis of one of such advertisements in more detail and it tries to show in a sophisticated way whether the advertisement is true or not. Dehghani et al. (2016) results show that entertainment, informativeness and customization are the strongest positive drivers. On the other hand, the advertising value affects both brand awareness and purchase intention of consumers accordingly.

One of the Coca Cola Company advertisements makes efforts to persuade customers that they would not recognize the taste difference between a Coca Cola and Coca Cola Zero. Therefore, the authors of this chapter have performed the testing of taste being perceived by the customer who perceives it (the quality perceived by the customer). The test results of the taste have become the subject of statistical elaboration aiming to approve or disprove whether the customers perceive the taste of Coca Cola and Coca Cola Zero in the same way. Thus the customer does not recognize the difference between both products, which may result in customer dissatisfaction. Another customer dissatisfaction has been given by Zeelenberg and Pieters (2004).

2. Theoretical bases

The taste of products has been evaluated by the customer verbally in analysis, and the verbal evaluation is transformed into the scoring (Blecharz and Stverkova, 2011; Blecharz, 2015). As for an easy understanding for the customer, five variants of answers have been chosen to make evaluation easier. Evaluation of the taste can be as follows:

- Excellent
- Rather good
- Average
- Rather bad
- Absolutely unsatisfactory

To each answer a score belongs, when "excellent" means 5 points, "rather good" is rated 4 points, ... to "absolutely unsatisfactory" rated with 1 point. The results are elaborated by simple statistics, i.e., the arithmetic mean is calculated. The arithmetic mean answers the question; which taste is perceived better.

However, to make it clear that there is *really* a difference in taste perception, it is not enough only to compare the arithmetic means. It is necessary to compare all the data, or the Coca Cola group data and the data for the Coca Cola Zero group. This can help to determine whether the difference in the taste perception is statistically significant or whether it is only the interference that is causing the difference in results of means.

For this purpose, testing the hypothesis by the ANOVA method is used, same as McDougall et al. (1994). There are two hypotheses determined: the zero and the alternative hypotheses

http://dx.doi.org/10.5772/66786

H_0: There is no difference between the two sets of data.

H_A: There is a difference between the two sets of data.

The hypotheses are evaluated by using the ANOVA method (analysis of variance). In this case, that difference in taste perception really exist, the so-called single-factor ANOVA is considered involving two groups of data (2 variability sources). Different marking of variables for the mathematical calculation are determined (processed by Ross (1989) and Roy (2001))[1]:

A = the examined factor (Coca Cola product),

A_1 = the first factor level (i.e., the first product group)

A_2 = the second factor level (i.e., the second product group)

A_i = the sum of results for the A_i-level,

$\bar{A}_i$= the average of results at the A_i, *tj.* A_i/n_{Ai}-level,

T = the sum of all results/surveying,

$\overline{T}$= the average of all results/surveying,

$_{Ai}$ = the number of surveying for the A_i - level,

N = the total number of surveying,

Y = the measured (determined) value,

S_T = the total sum of squares,

S_A = the sum of squares for the factor,

f = the number of degrees of freedom (= number of results -1),

C.F. = correction factor (T^2/N),

S_e = the sum of squares for the error,

V_i = variance,

F = F statistics used for F-test.

For a better understanding of the calculation, it is appropriate to demonstrate a simple example instead of formula listing.

Example 1

The task is to find out whether two variants of products differ by taste. **Table 1** shows respondents' evaluation.

---

[1]There is an alternative marking of variables by various authors, e.g., S_T as SS_T, S_A as SS_A, f as v, V as MS, etc.

A_1 (the first data group)	A_2 (the second data group)
5	3
3	3
4	5

Table 1. The data for the calculation model.

The process involves three steps:

a. the hypotheses H_0 and H_A are determined,

b. the single-factor ANOVA method is calculated, where the factor is of 2 levels,

c. the F-test is provided and the conclusion regarding the validity of hypotheses is stated.

a. The hypotheses:

H_0: There is no significant difference in the product taste perception.

H_A: The difference in the products taste perception is significant.

b. The ANOVA method (the calculation):

$\text{C.F.} = T^2/N = [(Y_1 + Y_2 + \ldots]^2/N = [(5 + 3 + \ldots + 4]^2/6 = 80.67$

$S_T = (5)^2 + (3)^2 + \ldots + (4)^2 - \text{C.F.} = 84 - 80.67 = 3.33$

$S_A = (A_1)^2/n_{A1}+(A_2)^2/n_{A2} - \text{C.F.} = (5 + 3 + 4)^2/3 + (3 + 3 + 4)^2/3 = 48 + 33.33 - \text{C.F.} = 81.33 - 80.67 = 0.663$

(*Note*: Remember that A_1 is the sum of results of factor A at the first level, n_{A1} is the number of those results.)

$S_e = S_T - S_A = 3.33 - 0.663 = 2.667$

$f_T = 6 - 1 = 5$

$f_A = 2 - 1 = 1$

$f_e = 5 - 1 = 4$

$V_A = S_A/f_A = 0.663/1 = 0.663$

$V_e = S_e/f_e = 2.667/4 = 0.667$

$F_A = V_A/V_e = 0.663/0.667 = 0.994$

$F_e = V_e/V_e = 1/1 = 1$ (not in the ANOVA table)

The results are sorted to the standard tabular format—the ANOVA table (see **Table 2**).

(a) The hypotheses testing

When testing hypotheses, the F values are compared. If the calculated F value is greater than the tabular (critical) one ($F_{\text{cal}} > F_{\text{crit}}$), the alternative hypothesis is preferred.

Variability source	f	S	V	F
factor (*A*)	1	0.663	0.663	0.994
Error (*e*)	4	2.667	0.667	
Total (*T*)	5	3.33		

Table 2. The ANOVA method for 1 factor, 2 levels.

Table 2 shows that the degree of freedom of the numerator for *F* is 1 (factor *A* = the row, *f* = the column) and the degree of freedom of the denominator is 4 (*e* = the row, *f* = the column).

The tabular (critical) *F* for 95% of the coefficient the reliability (CL), or 5% of the value alpha[2] significance level, is $F_{0.05}$ (14) = 7.7. See **Table 3** where the searched *F* is marked bold.

f_1f_2	1	2	3	4	5	6	7	8	9	10
1	161.45	199.50	215.71	224.58	230.16	233.99	236.77	238.88	240.54	241.88
2	18.513	19.000	19.614	19.247	19.296	19.330	19.353	19.371	19.385	19.396
3	10.128	9.5521	9.2766	9.1172	9.0135	8.9406	8.8868	8.8452	8.8123	8.7855
4	**7.7086**	6.9443	6.5914	6.3883	6.2560	6.1631	6.0942	6.0410	5.9988	5.9644
5	6.6079	5.7861	5.4095	5.1922	5.0503	4.9503	4.8759	4.8183	4.7725	4.7351
6	5.9874	5.1433	4.7571	4.5337	4.3874	4.2830	4.2066	4.1468	4.0990	4.0600
7	5.5914	4.7374	4.3468	4.1203	3.9715	3.8660	3.7870	3.7257	3.6767	3.6365
8	5.3177	4.4590	4.0662	3.8378	3.6875	3.5806	3.5005	3.4381	3.3881	3.3472
9	5.1174	4.2565	3.8626	3.6331	3.4817	3.3738	3.2927	3.2296	3.1789	3.1373
10	4.9646	4.1028	3.7083	3.4780	3.3258	3.2172	3.1355	3.0717	3.0204	2.9782
11	4.8443	3.9823	3.5874	3.3567	3.2039	3.0946	3.0123	2.9480	2.8962	2.8536
12	4.7472	3.8853	3.4903	3.2592	3.1059	2.9961	2.9134	2.8486	2.7964	2.7534
13	4.6672	3.8056	3.4105	3.1791	3.0254	2.9153	2.8321	2.7669	2.7144	2.6710
etc.										

Table 3. The table $F_{0.05}$ (f_1, f_2), 95% reliability.

The calculated *F* value is smaller than the tabular (critical) *F* value, i.e., $F_{cal} = 0.994 < F_{crit} = 7.7086$, therefore is *the zero hypothesis* is valid, *i.e., there is no difference in the taste perception.*

If the calculated *F* value would be greater than the tabular (critical) *F* value, i.e., ($F_{cal} > F_{crit.}$), the alternative hypothesis would be valid.

In the case study performed the ANOVA calculations are provided only by the Microsoft Excel 2013.

The symbols being used in MS Excel 2013 look as follows:

The sum of squares: SS,

[2]Not: the reliability is mostly (=CL) 95%, i.e., the significance level alpha 5%.

The dispersion: MS,

The degrees of freedom: in Excel signed as "difference" (better to change into "f"),

The *P* value: the probability,

The error...called as "all selections",

The factor: called as "inter elections".

Contrary to the previous manual calculation *P* value is mentioned additionally. The *P* value is the probability whether the monitored value difference between the particular sets of products data is given due to the interference (the sum of many random effects) than by the factor (product) (see **Table 4**).

The *P* value is the second possibility to make decision whether the zero/alternative hypothesis would be accepted or not.

The ANOVA

Variability source	*SS*	*Difference*	*MS*	*F*	*P value*	*F crit*
Inter selections	0.666667	1	0.666667	1	0.373901	7.708647
All selections	2.666667	4	0.666667			
Total	3.333333	5				

Table 4. The calculation using MS Excel.

- ***P* value < 5%** - the zero hypothesis H_0 **would be rejected**, i.e., everything is given by interference and the alternative hypothesis H_A would be accepted (=the factor impact exists).
- ***P* value ≥ 5%** - the zero hypothesis H_0 would be accepted[3] and the factor impact is considered unsubstantiated.

In the example, the value is $P = 0.3739$ or 37.39% and therefore the zero hypothesis would be accepted (also see the *F* comparison).

3. The case study: Coca Cola taste and Coca Cola Zero taste

In 2015, the Coca Cola Company introduced an advertisement in Czech Republic. The advertisement showed people drinking Coca Cola Zero but the product labels were hidden and they were thinking they were drinking Coca Cola. That is why the authors decided to provide real research to find out whether the customers would really not recognize the difference in the taste.

[3] Wonnacot and Wonnacot (1992) recommends to use the term "we accept the hypothesis", than the term "we do not reject the hypothesis".

http://dx.doi.org/10.5772/66786

In other words, they did not recognize any difference in the taste between both the soft drinks. In case, the customers would not recognize the difference in taste the advertisement is true. If there would be identified any difference in the taste perception, then the advertisement is misleading. The research was performed in November 2015. It was addressed to an overall 300 customers of various places in the Czech Republic. The selection of respondents had been performed randomly. Each respondent was tasting about 0.05 liter of the soft drink, cooled to the suitable temperature, and the soft drinks were labeled A_1 and A_2. For purposes of the research the sign A_1 has represented the Coca Cola and the sign A_2 represented the Coca Cola Zero.

They were defined with these hypotheses:

H_0: There is no difference in taste perception of Coca Cola and Coca Cola Zero.

H_A: There is a significant difference in taste perception of Coca Cola and Coca Cola Zero.

Now, let us look at the results.

The arithmetic mean of all 300 results of taste evaluation:

Coca Cola: the average = 3.68

Coca Cola Zero: the average = 3.03

The arithmetic mean shows that the taste of sweet Coca Cola is perceived better than the taste of Coca Cola Zero. Is this difference statistically significant? It means—which hypothesis is valid—the zero or alternative? We will do the assessment based on **Table 5**. **Table 5** shows the results.

To calculated ANOVA, the coefficient the reliability 0.95 (alpha = 0.05) was chosen. **Table 5** shows that the critical F is lower than the calculated F ($F_{cal} = 47.43 > F_{crit} = 3.86$). And, at the same time, the P value is practically equals nearly zero. Based on these results such conclusion can be stated that the alternative hypothesis is valid, i.e., that the customers definitely distinguish the taste of both soft drinks and that they prefer the taste of the classic sweet Coca Cola.

Someone could claim that it can be a statistical discrepancy (so-called I. sort of error), i.e., there exist the probability that the zero hypothesis would be rejected wrongly in 5% of cases even it is true. However, these doubts can be very easily disproved by determining the limited alpha value for our case. The alpha value is determined by MS Excel where the F values calculated and critical are equal. Alpha value is 1.4502×10^{-11} (!).

Variability source	SS	Difference	MS	F	P value	F crit
Inter selections	62.72667	1	62.72667	47.42487	1.45E–11	3.857056
All selections	790.9467	598	1.322653			
Total	853.6733	599				

Table 5. The ANOVA: the taste perception of Coca Cola and Coca Cola Zero.

So, it can be said with 100% (99.99999999999%) probability that the taste of Coca Cola and Coca Cola Zero perceived by the customer is different. The Coca Cola advertisement is untrue and misleading.

4. Conclusions

Using the analogical way, it would be possible to evaluate the truth of other advertisements also with other producers. According to the large majority of the supply over demand, the producers try to persuade the customers to purchase the product often by using an untrue, misleading, or false advertisements.

There is a question where is the limit between the truth and untruth, the limit between the fair and unfair practices. Some advertisements are based on the obvious exaggeration; e.g., drinking a glass of Red Bull drink we would get wings. It would be here all right, the advertisement does not say the truth, but it says with an exaggeration that we will be fresh "like we would have the wings".

Nevertheless, in the situation, when the advertisement argues that the product has a real feature, it cannot appear to be the untrue information. If it will be in the advertisement mentioned that the butter has 82% of fat and in fact it has only 70%, perhaps everyone will say that this is a false advertisement. If in this advertisement, if it appear that the information that you will not recognize the difference in the taste of two products and in fact it will be different, this will be the false advertisement again.

The untrue information in the advertisement presents the unfair practice of the "consumer's deception". The advertisements containing the untrue information should be forbidden and strictly penalized everywhere in the world. Only in this way will the consumer protection make sense.

Author details

Pavel Blecharz and Hana Stverkova*

*Address all correspondence to: hana.stverkova@vsb.cz

Department of Business Administration, Faculty of Economics, VSB-TU Ostrava, Ostrava, The Czech Republic

References

Blecharz, P., 2015. *Kvalita a zakaznik*. Praha: Ekopress.

Blecharz, P., Stverkova, H., 2011. Product quality and customer benefit. *Communications in Computer and Information Science*, Vol. 208, pp. 382–388. DOI: 10.1007/978-3-642-23023-3_58

Chang, Young Il, 2009. Understanding internet banking in China focused on process quality, outcome quality, customer satisfaction, reuse and word of mouth. *Journal of Information Technology Applications & Management*, Vol. 16, Iss. 3, pp. 45–58.

Dehghani, M., Niaki, M.K., Ramezani, I., Sali, R., 2016. Evaluating the influence of YouTube advertising for attraction of young customers. *Computers in Human Behavior*, Vol. 59, pp. 165–172. DOI: 10.1016/j.chb.2016.01.037

El Ouardighi, F., Feichtinger, G., Grass, D., Hartl, R., Kort, P.M., 2016. Autonomous and advertising-dependent 'word of mouth' under costly dynamic pricing. *European Journal of Operational Research*, Vol. 251, Iss. 3, pp. 860–872. DOI: 10.1016/j.ejor.2015.11.035

McDougall, P.P., Covin, J.G., Robinson, R.B., Herron, L., 1994. The effects of industry growth and strategic breadth on new venture performance and strategy content. *Strategic Management Journal*, Vol. 15, Iss. 7, pp. 537–554. DOI: 10.1002/smj.4250150704

Ross, P.J., 1989. *Taguchi Techniques for Quality Engineering*. Singapore: McGraw Hill Book Company.

Roy, R.K., 2001. *Design of Experiments Using the Taguchi Approach*. New York: John Wiley&Sons.

Wonnacot and Wonnacot' (1992). *Introductory Statistics for Business and Economics*. Praha: Victoria Publishing.

Shen, G.C.C., Chiou, J.S., Hsiao, C.H., Wang, C.H., Li, H.N., 2016. Effective marketing communication via social networking site: the moderating role of the social tie. *Journal of Business Research*, Vol. 69, Iss. 6, pp. 2265–2270. DOI:10.1016/j.jbusres.2015.12.040

Zeelenberg, M., Pieters, R., 2004. Beyond valence in customer dissatisfaction: a review and new findings on behavioral responses to regret and disappointment in failed services. *Journal of Business Research*, Vol. 57, Iss. 4, pp. 445–455. DOI: 10.1016/S0148-2963(02)00278-3

Empirical Study on the Financial Development to Promote the Urbanization Process in China: A Case of Hubei Province

Fangchun Peng and Yu Lu

Additional information is available at the end of the chapter

http://dx.doi.org/10.5772/66787

Abstract

As the central province, Hubei is a typical area of urbanization in China. Analysis of Hubei economic development has a certain representation in the process of urbanization. Since 1990s, research on 25 years of data and practices in Hubei shows that three major finance industries in Hubei, including banking, securities, and insurance, have played an important role in the process of urbanization. In the aspects of agricultural industries, urban construction, and industrial structure, the development of finance has promoted the process in agricultural industries, population urbanization, and industrialization.

Keywords: financial development, urbanization, statistical description, empirical study

1. Introduction

1.1. Significance of the topic

China is actively promoting the reform and opening-up policy and modernization, which including urbanization, information, new-type industrialization, and agricultural modernization. As one of the new "four modernization," urbanization has become an important and indispensable component of China's socialist modernization process. For this purpose, China issued the "National plan on New urbanization", which pointed out that China's urbanization rate of the population should reach 60% by 2020. The urbanization rate in 2014 was 54.77%, and the annual urbanization rate should be increased by 1 percentage point or so in future. In fact, China's urbanization rate presented a slowing trend, so it is difficult to accomplish the

No.E44: Financial Market and the Macro economy

urbanization index of "Thirteen Five Plan". Hubei province being as an important economic hub of the Yangtze River and as a more developed economic province of China's central area, which plays an important role in promoting the urbanization construction process.

Financial development is a very important driving force in the process of urbanization, especially for financial institutions that can provide the financial support to the process of urbanization. Meanwhile, urbanization can promote the financial system to be more perfect in return, so there is a close relationship between financial development and urbanization. Therefore, the study of financial development based on Hubei province has a great practical significance in promoting the process of urbanization.

1.2. Domestic and foreign research literature review

1.2.1. Foreign research

A famous American scholar Kuznets [1] described urbanization as a process of migration from rural to urban. Goldsmith [2] proposed the concepts of financial structure and financial interrelations ratio (FIR). Stopher [3] believed that financial development can provide strong financial support to the development of land resources in the process of urbanization.

1.2.2. Domestic research

Yinli [4] studied the relationship between the process in the financial industry and urbanization, and she thought that there was reciprocal causation of each other, the latter can be driven by the deep development of the former, and the latter also can be further developed to strengthen the former. Guo Jiang Shang [5] found that both the development of the financial industry and the process of industrialization can promote the process of urbanization, meanwhile development of the financial sector would promote the process of industrialization, and the financial development of urbanization utility is greater than utility of the industrialization.

A study on the interaction between financial development and urbanization of an area is done. Zhiwei [6] uses the panel data of 17 prefecture level cities in Henan Province as an example to study the influences of financial development on urbanization in the period 2001–2012, and he found that financial development significantly inhibited the advance of urban employment, but it significantly promoted the household population urbanization process in the short term. So it is not obvious for it to support the lifestyle urbanization and urban construction of urbanization.

1.2.3. Briefly reviewed

Both foreign and domestic scholars believe that there was a reciprocal causal relationship between financial development and urbanization, but in the actual situation of a certain region of our country, on account of the degree of economic development and urbanization in each region are not identical, and the data selected to be studied are also different, so the conclusion will also have differences. Viewed from present research current situation, whether financial development or urbanization development has a more mature theoretical system, but most of the research is done at the national level, and the current analysis of the Hubei

http://dx.doi.org/10.5772/66787

area alone is relatively scarce. Thus the author takes local areas to study the specific relationship between them, in order to make up for some deficiencies of the research field.

2. Research design and methods

The paper with the following three methods helps to expand research in three steps:

1. *Statistical analysis method*: to analyze development and urbanization in Hubei financial key indicators, analyze the banking, securities, and insurance industry developments and financial indicators of FIR reflecting the financial situation, to use a population urbanization rate reflecting the process of urbanization.

2. *Comparative analysis method*: to compare the differences of urbanization process of Hubei and developed identity (Zhejiang) and entire country, to discuss the lagged problems of urbanization in Hubei.

3. *Production function analysis method*: to study the situation of financial development promoting the process of urbanization in Hubei, including three aspects: first, agricultural loans reflecting that financial support the urbanization of agricultural and industrialization level; second, infrastructure investment reflecting the financial support of urbanization on urban construction level; third, financial investment in the second and third industries reflecting the urbanization of the industrial structure optimization level.

3. The general statistical analysis of financial development and urbanization process in Hubei

3.1. Financial development in Hubei

Since the establishment of the financial markets in 1990, China's financial industry has become the core strength of the national economy. This paper respectively selects the data of 25 years from 1990 to 2014, analyzing the current situation of Hubei's financial development in three aspects, including banking, securities, and insurance industry.

3.1.1. Banking

Overall, the total deposits and loans of banking institutions in Hubei show a rapid upward trend, from 406.56 billion and 732.77 billion in 1990 to 36494.82 billion and 25289.82 billion in 2014, both have a substantial growth. GDP growth is particularly rapid. From **Figure 1** bar chart, the FIR rose from 0.4932 to 1.3335 in nearly 25 years. There appears to be slight fluctuations between 2007 and 2008 due to the impact of the financial crisis, but the trend is upward, which indicates that there is progressive development of the financial market in Hubei. This paper uses the FIR which was proposed by DE Smith to reflect the ratio between one region that comprised values of financial assets and the country's or the region's total economic activity; in this way, it judges the scale of financial market and the ability to absorb the savings in this region. In this paper, the financial related rate is concluded by calculating the proportion of deposit balance in GDP (**Figure 1**).

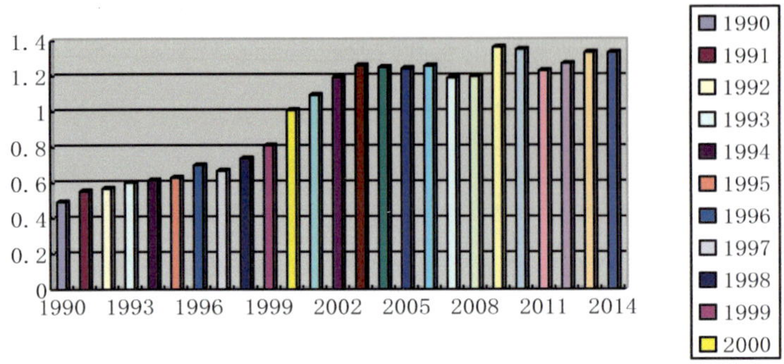

Figure 1. Financial related rate (FIR) of Hubei in 1990–2014. *Source*: Based on Hubei's 1990–2014 Statistical Yearbook data.

3.1.2. Securities industry

In recent years, the rapid development of China's securities market making increasingly more enterprises have more opportunity to choose the way of direct financing to raise funds. It can be seen from **Figure 2**, during the period of 1992–2014, the number of listed companies in Hubei has been increasing under favorable market situation. While in the part of corporate bond issuance, it can be seen through the table data that there are only a handful of enterprises to issue bonds, which indicates that the development of bond market is lagging behind in Hubei.

Overall, listed companies' amounts of share funding (SA) have fluctuated a lot in Hubei, there were even no domestic listed companies in some years, and corporate bond issuance amount (BA) was negligible. Security industry's development of Hubei has some influence on the urbanization process, but the influence is relatively weak compared with the banking sector. The paper will use the ratio between the amount of BA + SA and the total amount of assets to replace the financial structure, in order to reflect the degree of development of financial system in Hubei.

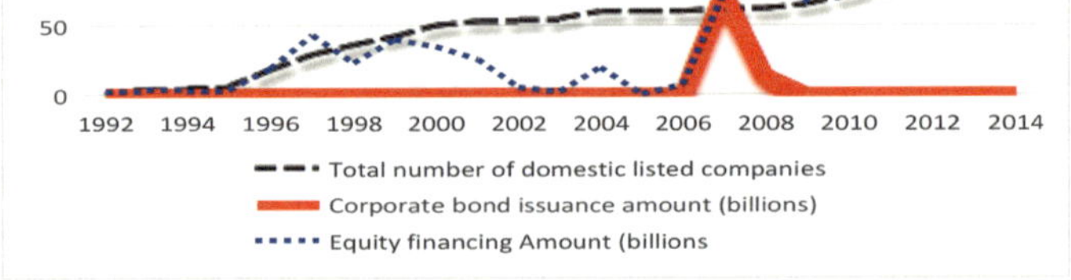

Figure 2. Main indicators of securities of Hubei in 1992–2014. *Source*: China Securities Regulatory Commission and deep statistical data of Flush.

http://dx.doi.org/10.5772/66787

3.1.3. Insurance

With the gradual development of the economy, the degree of market opening of China's financial deepening, and people's awareness of insurance has gradually increased, which brought a lot of space for the development of insurance industry in Hubei. According to the statistics of Hubei, there were 67 insurance companies in 2014, annual premium income reached 700.23 billion yuan (RMB), which increased 19.2% compared with the previous year. Life insurance companies and property insurance companies each had income of 21.937 billion yuan and 48.085 billion yuan, up 20.6% and 18.6%. It also increased more than 20% apparently in the payment of the indemnity (**Figure 2**).

Data from **Table 1** show that the insurance industry developed rapidly in Hubei since the twenty-first century. Premium income increased to 70 billion in the past 15 years, turned more than 10 times. The average annual growth rate of premium growth is about 20%, even more than 63% in 2008. It also reflects the rapid development of the financial sector and its role in promoting the development of urbanization to a certain extent.

Year	Premium income (billions)	Growth rate (%)
2000	60.67	6.9
2001	70.06	15.6
2002	91.38	30.3
2003	106.26	16.3
2004	122.54	15.3
2005	144.48	17.9
2006	160.80	10.8
2007	193.73	20.5
2008	317.15	63.7
2009	372.42	17.4
2010	500.33	34.3
2011	549.77	10.3
2012	595.41	8.3
2013	712.12	19.6
2014	700.23	19.2

Source: The Statistical Yearbook of Hubei.

Table 1. Main indicators of Hubei's Owners Insurance in 2000–2014.

3.2. The urbanization process of Hubei

3.2.1. Urbanized history in Hubei

Chinese urbanization process has become the epitome of an era, and Hubei is a typical province in central China. Overall, the process of urbanization in Hubei is consistent with the other

China's major provinces, roughly divided into three stages: (1) rapid development stage in 1949–1957; (2) in the last 1960s and 70s, the level of urbanization had a recession due to some historical causes; (3) since the reform and opening up in 1978, due to the effective implementation of government policies, urbanization has gradually went into the normalization process, showing the trend of rapid growth, which can be seen from the joint action of performance of government and private voluntary's promotion and the flow of rural population to the cities every year.

3.2.2. *Statistical analysis of urbanization in Hubei*

Since the new century, Hubei's economic and finance have achieved rapid development, and urbanization development has entered a new stage.

It can be apparently seen from **Figure 3** that the urbanization process has made tremendous progress in 1990–2014 and the urbanization rate (PUR) increased from 29 to 56%, an increase of 93%. Both in 1999 and 2000, the growth rate is prominent. The urbanization level is usually divided into three parts, namely the early, mid and late. When the urbanization rate is equal to or less than 30%, it is the initial stage. When the urbanization rate is in the range of 30–70%, it is the mid-stage. When the urbanization level is greater than or equal to 70%, it is the late stage. Up to 2014, the level of urbanization in Hubei has reached about 56%, and based on the division we can consider that Hubei has entered the middle stage. The urbanization rate (PUR) referred in this paper is the proportion of non-agricultural population in the total population (**Figure 3**).

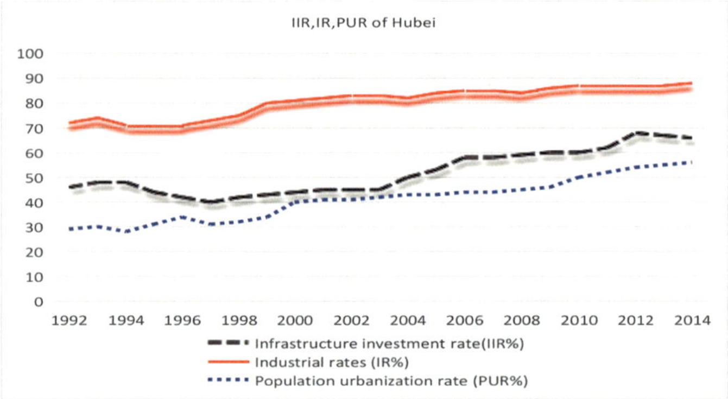

Figure 3. IIR, IR, PUR of Hubei in 1990–2014. *Source*: The Statistical Yearbook of Hubei.

Under the influence of reform and opening-up policy and development of the Yangtze River economic belt and other favorable factors, the development trend of urbanization in Hubei is rapid, and the urbanization rate is higher than the national average level. Although a lot of progress has been made, what can be seen in **Table 2**, Hubei still falls behind by 10 percent-

http://dx.doi.org/10.5772/66787

age point compared with the eastern Zhejiang province, and the data illustrate that it still has many problems in the process of development of urbanization in Hubei. For example, there is the problem of uneven regional development in Hubei. Except for a few cities which have rapid economic development, like Wuhan and Yichang, due to the smaller town size, weaker economic strength and lack of pillar industries, the urbanization development of most other regions is relatively lagged.

Year	2010	2011	2012	2014
Hubei	49.72	51.83	53.50	55.67
Zhejiang	61.62	62.30	63.20	64.87
Nation	49.95	51.27	52.57	54.77

Source: Consolidation of statistics data from National Bureau.

Table 2. Hubei Province, Zhejiang Province and the nation's urbanization rate in 2010–2014.

4. The empirical analysis of financial development to promote the urbanization process in Hubei Province

This paper analyzes the following three aspects of financial development's promoting the urbanization process: (1) using the agriculture loan amount to show the supports of agriculture industry by the financial institutions. (2) Using the investment to the basic constructions to show the supports of the construction of cities and towns by finance. (3) Using the investment to the second and third industry and the second and third industrial added-value to show the supports of optimization and upgrade of the industrial structure by finance.

4.1. Hubei financial supports for the development of the agricultural industry

Agricultural modernization is often developed along with the development of urbanization. The development and upgrading of organization, industrialization and marketization are the main features of agricultural modernization. In this process, the agricultural demand of modern science and technology and economic management methods, which cannot be separated from the support of financial capital. Therefore, the financial sector of government should guide and allocate financial capital, in order to play a positive role in promoting the development of modern agriculture (**Figure 4**).

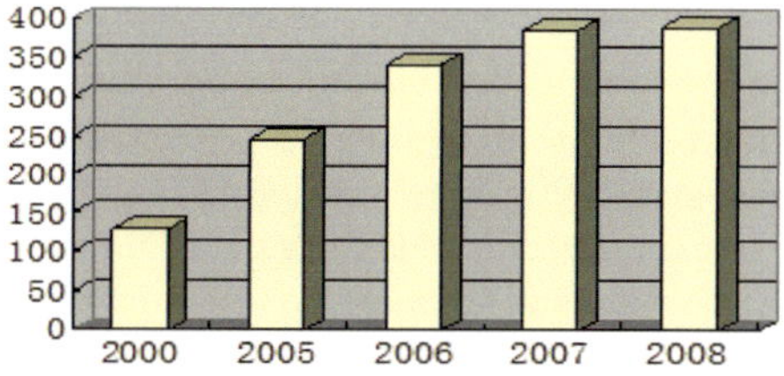

Figure 4. Hubei financial support for the agricultural funds during 2000–2008.

Since 2005, Hubei has issued a series of measures in succession including the use of risk prevention cash to promote the whole province's reform of rural credit cooperatives. According to statistics, 38 branches of them are smoothly combined and transformed into rural commercial banks or cooperative bank. By the end of 2014, compared to early 2014, great progress are made by the data obtained, in which loans reached 300 billion yuan, an increase of 53.3 million yuan. Agricultural loan balance is 225.6 billion yuan, an increase of 38.5 billion yuan, accounting for more than 80% of the total number of financial institutions in Hubei.

According to data of the Agricultural Bank of Hubei branch at the end of 2014, the Agricultural Bank had put 972.56 billion yuan loans to the poor areas. Since the reform of poverty alleviation and discount in our country in 2008, agricultural credit has developed rapidly. As far as the Agricultural Commercial Bank (ACB) is concerned, it has given out 69.96 billion yuan loans for poverty alleviation and discount. At the end of 2014, the loan balance is about 41.5 billion yuan, an increase of about 78% compared with the beginning of that year. About 269.84 billion yuan were put into poor areas of energy, electricity, water and transportation and other infrastructure areas.

So it can be illustrated that rural credit cooperatives and ACB and other financial institutions are playing a positive role in the development of agricultural industry in Hubei.

4.2. The supports of Hubei financial development to the construction of cities and towns

Urban construction involves many aspects, not only including airports, bridges, water resources, urban drainage and gas supply, providing intangible products or serves in science, education, culture and hygiene, but also every side of the lives of residents. So it is important material base of national production and life. In this paper, we mainly study the representative investment of infrastructure construction.

In the following table we can see that in the process of urbanization, the investment in infrastructure construction is in an increasing trend, the infrastructure investment rate(IIR) increased by 0.39 in 1990 to 0.66 in 2014, accounting for more than half of the total investments. The step-down growth rate of investment in the past two years may be related to the gradual saturation of infrastructure, but in general, infrastructure investment is in a rising trend. In the following the proportion of the infrastructure investment in total fixed investments will be set as infrastructure rate (PUR) and it is one of the indicators of measuring the level of urbanization (**Figure 3**).

4.3. The support of Hubei financial development to the optimization of industrial structure

From **Figure 3** we can intuitively see that during the period from 1990 to 2014, there had a rapid growth of the second and third industry investment in Hubei, of which the proportion of the third industry investment accounted for the largest. This showed that Hubei has been committed to optimizing the allocation of funds, the industrial structure is more reasonable through the transfer of funds between the three sectors and it had laid a good foundation for the development of new urbanization.

http://dx.doi.org/10.5772/66787

At present, more and more of the rural labor has turned into the second and third industry which bring them more abundant labor force and higher economic growth and this is the same as the trend of increases of the second and third industry (**Figure 3**).

We can draw a conclusion that the second and third industry added value are increasing steadily and industrial ratio also showed a rising trend. In this paper, the second and third industry divide GDP industrial ratio (IR) are used to show the level of the industrial structure optimize in Hubei, which is representative of one of the indicators of urbanization development. The third industry based on the service field, its development can promote employment furthest, and absorb more labor force from the rural labor force. That will undoubtedly sustain additional fund, only give full play to the function of optimizing the allocation of funds of government finance and financial system can better promote the adjustment of industrial structure in Hubei.

5. Conclusions and suggestions

5.1. Conclusions

Statistical analysis shows that the financial development of Hubei, urbanization and the role of the latter to promote the former. According to the results of statistical analysis:

1. From the perspective of financial development: during 1990–2014, Hubei's bank industry deposit and loan balances and GDP has a great growth. In addition, the securities industry is also gradually improving and developing, direct financing channels are also increasingly becoming more and more enterprise's financing options. The amount of financing of stocks and bonds have a certain growth, the number of listed corporation has reached 84, ranking first in the country's middle area. The insurance industry's income has increased rapidly, turning over a few times in 14 years from the beginning of the twenty-first century till now. Overall, the development trend of the financial industry in Hubei is more positive, there is a lot of room for development in the future.

2. From the perspective of the process of urbanization: Hubei has made great progress in the process of urbanization, the urbanization rate has increased to 55.67%, and the process of urbanization is in a high-speed development period. In addition, the industrial structure of the three major industries are gradually optimized and upgraded, due to the development of agricultural modernization, resulting in more rural surplus labor force, and these labor force transfers to the second and third industry gradually. From the current view of the process of urbanization in Hubei, already exceeded the national average level, but compared with the more developed eastern provinces, there are still a lot of differences.

3. From the perspective of the role of financial development to promote the process of urbanization. In the past 10 years, the amount of investment in the three major industries increased year by year, and the three industries have a certain degree of optimization. Its support for the second industry and the third industry is higher than the first industry,

which will make the industrial structure more reasonable. The two industries accounted for the proportion of the national economy is also gradually improving.

4. From the view of the function route of financial development on promoting the process of urbanization. Hubei's finance has promoted the process of urbanization in agricultural industry, urban construction and industrial structure optimization, and this specific path is driving wheel: financial development does not only promote urbanization, but also promote the industry of urbanization.

5.2. Suggestions

1. To accelerate the reform of the financial system in order to improve the level of financial services. Financial system reform may involve the process of urbanization in the financing and allocation of funds. According to the statistical yearbook data of Hubei, the resource of current funding for the construction of the town of Hubei is mainly from financial market funds, the government's financial support is very small. Therefore, on the one hand, the financial market mechanism should be improved to push the market allocate the financial resources efficiently to the urbanization field; On the other hand, the government should give more policy support, and pay attention to the advantages of "the function of policy finance", to strengthen the government's policy support. Using the policy finance of low interest rates and large financing welfare and high executive ability to prompt ACB, rural credit cooperatives (RCC) and other financial intermediaries to lower the loan threshold, the local government may provide more convenient financial services for the urbanization construction.

2. To establish a multi-level capital market system in order to broaden the financing channels. Urbanization is complex system engineering, which needs more diversified financial support, so it is necessary to rely on a multiple market including the first, second, third and fourth edition markets in order to meet the demand of funds. We mainly do well in two aspects: first, we should establish a multi-level capital market system, expand the total amount of financial capital, and improve the development of financial market; second, we must introduce private capital to participate in the financial market activities, give full play to the positive role in the private financial capital in the process of urbanization.

3. To optimize the industrial structure in order to promote the process of urbanization. Upgrading of the industrial structure is an important aspect of the development of urbanization; with the progress of urbanization process, the industrial form is also transformed from labor-intensive industry to capital- and technology-intensive industrial fields. In the process of development, we should use financial support to meet the needs of different types of enterprises, especially the new type of enterprises. We also have to increase the support to industrial parks, low carbon industry, and the characteristic industry, and relax the financing of SMEs to promote their growth and development, so as to achieve a diversified industrial development.

http://dx.doi.org/10.5772/66787

Author details

Fangchun Peng* and Yu Lu

*Address all correspondence to: 1679786130@qq.com

School of Economics and Management, Hubei University of Technology, Wuhan, PR China

References

[1] Simon Kuznets. Modern Economic Growth: Rate, Structure and Spread, New Haven, CT: Yale University Press, 1966.

[2] Raymond W. Goldsmith. Financial Structure and Financial Development [M] Zhou Shuo translation. Shanghai: Shanghai Joint Publishing, 1997.

[3] Peter R. Stopher. Intercity Passenger Forecasting: the Use of Current Travel Forecasting Producers [M]. Transportation Research Forum, 1993.

[4] Meng Yinli. Analysis of financial deepening, economic growth and urbanization effects [J]. Quantitative and Technical Economics Analysis, 2003 (4) P138-140.

[5] Guo Jiangshan. The dynamic analysis of VAR model of urbanization, industrialization and financial development–a case study of Hebei Province [N]. Journal of Hebei Normal University, 2011, 7(4) P69-74.

[6] Chen Zhiwei. An Empirical Study on the impact of financial development on Urbanization–a case study of Henan [J]. Economic Survey, November 2014 sixth P7-13.

Chapter 11

Comparison of the Bilateral Trade Flows of the Visegrad Countries with China

Lenka Fojtíková, Michaela Staníčková and Lukáš Melecký

Additional information is available at the end of the chapter

http://dx.doi.org/10.5772/66788

Abstract

China's economic miracle resulted in the fact that it is currently one of the largest economy and exporter in the world. Other states try to develop trade and investment relations with China because they see a big potential for their products in the Chinese market with an increasing middle-income group. This chapter is focused on the trade relations that are carried out among some Member States of the European Union and China. The object of the chapter is to show the main facts and trends in the foreign trade of the Visegrad countries with China and to compare their trade structure in the period 1995–2014. The description of the institutional framework of the trade relations among the listed countries and the analysis of the data about their bilateral trade showed that the V4 countries achieved a negative trade balance with China in the previous period, and the possibility to change it is practically very low because of the low level of their trade complementarity. From this aspect, another form of trade relations will apparently have to be developed, although the economic structure and other factors can play an important role in this case.

Keywords: merchandise trade, commercial services trade, Visegrad countries, trade structure, trade balance, trade complementarity, trade openness

1. Introduction

The economies of many countries in the world, developed as well as developing countries, are currently highly dependent on foreign trade. The liberalisation of global trade and a free movement of capital in the previous decades enabled to move production from one part of the world to other parts with the objective of decreasing the production costs of firms and increasing their market share and gains. From this aspect, China has been the favourite destination for many producers from the USA, Germany, Japan, etc., especially because of cheap labour force. On the one hand, a positive situation in the world economy represented by the economic growth of developed countries existed and simultaneously with this, the world trade liberalisation through the GATT proceeded in the 1980s. At the same time, China developed a strategy based on the policy of an 'open door' that has been applied since the end of the 1970s. Both facts mainly contributed to the current situation in the world, in which China is the largest world economy and also the largest producer and exporter of merchandise in the world. Although China has already changed its development strategy since 2010 from an export-led growth strategy to a strategy focused on the domestic consumption of more than 1.3 billion people living in China, its trade and investment activities in the world are growing all the time. Together with this, in another part of the world, the European Union (EU) tries to support its economic activity by following a new investment plan for Europe and investing 315 billion euros during three consecutive years in infrastructure, research and innovation and other productive sectors of economy in order to stimulate growth and employment. The Chinese government confirmed its interest to participate in this plan and to invest in Europe. Thus, China's foreign direct investment can also influence the structure of trade of the member states of the EU with China in the future. The main aim of this chapter is to show the main facts and trends in the foreign trade of the Visegrad countries with China and to compare their trade structure in the period 1995–2014. The structure of the chapter is as follows: first, the authors describe the theoretical environment of international trade with focus on trade policy. Consequently, the empirical part of the chapter includes an analysis of the Visegrad countries' foreign trade with China and compares it in the long-term perspective. In the chapter, the research methods of description, data analysis and comparison were utilised. The official documents, research papers and statistics of the UNCTAD and the World Trade Organization (WTO) were used as the main sources of information.

2. Trade policy as the background for the development of trade relations among countries

The level and scope of foreign trade of the individual countries is influenced by many factors. However, one of the most important factors is trade policy. Trade policy is a part of the general economic policy of a state. Fojtíková [1] (pp. 4–5) defines trade policy as a complex of rules and measures that the state carries out in the area of foreign trade using trade policy instruments. Lipková [2] states that the trade policy of a state is strongly influenced by the external policy of that state and its membership in different international political or military organisations, such as NATO, etc. This fact is also the reason for conflicts for achieving the internal/economic and external/political objectives of a state, especially at the time of economic crises and recessions. Besides the political objectives of a state, trade policy is also influenced by the entire economic policy of that state, its economic development and geographical size.

States use different trade policy instruments in order to achieve their economic and political objectives. The amount and type of the used instruments determine the character of the trade policy of a state. On the whole, if a state prefers to use more autonomous trade policy instruments, such as tariffs and different non-tariff barriers to trade, to conventional instruments, i.e. different international agreements or conventions, the character of the trade policy will be more protectionist than liberal. In practice, all states use both these trade policy instruments, but on a different scale.

Foreign trade has a micro as well as macroeconomic foundation. The motivation of most companies is to expand to foreign markets for the purpose of increasing their market share and economic gains. From the point of view of macroeconomy, a state tries to support its exports in order to increase the economic growth, employment and, on the whole, the economic welfare of the country. Sometimes, the state support of some industries can take the form of unfair trade practices, which the other countries can protect themselves from by trade defence instruments.

2.1. Trade policy of the Visegrad countries as members of the European Union

The Visegrad countries is the name for the four post-communist states of Eastern Europe, that is the Czech Republic, Hungary, Poland and the Slovak Republic that created a regional integration group (the Visegrad group or 'V4') at the beginning of the 1990s in order to develop new forms of political, economic and cultural cooperation after the dissolution of the communist regime in Eastern Europe. The Czech Republic, Hungary, Poland and Slovakia have always been part of a single civilisation sharing cultural and intellectual values and common roots in diverse religious traditions[1] [4]. They significantly increased their external policy activities after their entrance into the European Union (EU) in May 2004 and the V4 group focuses on spreading stability in the wider region of Central Europe. However, the entrance of the Visegrad countries into the EU meant that the national competencies in the area of international trade were delegated from the national level to the EU institutions,[2] especially the European Commission, the Council of the EU and the European Parliament, in the frame of the EU exclusive competencies [6]. In practice, this means that the EU with its 28 member states functions as one body in the trade negotiations with third countries. The main aims and principles of the Common Commercial Policy of the EU are included in Title II, Article 206-207 of the Lisbon Treaty. The common principles are applied in the area of tariff rates (by creating the Common Customs Tariff), the conclusion of tariff and trade agreements, the achievement of uniformity in measures of liberalisation, export policy and measures to protect trade, such as using trade defence instruments in the case of unfair trade practices [7].

The Common Commercial Policy (CCP) of the EU has a long history and is connected with the beginning of the European integration process. First, the Treaty of Rome (1958) formulated the procedure of harmonisation of the member states of the European Communities in the trade area and set the main aims and common principles for carrying out the common policy in the area of external trade. Later, the Maastricht Treaty of the European Union (1993) and the Amsterdam Treaty (1999) newly defined the Common Commercial Policy with a larger number of areas in which the common principles were carried out and the Treaty of Nice

---

[1] However, this does not mean that economic, social and other disparities among the V4 countries do not exist. More about this issue was written forexample by Melecký and Poledníková [3].

[2] The character and development of the trade policy of the individual V4 countries before their entrance into the EU was described for example by Fojtíková [5].

(2003) defined the procedural issues and the main decision method in the Council of the EU in negotiations on trade issues was set to be the qualified majority of votes (QMV). So far the last EU treaty, called the Lisbon Treaty (2009), brought other changes in the CCP of the EU. Namely, the Common Commercial Policy was placed under the External Action of the EU, part V of the Lisbon Treaty, which means that the CCP is carried out in the frame of the principles and aims external action of the EU. This means that the external political objects of the Union are determining for carrying out its trade policy objectives, although these political priorities could bring some economic losses.[3] The uniform principles are currently applied in all parts of the CCP, i.e. relating to trade in goods and services, the commercial aspects of intellectual property and also newly to foreign direct investment (FDI). The European Parliament also obtained more decision competencies than earlier and together with the Council of the EU acted *"in accordance with the ordinary legislative procedure, adopting the measures defining the framework for implementing the common commercial policy"*, according to Article 207 of the Lisbon Treaty [6]. However, the Council of the EU also acts unanimously in some cases, such as (1) trade in cultural and audiovisual services, where these agreements risk prejudicing the Union's cultural and linguistic diversity and (2) in the field of trade in social, education and health services, where these agreements risk seriously disturbing the national organisation of such services and prejudicing the responsibility of member states to deliver them [6] and other areas that belong to the shared competencies of the EU institutions with the national governments and their parliaments.

This means that the number of votes in the Council of the EU of the individual member states is important for the ability to influence the final decision when the Council of the EU decides by QMV. Together, the Visegrad group has 58 votes from the total of 352 votes in this structure: Poland 27, the Czech Republic and Hungary individually 12 and the Slovak Republic 7 votes [9]. However, the economic structure of the Visegrad countries is different (see **Figure 1**) and thus their trade priorities and interests can be also different.[4] In addition, the governments of the EU member countries try to support their exporters by different measures in the frame of their own export policy and strategies.

Although the main aim of the Union, according to Article 206 of the Lisbon Treaty, is *"to contribute in the common interest to the harmonious development of world trade, the progressive abolition of restrictions on international trade and on foreign direct investment, and the lowering of customs and other barriers"* [6], the EU develops its trade relations with non-EU member states in the frame of two different regimes. The preferential regime represents a more favourable treatment in the access to the EU internal market, while the non-preferential regime is based on the most

---

[3] The practical case of this situation is the economic sanctions that the EU has been applying against Russia since 2014 due to the Russian annexation of Crimea in Ukraine in 2014. Although Russia belonged to the three main trade partners of the EU with a share in the total EU trade of 9.5% in 2013 and some EU member countries had up to a 100-percent dependence on the Russian imports of oil and gas, the political decision of the EU was determining. The EU implemented the sanctions gradually in the frame of three waves and Russia also accepted retaliation measures and imposed embargo on the imports of some products from the EU [8]. The economic sanctions between the EU and Russia have been applied by both sides until now..

[4] For example, Poland inhibited the EU negotiation with Russia about the renewal of the Partnership and Cooperation Agreement in 2006. The negotiations could not be officially started until the time when the European Commission obtained mandate from all EU member states[11].

favoured nation clause for all members of the World Trade Organization (WTO).[5] Differences between these two trade regimes are obvious from the level of applied tariffs (preferential and conventional tariffs, i.e. the MFN tariff) as well as non-tariff protection. The simple average applied MFN tariff rate, including the *ad valorem* equivalents (AVEs) of non-*ad valorem* tariff rates, was 6.4% in 2014. Based on the relevant WTO definition, the average applied rate for agriculture was 14.4% and for non-agricultural products 4.3% [13]. The level of the preferential tariff is usually lower than the level of the MFN tariff. It is also possible to export almost a quarter of all tariff lines to the EU duty-free. On the other hand, the MFN tariff can reach the value in the range of 0–635.4%, most of them in the area of agricultural products [13].

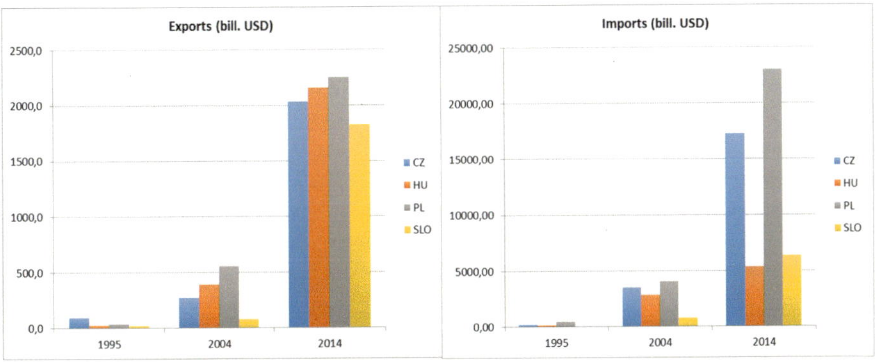

Figure 1. Value of the V4 countries trade with China in 1994, 2004 and 2014 (billion USD). Source: UNCTAD [10]; own processing.

Preferential agreements include the provision about reciprocal preferences or unilateral preferential treatment. Reciprocal preferences are included in regional and preferential agreements, on whose background free trade areas or customs unions are created. By the end of 2014, the EU notified the WTO of 37 regional trade agreements in force in the WTO, among which 23 agreements covered goods and 14 covered goods and services [13]. Countries with which the EU has preferential trade agreements include different parts of the world, for example Mexico, Chile and Peru in Latin America; Algeria, Egypt, South Africa, Tunisia, etc. in Africa; Albania, Montenegro, Switzerland, etc. in Europe, but also South Korea in Asia and many others. The level of preferential treatment is different in dependence on the type of agreement. The Association Agreements including 'deep and comprehensive free trade agreements' currently represent a modern form of trade agreements that comprehensively cover all areas of trade. This means that except for the removal of import and export duties and the removal of obstacles to trade in services, it also includes trade-related policies such as public procurement, competition, intellectual property; and approximates the trade and trade-related policies of these partners in line with the EU *acquis*, in areas such as sanitary

[5] Currently, trade relations based on the MFN clause are developed among 162 countries from the whole world that are the members of the WTO [12].

and phyto-sanitary measures, technical requirements and standards, customs procedures and trade facilitation. There are also tens of countries with which the EU negotiates or has a preferential agreement that is not yet applied, such as the negotiations for a Comprehensive Economic and Trade Agreement (CETA) with Canada, the negotiations about the Transatlantic Trade and Investment Partnership (TTIP) with the USA and many others [14].

The EU also grants unilateral preferences via the Generalised Scheme of Preferences (GSP),[6] but only to a selected number of countries. The EU currently grants more favourable treatment to 92 countries that are classified by the World Bank as low-income countries, such as Botswana, Cameroon, Colombia, Honduras, Vietnam, etc. Namely, the GSP includes three types of arrangements with a different level of preferences: (1) standard GSP, which provides tariff preferences to beneficiary countries; (2) the GSP+, which offers additional tariff reductions to 'vulnerable'[7] countries that ratify and implement core international conventions in the fields of human rights, labour standards, environmental protection and good governance and (3) Everything But Arms, which offers duty-free and quota-free access for all products except arms and ammunitions from the 49 least developed countries (LDC)[8] [15]. Regarding other unilateral measures outside the GSP, tariff preferences were granted as the autonomous trade measures to six countries in the Western Balkans in 2000 and subsequently renewed in 2005 and 2011 until the end of 2016.[9] The main intention of unilateral preferences is to allow developing countries, i.e. countries most in need, easy exports to the EU and to contribute to their economic growth.

2.2. Trade policy of China

China has been the leading exporter and importer in the world. In 2014, China shared by more than 12% in the world merchandise exports and by about 6% in commercial services exports. China's position is dominant especially in manufactures, in which China's share in world manufactures exports reached 18%. On the import side, fuels and mining are significant lines, in which China achieved a share of 13% in world imports. China's share in the world commercial services trade is less significant than in the area of merchandise trade, but its position is also significant; it takes the third position among the leading exporters and importers in the world [16].[10] China reached this position after a long-term process of domestic reforms that started by the end of the 1980s. The economic reforms included the policy of 'open door' and trade liberalisation. This strategy brought China the expansion of production and employment, which has been showing in double-digit economic growth for more than two decades, and a positive

---

[6] The current scheme was established by Regulation (EU) No 978/2012 under which preferences started to be applied on 1 January 2014 and will be effective for 10 years.

[7] A vulnerable country means a country: (1) which is not classified by the World Bank as a high-income or upper-middle income country during three consecutive years; (2) whose imports into the EU are heavily concentrated in a few products; and (3) with a low level of imports into the EU [15].

[8] Most of these countries are from Africa (34) and Asia (9), the rest are from Australia and the Pacific (5) and Caribbean (1).

[9] Specific unilateral preferences were also granted to Moldova and Ukraine; these preferences were applied until the end of 2015.

[10] When we consider the EU-28 as one unit, then China is in the third position among the world exporters and importers of commercial services. However, when we consider the EU member states individually, then China takes the fifth position on the export side and the second position on the import side [16].

trade balance. Because of this fact, China has the largest reserves of foreign exchange and gold in the world, amounting to more than 3 trillion USD in 2015. Nowadays, China is not only the leading exporter in the world, but also an investor's country. Overall, China became the largest FDI recipient in the world in 2014 with 129 billion USD, mainly because of an increase in FDI in the services sector. On the other hand, China's FDI outflows reached116 billion USD at the same time [17].[11] As Baláž et al. [18] states, looking for new markets is a natural result of the Chinese growth model oriented to export. The huge economic expansion of China that has been caused especially by a very high share of investment in the GDP since 2000, accompanied by export growth to the world markets, significantly increased the incomes of companies and also their workers, which positively contributed to the total domestic consumption.

The entrance of China into the WTO in 2001 also contributed to the positive economic development of China. The results of the empirical analysis carried out by [19] confirmed that the entrance of China into the WTO had a positive influence on FDI inflows (including into the financial market) regardless of the fact that many state regulators and structural obstacles to foreign banks existed in China for the whole time. However, the benefits from China joining the WTO are on both sides. This means that China obtained access to the MFN clause that enables it to trade with more than 160 members of the WTO on a non-discrimination base, and it can also participate in forming multilateral trade liberalisation. On the other hand, trade with China is more transparent than before, because China accepted multilateral trade commitments on removing barriers to trade, the liberalisation of some economic sectors and general rules of the WTO. As Fojtíková [20] states, China has been progressively lowering its MFN tariff and reduced non-tariff barriers to trade since 2001. In 2013, the average MFN tariff was 9.4%. There is a higher tariff for agricultural products at 14.8% than for non-agricultural products (8.6%). A duty-free tariff is applied to about 10% of tariff lines. However, the preferential tariff applied to 37 LDC was 5% in 2013 and to the ASEAN countries it is even lower [21]. However, non-tariff barriers currently represent a more significant problem than tariffs in general. It is also influenced by the different economic system of China (called the 'socialist market economy') than the pure market economy. The significant share of China in world trade and a different economic system also contribute to the fact that China is very often a participant of trade disputes in the WTO.

China also develops its trade relations through bilateral and regional trade agreements. China has concluded 12 free trade areas with over 20 states and regions, such as Pakistan, Chile, New Zealand, Singapore, Peru and Costa Rica. China's trade relations with Hong Kong and Macao are developed through the Closer Economic Partnership Arrangements (CEPA) that was signed in 2003. China has also signed free trade agreements with Switzerland and Iceland, but they are yet to enter into force. China has been a member of the Asia-Pacific Economic Cooperation (APEC) forum since 1991 and a member of the Asia-Europe Meeting (ASEM) since its inception, in 1996. It acceded to the Asia-Pacific Trade Agreement in 2001. China's free trade agreement with the Association of Southeast Asian Nations (ASEAN) came into effect in 2005. China is currently negotiating free trade areas with the Gulf Cooperation Council (GCC) countries, Australia, Norway, Korea and Japan. Negotiations for a Regional

[11] The data cover only Mainland China, i.e. without Hong Kong and Taiwan.

Comprehensive Economic Partnership (RCEP) between ASEAN members, Australia, China, India, Korea, the Republic of, Japan, and New Zealand were launched in 2012. All of these agreements and negotiations are based on reciprocal arrangement. The unilateral preferential treatment is applied by China to the 40 least developed countries. Except for trade agreements, China had concluded 131 bilateral investment protection and promotion agreements, for example with Japan, Korea, etc. [21].

China's trade policy is managed by the National People's Congress of the People's Republic of China (NPC) that is the main body of state power in China. The president promulgates legislation adopted by the NPC or its Standing Committee, but does not have the power to veto it. He is responsible for ratifying or abrogating bilateral, regional or international treaties and agreements. The National Development and Reform Commission (NDRC) is in charge of devising the overall national economic and social development policy. The Ministry of Commerce (MOFCOM) has the main responsibility for policy coordination and the implementation of all trade-related issues. MOFCOM is in charge of, *inter alia*: formulating strategies, guidelines and policies related to foreign trade and international economic cooperation; drafting laws and regulations governing foreign trade and investment; studying and putting forward proposals on harmonising the domestic legislation on trade and economic affairs; and bringing domestic laws into conformity with multilateral and bilateral treaties and agreements. Other ministers also cooperate in trade policy formulation and implementation. China's overall main trade policy objective is to accelerate its opening up to the outside world [21].

3. Trade relations of the EU with China: institutional framework

China is the second trading partner of the EU behind the USA and the EU is China's biggest trading partner. China and the EU currently trade well over 1 billion USD a day. However, bilateral trade in services only amounts to 1/10 of the total trade in goods. Investment flows are also a vast untapped potential; China accounts for just 2–3% of overall European investments abroad, whereas Chinese investments in Europe are rising, but from an even lower base [14]. From this point of view, the international agreement constitutes the essential institutional framework for the development of trade and investment relations between both sides. The EU currently develops economic relations with China on the basis of the EU-China trade and cooperation agreement that was signed in 1985. However, the EU and China are global players, who develop their strategic partnership in order to increase cooperation with each other on key international and regional issues, such as foreign affairs, security matters and international challenges such as climate change and global economic governance. The cooperation is developed in the frame of the EU-China 2020 Strategic Agenda for Cooperation that was agreed on at the EU-China summit in 2013 [22]. The 17th EU-China summit was held in June 2015. Bilateral cooperation between the EU and China proceeds at a three-level dialogue, namely a High-Level Strategic Dialogue, High-Level Economic and Trade Dialogue and High-Level People-to-People Dialogue. The results of this work are discussed during annual summits. In the frame of trade dialogue, China announced its intention to contribute

http://dx.doi.org/10.5772/66788

to the Investment Plan for Europe that the European Commission published in 2014 in order to increase the GDP of the EU and create up to 1.3 million new jobs by2017.

Although a new trade agreement between the EU and China is not recently negotiated, the idea about a comprehensive EU-China Investment Agreement was launched at the 16th EU-China summit in 2012. The agreement should provide for the progressive liberalisation of investment and the elimination of restrictions for investors to each other's market. The object of this agreement is provide a simpler and more secure legal framework to investors on both sides by securing predictable long-term access to EU and Chinese markets respectively and providing strong protection to investors and their investments [14]. The first round of negotiations for an EU–China investment agreement took place in Beijing on 21–23 January 2014. In January 2016, the EU and China negotiators reached clear conclusions on an ambitious and comprehensive scope of the upcoming EU–China investment agreement and moved into a phase of specific text-based negotiations. The negotiators will continue working intensively during 2016 in order to hammer out the details of the agreement.

Although the trade and investment relations of the EU member states with the third countries are first given by the institutional framework of the EU agreements, the EU member states have created their own export strategies and signed inter-governmental agreements with non-EU states about economic cooperation. This means that an eligible institutional framework on the EU level as well as good political relations among countries and their representatives are both important for developing trade relations. Historical relations among the V4 countries and China started in 1949 when the former Czechoslovakia (the Czech Republic and Slovak Republic are individual states since 1993), Poland and Hungary as part of the Eastern Bloc recognised the establishment of the People's Republic of China. Relationships between China and these countries cooled down after the deterioration of the USSR and China relations during 1950s and especially 1960s, because the countries in Eastern Europe were Moscow satellites. A new era in the development of political and economic cooperation among China and the Czech Republic, the Slovak Republic, Hungary and Poland started after the fall of communism in Eastern Europe. However, the external trade policy of these post–communist states was oriented especially on Western Europe, which also had an impact on changes in the territorial structure of the foreign trade of these countries.

The entrance of the Visegrad countries into the EU in 2004 signified another milestone in the development of China's bilateral relations with the Czech Republic, Hungary, Poland and Slovakia. The current economic cooperation of China with the V4 countries is based on the '16+1' cooperation platform that was introduced by Chinese Prime Minister Wen Jiabao during his visit to Poland in 2012. This platform has an economic dimension and serves as a mutual framework of cooperation between 16 Central and Eastern European (CEE) states and China. The priorities of China's strategy include twelve areas, including setting up a special credit line of 10 billion USD to facilitate the bilateral cooperation in IT industry, infrastructure, construction and green economy, creating a special fund for investment cooperation in the value of 500 billion USD to support bilateral trade, setting up special economic zones in each of the CEE countries within the next 5 years, setting up an association to promote tourism in China and the CEE countries, setting up a research fund to promote the studies of bilateral relations between China and the CEE countries and many others [23]. The leaders of the CEE countries and China have already met four times since 2012. After the summit in Warsaw, summits in Bucharest, Belgrade and the Chinese city of Suzhou took place until 2015. During these summits, the leaders of the CEE countries and China reached a wide number of agreements in the area of energy, infrastructure, trade and tourism and scientific and technical cooperation [24]. The Visegrad countries find a big export opportunity in China's market, but also the source of FDI for increasing their domestic production and employment. For this reason, developing relations with China is one of the main objectives of the external relation strategies of these countries. On the other hand, deepening the economic cooperation between China and the CEE countries is one of the steps to be achieved by the Chinese national strategy of building the 'New Silk Road Economic Belt' announced by president Xi Jinping in 2013.

4. Comparison of foreign trade of the V4 countries with China

Before we start to compare the foreign trade of the V4 countries with China, it is important to show the basic presumptions of these countries for trade. From this point of view, the number of the population and geographical size of a country, as well as the economic level, play the most important role, although the intensive factors of productivity, such as innovation and new technology, can currently substitute these extensive factors. **Table 1** shows the basic data from 2014. The highest population is in Poland and the lowest population is in Slovakia. The Czech Republic and Hungary are the most similar from this aspect. The characteristics of the V4 countries in the area of geographical size are similar to population. On the whole, the share of the total V4 population accounts for less than 5% in the total Chinese population and less than 6% in the Chinese area. However, the economic level measured at constant prices is the highest in the Czech Republic and China achieves only about a quarter of this value (see **Table 1**). In the nominal expression of GDP, the highest economy is still China and the V4 countries achieved only 14.5% of its level in 2014.

	Population (in thousands)	Area (in km^2)	GDP/cap. (in USD)	GDP (in USD)	Trade openness[1] (in %)
China	1,369,435.7	9,596,960	3799.4	5,295,557.6	46.4
Czech Republic	10,542.7	78,867	14,625.8	157,088.2	160.7
Hungary	9889.5	93,028	11,803.0	117,241.8	173.6
Poland	38,620.0	312,685	11,238.8	429,551.2	91.8
Slovakia	5422.9	49,035	12,196.1	66,519.3	179.5

[1]Trade openness expresses the share of the total trade in goods and services in GDP.
UNCTAD [10]; own processing.

Table 1. The main characteristics of China and the V4 countries, 2014.

Slovakia recorded the highest level of trade openness among the monitored economies, achieving almost 180% of GDP in 2014. However, the other small economies, i.e. the Czech Republic and Hungary, are also open economies. The basic characteristics of the V4 countries and China

http://dx.doi.org/10.5772/66788

confirm the fact that the higher the country is (from the point of view of population and land), the lower is the openness of its economy recorded. The structure of trade openness is also different in the individual countries (see **Table 2**). It is given by the structure of their economies. Although services take the largest share in the GDP and employment in all considered countries, excluding China, the main area of trade is good for all of them. There are also some differences among the countries, for example Slovakia recorded only a 1.8% share of services in its total trade in 2014, but Poland and Hungary recorded a more than 16% share of services in their total trade at the same time. Because merchandise trade plays the most important role for the V4 countries and China, the analysis of the V4 trade with China will be focused only on this area.

	Structure of economy (% of GDP)			Trade (mil. USD)		Share of services in the total trade (%)
	Agriculture	Industry	Services	Merchandise trade	Commercial services trade	
China	8.9	42.7	48.4	4,301,662	614,038	12.5
Czech Republic	2.7	38.2	59.2	326,335	48,380	12.9
Hungary	4.4	30.9	64.8	215,713	41,460	16.1
Poland	3.3	41.1	55.6	436,513	83,569	16.1
Slovakia	3.4	30.4	66.2	948,710	17,128	1.8

UNCTAD [10]; own processing.

Table 2. Trade characteristics of China and the V4 countries, 2014.

Although the intra-EU trade, i.e. trade among the individual member states of the EU, plays the most important role for most countries in the EU, China ranks among the main non-EU trade partners, such as the USA, Russia, etc. However, more than export, it is import from China that is dominant in these countries from the long-term perspective. The results of this fact are recorded in **Table 3**. While the share of China in the V4 countries exports increased in the range of 0.7 –1.7 percentage points (pp) between 1995 and 2014, the share of China in the V4 countries imports increased more significantly. The Czech Republic recorded the highest increase of the Chinese share in its total imports, i.e. by 10.5 pp. between 1995 and 2014. In contrast to this, Hungary recorded only 4.4 pp. increase of the share of China in its imports

	Export to China (%)			Import from China (%)		
	1995	2004	2014	1995	2004	2014
Czech Republic	0.44	0.41	1.17	0.83	5.26	11.35
Hungary	0.18	0.70	1.92	0.83	4.77	5.20
Poland	0.15	0.75	1.05	1.60	4.61	10.61
Slovakia	0.20	0.28	2.12	0.71	2.71	8.18

UNCTAD [10]; own processing.

Table 3. Share of China in the V4 countries trade in 1995, 2004 and 2014 (%).

at the same time. On the whole, exports and imports among the V4 countries and China increased significantly, especially in 2014, which is graphically shown in **Figure 1**.

In nominal expression, the development of the value exports and imports among the V4 countries and China is shown in **Figure 1**. Impressive results were achieved especially by Slovakia. The Slovak exports to China increased more than 107 times (from the value of 16.7 million USD in 1995 to 1.8 billion USD in 2014) and the imports 110 times between 1995 and 2014. On the export side, Hungary also recorded a positive result when its export to China increased almost 99 times between 1995 and 2014, while the value of imports increased only 42 times. However, the value of Hungarian imports remained higher than the value of exports. The significant increase of bilateral trade of the V4 countries with China occurred especially after the entrance of the V4 countries into the EU in 2004. At this time, the Czech Republic, Hungary, Poland and Slovakia became politically and economically stable markets for Chinese goods and investments. During a 10-year period (2004–2014) the trade volume of the V4 countries with China increased almost five times. Although Szikorová [25] states that the Central European region has became a Chinese bridge to the EU market, the four Visegrad countries have also increased their export to China; with the exception of direct exports to China, exports from the V4 countries to China have been strengthened via multinational companies, which in general proved to be more successful on the Chinese market.

However, the result of this development is a trade deficit that theV4 countries recorded in merchandise trade with China in the monitored period. The reasons for trade deficit can be different, from higher price competitiveness of the Chinese products on the world market, the existence of state owned firms, a good marketing of sale, to an imbalanced trade policy. While the Visegrad countries opened the way for cheap imports from China and carried out a liberal trade policy as a part of their transformation process in the 1990s, China did not grant them reciprocal market access. The large protective measures, including bureaucracy, violations of property rights, unclear legislation and the lack of transparency in tax laws started to be improved after China's entrance into the World Trade Organization in 2001. However, the trade deficit of the Visegrad countries with China was higher in 2014 than 2004. The highest trade deficit with China, amounting to about 20.7 billion USD, was recorded by Poland in 2014 (see **Figure 2**). Poland is also China's biggest trade partner in Central and Eastern Europe and is followed by the Czech Republic and Hungary [26]. The trade deficit of the Czech Republic reached 15.2 billion USD at the same time. The trade deficit of Slovakia was 4.5 billion USD in 2014 but was more than six times higher than in 2004. Hungary reached the lowest trade deficit with China during the whole monitored period in the value of about 3 billion USD [10].

From the sectorial point of view, no significant changes occurred in the structure of the V4 countries trade with China between 1995 and 2014. On the export side, the prevailing part of the V4 countries exports was carried out in the frame of the group of machinery and transports equipment (SITC 7), with the exception of Poland, in which the commodity structure of its exports to China was more variable (see **Table 4**). On the import side, the situation in all V4 countries was the same, but small changes occurred between 1995 and 2014. While manufacture goods (SITC 8) were a dominant group in the V4 countries imports in 1995, the group SITC 7 has been the main import group since 2004. After the entrance of the V4 countries

http://dx.doi.org/10.5772/66788

into the EU in 2004, the Chinese investments to this region were also recorded. Between 2005 and 2013, the total Chinese outward foreign direct investments reached 100 million USD in the Czech Republic, 4590 million USD in Hungary and 1600 million USD in Poland. In these countries, China invested especially in the transportation/automotive and energetic sector [26], i.e. in sectors that are, according to trade statistics, predominantly included in the SITC 7 commodity group.

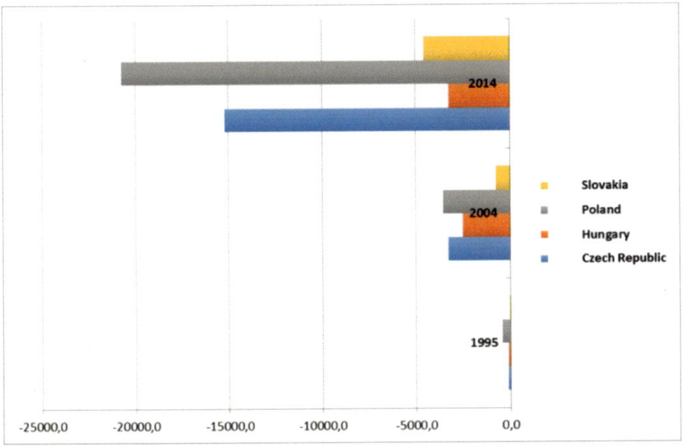

Figure 2. Trade balance of the V4 countries with China in 1995, 2004 and 2014 (million USD). Source: UNCTAD [10]; own processing.

	Exports			Imports		
	1995	2004	2014	1995	2004	2014
Czech Republic	SITC7	SITC7	SITC7	SITC8	SITC7	SITC7
Hungary	SITC7	SITC7	SITC7	SITC8	SITC7	SITC7
Poland	SITC5	SITC7	SITC6	SITC8	SITC7	SITC7
Slovakia	SITC7	SITC7	SITC7	SITC8	SITC7	SITC7

UNCTAD [10]; own processing.

Table 4. Main commodity groups in the V4 exports and imports with China in 1995, 2004 and 2014.

On the whole, trade complementarity among the V4 countries with China is very low with a tendency to decline as is shown in **Table 5**. The results of trade complementarity between the individual V4 countries and China were obtained from UNCTAD [10] and were calculated at a 3-digit level SITC, revision 3. On the whole, the value of the index of complementarity was nearer zero than one. For this reason, there is a very small probability of increasing mutual trade among the individual V4 countries and China and from this point of view a change in the development of the trade balance of Czechia, Hungary, Poland and Slovakia with

China will be very hard to achieve. However, it could be the subject of another analysis if the V4 countries achieve a revealed comparative advantage in the sector SITC 7 and if China's bilateral trade with the Visegrad countries has an intra-industry or inter-industry character. However, it is necessary to see the future of the V4 countries trade relations with China not only from the merchandise trade point of view but also from the point of view of trade in commercial services. In 2014, the Chinese government issued a number of policies to facilitate foreign direct investment in a broad range of services, including financial services, tourism, entertainment and healthcare. Services also currently take up a greater share of China's FDI than the manufacturing sector [27].

	1995	2004	2013
Czech Republic	0.50	0.44	0.40
Hungary	0.43	0.42	0.41
Poland	0.35	0.33	0.38
Slovakia	0.42	0.36	0.35

UNCTAD [10]; own processing.

Table 5. Trade complementarity of the V4 countries with China in 1995, 2004 and 2013.

5. Conclusion

The main aim of the chapter is to show the main facts and trends in the foreign trade of the Visegrad countries with China and to compare their trade structure in the period 1995–2014. The results of trade analysis showed that the V4 countries have a different level of trade openness, but for all of them China represents a very important trade partner. All V4 countries recorded, in nominal expression, an increase of their merchandise trade with China on the export, but also import side. It followed from the fact that China significantly increased its position on the V4 markets during the monitored period and became one of the V4 main trade partners from non-EU Member States (China ranks among the five main partners of all of them). However, the V4 countries trade balance with China was negative for the whole time. The highest deficits were achieved by Poland. The sectorial structure of the bilateral trade of the V4 countries with China did not record significant changes and the machinery and transport equipment represent the main tradable items on the export, but also import side. From this aspect, the trade complementarity of the V4 countries with China had a declining trend and is generally on a very low level. In the future, it will be necessary to complement the foreign trade of the V4 countries with China with foreign direct investments flows.

Acknowledgement

This chapter was created within the frame of project registration number CZ.1.07/2.3.00/20.0296 supported by the Education for Competitiveness Operational Programme.

Author details

Lenka Fojtíková*, Michaela Staníčková and Lukáš Melecký

*Address all correspondence to: lenka.fojtikova@vsb.cz

VŠB-Technical University of Ostrava, Faculty of Economics, Department of European Integration, Ostrava, Czech Republic

References

[1] Fojtíková, L. (2009). Zahraničněobchodnípolitika ČR. Historie a současnost (1945–2008). Praha: C. H. Beck. 246 p. [Foreign Trade Policy of the Czech Republic. History and the Present (1945-2008)].

[2] Lipková, L. a kol. (2011). Medzinárodné hospodárske vzťahy. Bratislava: Sprint dva. 432 p. [International Economic Relations].

[3] Melecký, L., Poledníková, E. (2013). Využitívícerozměrnýchstatistickýchmetod pro hod noceníregionálníchrozdílůzemíVisegrádskéčtyřky. Scientific Papers of the University of Pardubice. Series D, Faculty of Economics and Administration, Special Edition. Vol. 20, N. 26, pp. 140–150. [Use of Multivariate Statistical Methods for Visegrad Four Regional Disparities Assessment].

[4] Visegrad Group. (2016). About the Visegrad Group [Online]. [cit. 2016-03-06]. Available at: http://www.visegradgroup.eu/about.

[5] Fojtíková, L. (2011). Opatření zahraničněobchodnípolitiky a otevřenost ekonomiky v zemích Visegrádské skupiny. Ekonomická revue - CEREI. Vol. XIV, N. 1, pp. 51–66. [Measures of Foreign Trade Policy and Economic Openness in the Visegrad Countries Group].

[6] Official Journal of the European Union. (2007). Treaty of Lisbon. Amending the Treaty on European Union and the Treaty Establishing the European Community (2007/C 306/01) [Online]. [cit. 2016-03-05]. Available at: http://eur-lex.europa.eu/legal-content/EN/TXT/PDF/?uri=OJ:C:2007:306:FULL&from=EN.

[7] Fojtíková a kol. (2014). Postavení Evropské unie v podmínkách globalizované světové ekonomiky. Vol. 27. Ostrava: VŠB-TU. 390 p. [Position of the European Union in the Conditions of Globalized World Economy]

[8] Fojtíková, L., Vahalík, B. (2015). *ZahraničníobchodČeskérepubliky pod vlivemgobalizačníchjevů a procesů*. Vol. 39. Ostrava: VŠB-TU. SAEI. 196 p.

[9] Rada EU. (2016). Kvalifikovaná většina [Online]. [cit. 2016-03-06]. Available at: http://www.consilium.europa.eu/cs/council-eu/voting-system/qualified-majority/. [Qualified Majority]

[10] UNCTAD. (2016). UNCTADSTAT [Online]. [cit. 2016-03-10]. Available at: http://unctadstat.unctad.org/wds/ReportFolders/reportFolders.aspx?IF_ActivePath=P,15912.

[11] Fojtíková, L., Lebiedzik, M. (2008). Společné politiky Evropské unie. Historie a současnost se zaměřením na Českou republiku. Praha: C. H. Beck.179 p. [Common Policies of the European Union. History and the Present with Focus on the Czech Republic]

[12] WTO. (2016). Members and Observers [Online]. [cit. 2016-03-06]. Available at: https://www.wto.org/english/thewto_e/whatis_e/tif_e/org6_e.htm.

[13] WTO. (2015). Trade Policy Review. Report by the Secretariat. The European Union [Online]. [cit. 2016-03-06]. Available at: https://www.wto.org/english/tratop_e/tpr_e/s317_e.pdf.

[14] European Commission. (2016). Trade. Countries and Regions [Online]. [cit. 2016-03-07]. Available at: http://ec.europa.eu/trade/policy/countries-and-regions/.

[15] European Commission. (2015). The EU's Generalised Scheme of Preferences [Online]. [cit. 2016-03-06]. Available at: http://trade.ec.europa.eu/doclib/docs/2015/august/tradoc_153732.pdf.

[16] WTO. (2015). International Trade Statistics 2015 [Online]. [cit. 2016-03-06]. Available at: https://www.wto.org/english/res_e/statis_e/its2015_e/its15_world_trade_dev_e.pdf.

[17] UNCTAD. (2015). World Investment Report 2015 [Online]. [cit. 2016-03-06]. Available at: http://unctad.org/en/PublicationsLibrary/wir2015_en.pdf.

[18] aláž,P.,Szökeová,Zábojník,S.(2012).Čínskáekonomika.Novádimenziaglobalizáciesvetov éhohospodárstva. Bratislava: Sprint dva. 279 p.[Chinese Economy. A New Dimension of the World Economic Globalisation]

[19] Fojtíková, L., Kovářová, J. (2014). Influence of China's Entry into the WTO on Crossborder Financing. Proceedings of the 14th International Conference on Finance and Banking, pp. 74–79.

[20] Fojtíková, L. (2012). China's External Trade after Its Entrance into the WTO with the Impact in the EU. *Proceedings of the 1st International Conference on European Integration 2012*, pp. 56–65.

[21] WTO. (2015). Trade Profile. China [Online]. [cit. 2016-03-06]. Available at: http://stat.wto.org/CountryProfiles/CN_E.htm.

[22] European Commission. (2013). EU-China 2020 Strategic Agenda for Cooperation [Online]. [cit. 2016-03-07]. Available at: http://eeas.europa.eu/china/docs/20131123_agenda_2020__en.pdf.

[23] Drelich-Skulska, B., Bobowski, S., Jankowiak, A. H., Skulski, P. (2014). China's Trade Policy Towards Central and Eastern Europe in the 21st Century, Example of Poland. Folia Oeconomica Stetinensia. Vol. 14, N. 1, pp. 150–174.

[24] CCTV.com. (2016). *The 4th China-CEE summit opens in China's Suzhou* [Online]. [cit. 2016-05-27]. Available at: http://english.cntv.cn/2015/11/24/VIDE1448349240074452.shtml.

[25] Szikorova, N. (2012). Development of the Chinese-Slovac Economic Relations. *Journal of US-China Public Administration*. Vol. 9, N. 12, pp. 1368–1376. ISSN 1548-6591.

[26] Jaroch, E. (2016). China's Foreign Policy Towards CEE Countries: Determinants, Development and Problems [Online]. [cit. 2016-05-26]. Available at: http://peoplessquare.pl/2016/03/19/chinas-foreign-policy-towards-cee-countries-determinants-development-and-problems/.

[27] KPMG. Global China Practice. (2015). China Outlook 2015 [Online]. [cit. 2016-05-27]. Available at: http://www.kpmg.com/ES/es/Internacionalizacion-KPMG/Documents/China-Outlook-2015.pdf.

Chapter 12

Sustainable Consumption in the Luxury Industry: Towards a New Paradigm in China's High-End Demand

Patrizia Gazzola, Enrica Pavione and Roberta Pezzetti

Additional information is available at the end of the chapter

http://dx.doi.org/10.5772/64477

Abstract

In the present competitive global environment, many drivers should motivate the growing attention paid to sustainable consumption in the luxury industry. From the demand side, in Western mature markets wealthy consumers show a growing awareness of environmental and social issues and, therefore, seek new forms of luxury which show respect for both natural resources and human beings, yet standing by traditional factors such as quality, rarity, creativity, originality and craftsmanship of goods. On the firm's side, sustainable consumption offers luxury firms a particularly suitable platform to enrich the value-set of products as brands identity and brand image. Starting from a review of the literature on the concept of sustainable consumption, the paper provides an analysis of the main drivers that are leading to the emergence of "sustainable luxury". The aim of the paper is to investigate the opportunity for the development of this new competitive paradigm within the Chinese luxury market, by analyzing the distinct feature of Chinese high-end demand. The paper also taking into account the growing role played by the Chinese central government in creating the conditions for sustainable consumption or "circular economy".

Keywords: sustainable consumption, sustainable luxury, purchasing behaviour's drivers, Chinese luxury demand

1. Introduction: Emerging sustainable consumption within the framework of sustainable development

The concept of sustainable consumption, emerged from the Rio Earth Summit in 1992, has become one of the most important goals of the United Nations within the framework of sustainable development. Although sustainability has both social and economic dimensions, the increasing focus on sustainable consumption is due to the fact that humans of the world consume goods and services on a regular basis which contributes towards environmental change deprivation of resources, both renewable and non-renewable, and causes waste and pollution (Mathews, 2012; Hörndahl and Dervisevic, 2015). This concept is connected with the notion of reshaping consumerism and it is not concerned with consuming less but about consuming differently and in a more efficiently way (Beaton and Perera, 2012).

According with Mathews (2012) the measures to promote sustainable consumption are few in comparison to sustainable production initiatives, and have generally been poorly implemented. Akenji (2014) argues that consumers are not getting clear information about how they should differentiate sustainable from non-sustainable products. However, with the recently adopted 10-Year Framework of Programmes at the Rio+20 Summit, governments will be looking to hasten their search for policies that boost sustainable consumption.

There are different values that can affect consumer's behaviour (Stern, 2011; Holbrook, 1999): cultural, attitudinal, contextual and personal values as well as a habit or a routine; these are all values that may affect the decision of a consumer's purchase (Maniatis, 2015). The context of sustainable consumption is both sociological and personal given that the consumer needs to consider enhancements in self-lifestyle and of lifestyles of others in the community (Spaargaren, 2003). A research conducted in China by Xu et al. (2012) shows that sustainable labels are significant enablers for consumers willing to pay more for sustainable products. Consumers trust sustainable brands because they think of sustainable features in the product (Thogersen et al., 2012) and, in many cases, they are ready to pay a premium price for a product marked with credible sustainable labelling provided they understand clearly the economic and ecological benefits of the product and are able to trace these benefits to tangible evidence (Owusu and Anifori, 2013; Xia and Zeng, 2006; Xu et al., 2012). An increasing number of consumers tend to trust their own consciousness about health and the environment, and the certifications and labelling of the product constituents in making decisions for purchasing (Kai et al., 2013). Therefore, consumers tend to mix their sustainable knowledge and attitudes with sustainable brand awareness while choosing a 'green' product (Matthes et al., 2013; Zhao et al., 2014).

The concept of sustainable consumption may be applied in many different industries; in the luxury industry, in particular, is becoming an important phenomena that affect both strategic choices and business models of luxury firms, together with brand management practices (Mathews, 2012). In particular, the luxury industry is experiencing a new paradigm based on of sustainable luxury; this concept is intended to impact strongly on the competitive strategies and the purchasing decisions of consumers in both mature economies and emerging ones, like China.

In the present chapter, the approach to sustainability in the luxury industry is not analysed as a mere brand marketing tool, simply connected with sustainability, but it is considered as an integral element of the company's value proposition and a driver of competitive advantage. According to the perspective adopted in this chapter, the concerns of ethical nature extend to the entire value chain and to all the stakeholders, integrating in a strategic vision of long-term budgetary and social success. This perspective, in a phase of growing affirmation, reflects on the need for luxury-sector companies to redirect their strategic approach towards

http://dx.doi.org/10.5772/64477

transforming their social responsibility and sustainability into an 'endogenous factor' and a competitive opportunity to benefit both the individual companies and the overall sector.

2. The new paradigm of 'sustainable luxury'

The luxury industry is going to embrace sustainability into the competitive strategy of firms. It is a slow but steady evolution. In the 1980s and 1990s, there were the anti-fur campaigns; many brands and luxury retailers have since eliminated the use of fur in their products or taken measures to ensure animal welfare conditions in their fur supply chains. Then, beginning in the late 1990s, numerous sweatshop scandals pressured to implement factory-compliance monitoring programs. In the past several years, the luxury industry has faced intensifying criticism about its environmental footprint. Currently it is establishing the concept of 'sustainable luxury', a term coined to refer to the commitment of luxury companies responsible for their production to both the society and the environment.

In the last years, many drivers should motivate luxury companies to engage in more sustainable practices in response to a variety of social, environmental and economic pressures (Pavione et al., 2016; Colombo and Gazzola, 2015a; Pavione and Pezzetti, 2015). On one hand, the evolution of the global demand: from the unreachable dream connected to the possession of a particular product, the concept of luxury tends to free itself from the economic value of a product and from the individual's spending capacity to be connected instead to a more intrinsically ethical/social idea of value, to a lifestyle connected to emotional and experiential values. High-end consumers, both in mature markets and emerging economies, show a growing interest for new forms of luxury that show respect for natural resources and human beings, yet standing by traditional factors such as quality, rarity, creativity, originality, craftsmanship and savoir-faire. In this framework, sustainable luxury comes to life and the luxury industry is undergoing a process of self-analysis and redefinition of competitive strategies in the light of social responsibility and sustainable dimension (Guercini and Ranfagni, 2012). Revisiting products, services, communication strategies and managerial processes in the direction of sustainable development and developing new socially responsible business models are rapidly becoming the key dimensions to create long-lasting financial and non-financial value, along with strong relationships with wealthy customers. On the other hand, the increasing focus on sustainability by the luxury companies is linked to the need to preserve over time the natural resources and rare raw materials indispensable to ensure the survival of the luxury industry.

In the present scenario, in which wealthy consumers are becoming more concerned with environmental and social issues, luxury companies are increasingly asked to demonstrate their sustainability efforts and to base both brand identity and brand reputation on a set of values through which they should be known and publicly judged by both clients and the market; in this perspective, sustainable development strategies offer a particularly suitable platform to enrich the value-set of high-end products and luxury brands. From a mere marketing choice, sustainable luxury need to be incorporated in the core strategy of luxury firms and become part of the firm's core business (Pavione and Pezzetti, 2014, 2015; Colombo and Gazzola, 2015b): recent empirical evidences show how the higher performing companies are able both to integrate sustainable development strategies in the processes of firm's governance and reconsider their business models, with the objective of capturing the opportunity for growth that a sustainable approach provides (Hoffmann and Coste-Maniere, 2012).

A particularly revealing aspect to emphasize concern over the different characteristics is the relationship that may assume between luxury and sustainability. On one hand, the prestige connotation of a luxury product may be reinforced, increasing the exclusivity of the brand and its perceived value; sustainability in this case is seen as an additional attribute to the pre-existing offer of a luxury product, in some way 'instrumental' to its reinforcement. Many famous brands, such as Gucci and Herm ès, moved in this direction. On the other hand, sustainability may be conceived of as an original source of luxury. In more recent times, in the realm of business experience, niche luxury products seem to revolve around the promotion of sustainable aspects of the production line (such as, for example, a particular valuable raw material), around which a luxury brand is built. In this type of experience, the sustainable resource for luxury does not increase the perceived value of a pre-existing product, but generates an exclusive property. This latter aspect appears particularly revealing and may lead to defining new business models from the existing natural resources and from their connection with the territorial realities (Banathy, 1996) and with local actors; this is the case for well-known Italian high-end brands like Brunello Cucinelli, Ermenegildo Zegna and, more recently, Prada. In this case, defining innovative inter-organizational solutions for creating proposals for sustainable offers may represent a driver to create and develop new luxury brands created under the sign of sustainability, from an ecological and social perspective.

An approach oriented to sustainability may further have a meaningful impact on the developmental processes of a new product and on defining the relationship between concept and competitive positioning. In the textile industry, for example, the development of production techniques and innovative products today represent the main fulcrum of the competitive advantage. An important part of the research and development in the field of new materials, finishing and the manufacturing processes is guided by a growing drive towards sustainability, which becomes an essential strategic element. In the same way, the innovation incorporated in the creative process assumes a fundamental role. Those seeking new design systems represent the point of departure for sustainable change management, often oriented to seek a blend between art and luxury and to the requirement to respond to the demand for ethical products and behaviour by the consumers (Pavione and Pezzetti, 2015). For example, in recent years, the realm of innovative design developed 'slow design', which emphasizes the centrality of the creative process rather than the product (Fuad-Luke, 2002).

3. Sustainability in the fashion luxury: emerging trends

Fashion, above all for the high-end segment, has historically developed business models over time that have given little space to sustainability and only recently luxury fashion firms have taken up the challenge of sustainability, mainly imposed by a profoundly changed context. If, on the surface, the concepts of fashion and of sustainability may appear antithetical, in reality, at least until the era of mass consumption, fashion was sustainable by definition, based on the artisanal processing of naturel resources. The business models of the fashion-clothing system still show a certain delay in showing themselves overall sensitive to the paradigm of sustainability, regarding other luxury sectors that have been most affected. The reasons refer to the fact that the world of fashion was traditionally based on a model of interaction with both customers and the market based nearly exclusively on image, evocation and communication, rather than on the production processes and development of sustainable assets.

An important evolution of the concept of sustainability in fashion began from the third millennium as a consequence of new trends emerging from the demand side. The need for a sustainable approach in the fashion market appears consistent with the changed characteristics of the high-end customer, increasingly interested in receiving the stamp of social approval in a globalized society and thus to express a request for differentiated products that, along with the traditional intrinsic values of fashion, also show attention to the quality of life. To the 'beautiful and well-made' product, the modern consumer tends to add a social and hedonist dimension, represented by the capacity of the product to provide wellness, to show a social value, made of links with the territory, intrinsic knowledge and propriety of the stakeholder community relations. The recent economic crisis has further increased the share of wealthy consumers who put more attention to the value dimension of the luxury fashion system.

In recent years, sustainability has been an element of brand differentiation and a source of long-term competitive advantage (Ricchetti and Frisa, 2011). The growing demand for sustainable fashion pushes companies to review their business models, developing strategic partnerships and new relationship modes with all of the actors involved, in a sort of 'green agreement' based on cooperation and sharing objectives. Specifically, the new business models emerging in luxury fashion are based on some critical success factors that compensate, in financial performance, for investments in sustainability, savings in terms of resources used (water and energy in particular), less waste of materials, reduction of cost of non-sustainability (deriving, for example from legal impositions), technological innovations that translate into the ability to introduce new products, adoption of new collaborative practices among the actors of the supply chain, and greater attention to the relationships with both the local communities and the customers. In testimony to the importance of the social dimension in companies' strategic vision, systems for measuring the performance of luxury companies tend to be increasingly based not only on the economic results, but also on the social and environmental consequences of firms' strategic choices.

Within this framework, in constant evolution, it is interesting to investigate whether and to what extent the sustainable consumption of luxury products represents a driver to purchasing decisions and consumer by the demand from China, destined to become in the medium term, the main market for the Western global luxury brands.

4. Distinct features and driving forces of China's luxury market

In the past decade, the fashion industry in China has undergone tremendous change and is continuing to expand. In the first decade of this millennium, the fashion market in China tripled and it is expected to increase threefold again in the second decennium (Boston Consulting Group, 2014). In 2014, China was still among the top nations in terms of total sales of luxury goods with 15 billion euro and it accounted for the fifth largest luxury goods market after the U.S (64.9 billion euro), Japan (18 billion euro), Italy (16.1 billion euro) and France (15.3 billion euro) (Bain & Company, 2014a, b). After a market slowdown experienced in 2014, with a decreased 2% year-on-year to 15 billion euro, in 2015, the Chinese demand for luxury goods was boosted by the improving economic conditions and the rising disposable incomes resulted in growing demand for luxury goods (Hurun Research Institute, 2015; KPMG, 2007). The growth of the Chinese demand for luxury products is also driven on one hand by the growing luxury companies' engagement in marketing campaigns conceived for the Chinese market and on the other hand by increasing investments in emerging e-commerce.

With a population of about 1.4 billion and a rapid economic growth, China offers the world potentially the largest consumer market for the fashion industry, both in the luxury and in the masstige segments. The economic growth, the demographic trends and the expected rise of the Chinese upper middle-class will further stimulate and expand the potential of China's fashion market. Luxury, which represents high-end merchandise life, set off a massive consumption boom in today's China. In addressing the growth rate of China's luxury market, Goldman Sachs predicted that China will soon become the world's fastest growing luxury market. In fact, China is not only an exceptionally promising new market, but also the current situation of the last untapped market. Because now that the European and American market is relatively saturated, Asians, in general, are keen on luxury goods, the Chinese luxury market naturally becomes a top priority for international brands to compete (Goldman Sachs, 2010).

Chinese luxury goods industry is increasing with immense business opportunities, as market capacity is enormous. According to Roberts, the further growth of luxury market in China is expected to be under-pinned by improving purchasing powers, an increasing number of upper middle-class Chinese and trading up from the mass of consumers (Roberts, 2015). In particular, the demand for luxury goods is strongly sustained by both the growing number of wealthy individuals (in 2013 China accounted for 1.09 million millionaires) and the magnitude of China's middle-class growth: according to Barton, Chen, Jin, the number of 'upper middle-class' Chinese will increase tremendously in the next years and by 2022, 54% of China's urban consumers will be regarded as part of it; in particular, the growth of the middle-class is predicted to be faster in smaller cities with huge economic potential and, consequently, an increasing huge demand for luxury goods will arise from smaller centres located in northern and western China powered by urban consumerism (Barton, Chen, Jin, 2013; Gouxin et al., 2012).

In this framework, China is becoming a great power of luxury consumption and in competition is rapidly growing as luxury brands becoming increasingly popular among Chinese consumers (Xiao Lu, 2008; Wang et al., 2000; Zhang and Sharon, 2003). But it should be noted that Chinese luxury market has only just begun. Consumer attitudes are still in the relatively early stages and the understanding of luxury is not mature enough, even though in most recent years Chinese luxury consumers are becoming more sophisticated. Traditionally, Chinese luxury consumers have been divided into two main categories (Seringhaus, 2002). One category is represented by wealthy consumers who prefer to avoid the crowds and

http://dx.doi.org/10.5772/64477

pursue personalized service; they frequently visit luxury retail stores, buy the latest and most popular products, generally, and will not consider the problem of the price. The second category is represented by office workers, the most typical ones hired by the foreign companies, who will spend a whole month wages to buy a commodity. Chinese people like to follow the brand and pay particular attention to status as they have been taking the 'face' very seriously. Once they have wealth, they began to pursue to display successfully their external performance, the luxury brand's philosophy gives them enough purchasing power, to reflect their success and wealth, and those close to the glittering trademark products are often the most popular in China.

This picture can be considered still valid in spite of different factors that, in the short and medium term, will transform the characteristics of the Chinese demand for luxury products. A first evolutionary factor in the Chinese luxury market is related to the dimension of the youth demand: unlike the Western countries, Chinese luxury consumers are very young, mostly under 40 years of age or less. Young people between 25 and 30 years old in China are rapidly developing luxury consumer groups, and the speed is much faster than the developed Western countries; increasing demand is also driven by the so called 'post-90s generation', that has high brand awareness and, in most cases, is today willing to splurge on luxury goods. The growing importance of this market segment explains the effort made by global luxury brands on making their brand image 'younger' and more fashionable to capture the next generations of trendy customers (Bain & Company, 2015). Another changing factor is related to the demand's composition. Traditionally, Chinese luxury consumers are mostly male: in 2000, women accounted for only 25% of the total consumption of the population but at present, due to the economic independence of women in society, the proportion of female consumers in the luxury market is rapidly growing. In 2014, the sales of woman's wear saw momentum, with an annual growth rate of 11%, while men's wear experienced a decrease in sale of 10%, followed by men's watches (13%); the growing female demand for luxury goods represents an opportunity for global luxury brands with a strong focus on women as target. Professional women who are independent financially are changing the Chinese luxury goods industry's customer bases, which were male-dominated in the past. Fashionable and wealthy urban women who are willing to treat themselves are very obsessed with luxury glamor. The growing demand for luxury products by Chinese women is a factor intended to drive the demand for sustainable luxury, given the attention traditionally paid on sustainability by this segment of demand.

Traditionally, Chinese consumers who buy luxury goods were not going to research the history behind the brand and enquire connotation but new trends are today emerging in the largest cities like Beijing and Shanghai. According to a recent research conducted by Bain & Company, young Chinese consumers of luxury goods are rapidly changing their attitude towards purchasing of luxury goods and paying growing attention to the true value of the purchased goods: exclusive designs, high level of quality, historical heritage of brands and the value of 'made in', brand reputation play an increasing importance in guiding the purchasing choices of Chinese wealthy consumers. In particular, the rising young upper middle-class is becoming more sophisticated and knowledgeable about luxury and increasingly looks for low-key, unique products instead of those with visible logos; as a result, luxury niche brands have an increasing popularity among Chinese consumers, as individuality is becoming key for many consumers whose purchasing decisions will more and more driven by emotions and self-expression. In particular, in the fashion luxury niche brands are rapidly becoming the most sought after (Roberts, 2015) as they are able to create a stronger sense of identity. As a result, brand loyalty (traditionally very low among Chinese consumers) is expected to improve, as younger consumers are becoming more loyal than older generations, but the Chinese consumer will remain far less faithful to brands than Western consumers. When markets become more mature, there will be greater differentiation between brands, and consumer individualism will naturally select brands, which they identify with (Atsmon, Dixit, Wu, 2011). At the same time, even though the luxury market is still dominated by foreign brands, Chinese consumers also show an increasing interest in selected home-grown luxury brands that are able to emphasize local craftsmanship and Chinese culture and that are rapidly gaining international attention and recognition: this is the case of Shag Xia and, more recently, Shiatzy Chen and Ne Tiger.

These new trends affect not only customer behaviour and purchasing choices, but also the distribution of luxury products, both in largest and smaller cities. Despite the increasing popularity of e-commerce in China non-grocery specialists, in particular department stores, continued to dominate the distribution of luxury goods (Roberts, 2015; Fung Business Intelligence Centre, 2015). Although there were an increasing number of consumers who were keen on online purchases of products such as non-luxury clothing and footwear, beauty and personal care, and even some luxury accessories, when purchasing luxury goods (such as clothing, leather goods, watches) online they still had concerns about authenticity. Chinese consumers are more interested in the largest city's department stores (**Table 1**) and shopping malls (**Table 2**) that give them the opportunity to walk around and pick different global luxury brands.

Department store	City	Place of origin	Selected high-end/luxury brand
SKP Beijing (Formerly known as Shin Kong Plaza)* 新光天地	Beijing	Taiwan	CÉLINE, VERSACE, PRADA, VALENTINO, GUCCI, MOSCHINO, FENDI, EMPORIO ARMANI, LANVIN, LOEWE, FURLA, BALENCIAGA
Reel Department Store 芮欧百货	Shanghai, Shenzhen	China	D&G, BALENCIAGA, Dior, GIVENCHY, RALPH LAUREN, GUCCI, Pringle, LANVIN, ALEXANDER WANG
Renhe Spring Department Store 仁和春天百货	Chengdu	China	YVES SAINT LAURENT, GUCCI, GIVENCHY, BALENCIAGA, dunhill, Cartier, BURBERRY, EMPORIO ARMANI
Charter Department Store 卓展百货	Shenyang, Changchun, Beijing, Harbin	China	dunhill, D&G DOLCE&GABBANA, KENZO, GUCCI, CÉLINE, GIVENCHY, EMPORIO ARMANI, BURBERRY, Dior, CERRUTI 1881, OMEGA, TIFFANY&Co.

Table 1. Selected high-end/luxury department stores in China, as of May 2015.

Enterprise	Selected shopping mall	Location	Selected high-end/luxury brand
Beijing Yintai Property Co., Ltd. 北京银泰置业有限公司	Yintai Centre 银泰中心	Beijing, Hefei	LV, CÉLINE, VALENTINO, LANVIN, GUCCI, Cartier, HERMÈS PARIS, GIORGIO ARMANI, D&G, BOTTEGA VENETA, Louboutin, Lane Crawford
Hang Lung Properties 恒隆地产有限公司	Plaza 66, Forum 66, Center 66, Olympia 66 恒隆广场	Shanghai, Shenyang, Wuxi, Tianjin, Dalian, Kunming, Wuhan	BURBERRY, HERMÈS PARIS, BOTTEGA VENETA, CÉLINE, LANVIN, GIVENCHY, Dior, PRADA, BALENCIAGA, GIORGIO ARMANI, LV
The Wharf (Holdings) Limited 九龙仓集团有限公司	Chengdu International Finance Square 成都国际金融中心	Chengdu	MOSCHINO, GUCCI, GIORGIO ARMANI, LOEWE, BALENCIAGA, VALENTINO, Salvatore Ferragamo, BOTTEGA VENETA
Swire Properties Limited 太古地产	Taikoo Hui 太古汇	Guangzhou	dunhill, CERRUTI 1881, GIVENCHY, LANVIN, BOTTEGA VENETA, Dior, PRADA, GUCCI, BURBERRY
China Resources Land Limited 华润置地有限公司	The MixC 万象城	Shenyang, Hangzhou, Shenzhen, Chengdu, Nanning, Zhengzhou, Chongqing, Wuxi, Hefei, Qingdao, Shanghai, Taizhou, Ganzhou, Xi'an, Wenzhou, Nantong, Taiyuan, Jinan, Xiamen, Liuzhou, Shijiazhuang, Fuzhou, Ningbo	LV, MOSCHINO, CERRUTI 1881, dunhill, LANVIN, VALENTINO, LOEWE, Dior, BURBERRY, GUCCI, Cartier

Source: Fung Business Intelligence Centre (2015).

Table 2. Selected high-end/luxury shopping malls in China, as of May 2015.

The increased focus on 'exclusivity' affects not only the product design but also store footprint; this implies the need for luxury companies to better control their brand image and oversee expansion by opening more directly-operated stores, buying back franchises from local partners and taking stakes in other China retail partners (Fung Business Intelligence Centre, 2015).

In addition to this growing consumer demand, the Chinese government is strongly promoting consumption to boost economic growth as an alternative to the current drivers of internal investment and export demand, the minimum wage has risen, public holidays have been created or extended, China's one-child policy stopped and retail markets have been deregulated. Therefore, there is a need to reshape consumerism for this rising 'upper middle-class' to prevent the depletion of resources and its polluting consequences. In particular, the Chinese government has ambitious goals for promoting sustainable consumerism in China and the government's policies for energy-use and carbon emission reductions may aid in the effort towards sustainable consumption. A historical review of environmental policy in China shows how changes in the political arena led to the acceleration of environmental protection in the country. In particular, the 12th Five-Year Plan (2011–2015) is considered by the current government to be the greenest strategy document in the country's history that reflects the government's goal to promote and support sustainable consumption or 'circular economy'. Three trillion yuan (£284 billion) has been pledged to be spent over the 5 years on environmental protection alone, which is double the amount spent during the previous period. It is not just the size of the funding, but the position in the Plan of both climate change and the environment, which shows the prominence of these issues for the government – and how it wants to be seen on the global political stage. China, although starting relatively recently, has steadily developed its environmental policy and it is now a priority for the central government. In particular, the plans for the 'Circular Economy' are most relevant for this study, and clearly show that the government is seeking to promote sustainable consumption through focused policies. The 'Circular Economy approach', as described by Article 2 of the Circular Economy Law of the People's Republic of China is "a generic term for the reducing, reusing and recycling activities conducted in the process of production, circulation and consumption". It was advocated as a national strategy in 2006, and promoted at the individual firm level, chained industries, regional level, and households (Zhu, 2008). The Circular Economy strategy adopted by the central government reflect the emerging 'three R's' business models adopted in recent years by several luxury players based on the principles of recycling, reuse and reduction. The areas of intervention to ensure sustainability in the fashion system exist on all the phases of the value chain (Ricchetti and Frisa, 2011):

- provision of raw materials (low ecological impact, deriving from recycling, renewable sources or fair trade initiatives);

- production (reducing energy consumption, reusing water resources in the manufacturing processes, adopting technologies inspired by ecological principles, recycling discards and excess production, etc;

- logistics (rationalizing transportation flow, improving vehicles and reducing packaging);

- promotion (eco-compatible preparation, forms of communication and encouraging reduced environmental impact);

- end of life (reuse, recycling and degree of biodegradability).

In the world of luxury fashion, in particular, these dimensions have assumed increased importance and the various combinations of the three R's translate into differentiation of market strategies and business models, in which the element of sustainability is configured as a critical success factor. These emerging trends open today new opportunities for boosting sustainable consumption of luxury products among wealthy consumers.

5. Conclusion: towards a sustainable luxury Chinese demand?

Consumption is an important engine of economic growth. Advocating sustainable consumption can guide consumers to buy environmentally friendly products, also in the luxury markets. Sustainable consumption is an important aspect of sustainable development and has attracted growing attention among both Chinese governments and managerial literature.

http://dx.doi.org/10.5772/64477

With a rapidly growing upper middle-class (Wang, 2010) and pressure from the central government for increasing domestic consumption, consumerism in China is currently heading along an unsustainable path with severe environmental consequences that are likely to affect not only China but also the rest of the world.

However, in recent years, two main drivers have been pushing in the direction of creating the framework for both sustainable development and sustainable consumption in China, that will affect in the medium term the Chinese demand for luxury products: from one hand the strategic commitment of many international giants that dominate the global luxury industry towards sustainability and sustainable strategies as integral elements of the value proposition and drivers of competitive advantages; on the other hand, the Chines central government pressures for boosting green development by creating the conditions for sustainable consumption or 'circular economy'.

As far as the first trend, Western leading luxury brands (in particular in the fashion industry) are increasingly committed to integrating sustainability into both business strategies and business models, starting with the sustainable management of the entire supply chain. Brand enhancement strategies are also increasingly focused on the conquest of a strong brand reputation and brand loyalty based on the sustainable dimension of the concept of luxury that they propose to high-end customers. In this framework, not only big luxury international groups like the French Kering but also emerging niche brands already have started to influence and 'educate' the western consumers and they today play an important role in creating the condition for influencing the purchasing attitude and behaviour of Chinese demand towards new form of sustainable luxury with regard to both the high-end and the masstige segment.

The second driver of evolution towards sustainable consumption in the luxury Chinese market considers the key role of the government. As stress before, Chinese consumers need to be educated on sustainable consumption in order to change their consumption behaviour from unsustainable to sustainable practices. Thus an information-rich learning environment, promoted and supported by the government, both at the central and local level, could motivate and enable sustainable consumption in the medium-term. China has been pushing for sustainable consumption in recent years, as the country has faced a series of 'urban diseases' after three decades of huge and rapid economic growth. These urban ills include traffic jams, limited ability to handle sewage and garbage, and polluted air, water and soil (Xinhua, 2016). In this framework, the Chinese central government understands the need to find a new greener path for supporting economic and social development and established several policies that encourage sustainable consumption behaviours (Qu et al., 2015). In particular China's central government is continually improving its energy and environmental policies and is extremely motivated both to try consumption-shaping policies that can be implemented locally and to search internationally for best practices: the 12th Five-Year Plan has ambitious targets for resource and environmental protection and it represents the greenest strategy document in the country's history. Again, recently ten Chinese ministries, including the National Development and Reform Commission (NDRC) and the Ministry of Finance, jointly issued guidelines on green consumption, with the aim to ensure that the country adopts a 'green and healthy' consumption mode by 2020. However greater efforts should be made to more influence people's behaviour towards sustainable consumption, as recommended by the Sustainable Consumption Roundtable (SCR) (2006).

In this framework, the luxury industry can today represent for China a 'joint laboratory' to innovate firms' strategic approaches and to boost sustainable consumption of luxury products in consideration to the impact that scarce resource depletion and environmental pollution have on both the survival of the luxury industry and the society. The dimension of the challenges implies a reinforced cooperation between all actors involved. On one side, luxury companies can act as a driving force to create the conditions for enabling a virtuous process towards sustainable luxury consumption demand among wealthy Chinese, due to the importance that Western luxury brands play in the purchasing attitude of consumers. On the other side the dimension of the challenges imply a strong government involvement to change society's attitudes towards consumption. To stimulate the desire of consumers for sustainable products will require innovative partnerships between all actors involved, both private and public.

These emerging trends open interesting research opportunities under the profile of the analysis of both strategic behaviour and the emerging in the Chinese luxury market, today barely investigated by managerial literature.

Author details

Patrizia Gazzola*, Enrica Pavione and Roberta Pezzetti

*Address all correspondence to: patrizia.gazzola@uninsubria.it

Department of Economics, University of Insubria, Monte Generoso, Varese, Italy

References

Akenji, L. (2014). Consumer scapegoatism and limits to green consumerism. Journal of Cleaner Production 63, 13–23.

Atsmon, Y., Dixit, C., Wu, C. (2011). *Tapping China's Luxury Goods Market*, McKinsey & Company, April, http://www.mckinsey.com/business-functions/marketing-and-sales/our-insights/tapping-chinas-luxury-goods-market

Bain & Company (2014a). *China Luxury Market Study*, January 20, Bain & Company Inc., Boston

Bain & Company (2014b). *Luxury Goods Worldwide Market Monitor*, Fall-Winter, Bain & Company Inc., Boston.

Bain & Company (2015). China Luxury Market Study. January 20, Bain & Company Inc., Boston.

Banathy, B.H. (1996). *Designing Social Systems in a Changing World*. New York: Plenum Press.

Barton, D., Chen, Y., Jin, A. (2013). *Mapping China's Middle Class*, McKinsey & Company, June, http://www.mckinsey.com/industries/retail/our-insights/mapping-chinas-middle-class.

Beaton, C. and Perera, O. (2012). *Global Outlook on Sustainable Consumption and Production Policies: Taking Action Together (Report 2012:1)*. France: United Nations Environment Program (UNEP).

The Boston Consulting Group (2014). *True Luxury Global Consumer Insight*, http://www.bcg.it/documents/file182491.pdf

Colombo, G. and Gazzola, P. (2015a). *Building CSR in the corporate strategy*, Strategica International Academic Conference, Third Edition, Bucharest, October 29–31, 2015, Local Versus Global, SNSPA, Bucharest, Romania.

Colombo, G. and Gazzola, P. (2015b). *From uncertainty to opportunity: How CSR develops dynamics capabilities*. 15th Annual Conference of European Academy of Management, Warsaw, 17–20 June.

Fuad-Luke, A. (2002). *Slow Design: A Paradigm Shift in Design Philosophy?*, Development by Design, Bangalore, India.

Fung Business Intelligence Centre. (2015). *Luxury Market in China*, May 22, Hong Kong.

Goldman Sachs, (2010), BRICs, Goldman Sachs, http://www2.goldmansachs.com/ideas/brics/index.html

Gouxin Li, Goufeng Li, Kambele Z. (2012). "Luxury Fashion Brand Consumers in China: Perceived Value, Fashion Lifestyle, and Willingness to Pay", in *Journal of Business Research*, vol. 65, 10, pp. 1516–1522.

Guercini, S. and Ranfagni, S. (2012). *Social and green sustainability and the Italian Mediterranean fashion brands*. EUROMED Marseille, 28–29 June.

Hoffmann, J. and Coste-Maniere I. (2012). *Luxury Strategy in Action*. Palgrave Macmillan, New York.

Holbrook M.B. (1999). *Consumer Value: A Framework for Analysis and Research*, New York: Routledge.

Hörndahl, M. and Dervisevic, S. (2015). *Shanghai's Development into Sustainable Consumption: An Insight from a Retail Apparel's Industry on Change in Consumer Behaviour*, Dissertation Thesis, VT2015KF13, Borås.

Hurun Research Institute, (2016). *12th Chinese Luxury Consumer Survey*, Hurun Institute, Shanghai

Kai, S.B., Chen, O.B., Chuan, C.S., Seong, L.C. and Kevin, L.L.T. (2013). *Determinants of willingness to pay for organic products*. Middle-East Journal of Scientific Research. 14 (9), 1171–1179.

KPMG (2007). *Luxury Brand in China*, KPMG Report, Hong Kong.

Maniatis, P. (2015). *Investigating factors influencing consumer decision-making while choosing green products*.Journal of Cleaner Production. 132, 1–14.

Mathews, C. (2012). *Towards a Framework for Sustainable Consumption in China. Diss. Imperial College London*: London: Faculty of Natural Sciences.

Matthes, J., Wonneberger, A. and Schmuck, D. (2013). *Consumers' green involvement and the persuasive effects of emotional versus functional ads.* Journal of Business Research. 1e9.67 (9), 1885–1893. http://dx.doi.org/10.1016/j.jbusres.2013.11.054 (accessed 03.01.14).

Owusu, V. and Anifori, M.O. (2013). *Consumer willingness to pay a premium for organic fruit and vegetable in Ghana.* International Food and Agribusiness Management Review. 16 (1), 67–86.

Pavione, E., Pezzetti, R. and Dall'Ava M. (2016). *Emerging competitive strategies in the global luxury industry in the perspective of sustainable development: the case of Kering Group*, MDKE, 4 (2), 241–261.

Pavione, E. and Pezzetti, R. (2015). *Responsible and sustainable luxury in the global market: new emerging strategies in the luxury sector*, In: Brătianu C. etal (Ed.) Strategica International Conference "Local versus Global", Bucharest, Romania, October 29–31: pp. 73–80.

Pavione, E. and Pezzetti, R. (2014). *Emerging competitive strategies in the luxury sector: exploiting the mass-market vs refocusing on the high-end segment.* 17th Toulon-Verona Conference "Excellence in Services", John Moores University, Liverpool.

Ricchetti, M. and Frisa, M.L. (Eds), (2011). *Il bello e il buono. Le ragioni della moda sostenibile.* Marsilio Editore, Venezia.

Roberts. F. (2015). Luxury Goods Industry in 2015, Euromonitor International, http://blog.euromonitor.com/2015/11/luxury-goods-industry-in-2015.html.

Qu, Y., Li, M., Jia, H. and Guo, L. (2015). *Developing more insights on sustainable consumption in China based on Q methodology.* Sustainability, 7 (10), 14211–14229.

Seringhaus, F.R. (2002). *Cross-cultural exploration of global brands and the internet.* In: 18th Annual IMP Conference, Groupe ESC Dijon Bourgogne Dijon, France, 1–29.

Spaargaren, G. (2003). *Sustainable consumption: A theoretical and environmental policy perspective.* Society & Natural Resources. 16, 687–701.

Stern, P.C. (2011), *Contributions of psychology to limiting climate change.* American Psychologist. May–Jun, 66 (4), 303–314.

Sustainable Consumption Roundtable (SCR) (2006) Towards sustainable consumption. Available at: http://www.sd-commission.org.uk/publications.php?id=367 (accessed on 15 March 2016).

Thogersen, J., Jorgensen, A., Sandager, S. (2012). *Consumer decision-making regarding a "green" everyday product.* Psychology Mark. 29 (4), 187–197.

Xia, W. and Zeng, Y., (2006). Consumer's attitudes and willingness-to-pay for green food in Beijing. http://dx.doi.org/10.2139/ssrn.2281861. Available at SSRN: http://ssrn.com/abstract1/42281861 (June 19) .

Xiao Lu, P. (2008), *Elite China*, John Wiley Hoboken, NJ.

http://dx.doi.org/10.5772/64477

Xinhua (2016) *China issues guidelines to boost green, sustainable consumption*. The Global Times 2016-3-1.

Xu, P., Zeng, Y., Fong, Q., Lone, T. and Liu, Y. (2012). *Chinese consumers' willingness to pay for green and eco-labeled seafood*. Food Control. 28, 74–82.

Wang, H.H. (2010). *The Chinese Dream: The Rise of the World's Largest Middle Class and What it Means to You*. Bestseller Press, Hong Kong.

Wang, C.L, Chen, Z.X., Chan, A.K. and Zheng, Z.C. (2000). *The influence of hedonic values on consumer behaviours: an empirical investigation in China*, Journal of Global Marketing, 14 (2), 169–186.

Zhao, H., Gao, Q., Wu, Y., Wang, Y. and Zhu, X. (2014). *What affects green consumer behavior in China? A case study from Qingdao*. Journal of Cleaner Production 63, 143–151.

Zhang, J. and Sharon, S. (2003). *Cultural values in advertisements to the Chinese X generation: Promoting modernity and individualism*, Journal of Advertising, 32 (1), 23–33.

Zhu, D. (2008). *Background, pattern and policy of China for developing circular economy*. Chinese Journal of Population, Resources and Environment. [Online] 6 (4). Available at http://www.indigodev.com/documents/CE_Zhu_Background.pdf.

Chapter 13

Multicriteria Decision Analysis of Health Insurance for Foreigners in the Czech Republic

Haochen Guo

Additional information is available at the end of the chapter

http://dx.doi.org/10.5772/66790

Abstract

Multicriteria decision making (MCDM) is the superclass of model in most readily understandable branch of basic leadership. It is a branch of a general class of operations research (OR) models which manages choice issues under the nearness of various choice criteria. MCDM techniques have advanced to oblige different sorts of utilizations. Many techniques have been produced, with even small varieties to existing strategies bringing about the production of new branches of examination. The aim of this chapter is to present selected MCDM methods and application in case of health insurance decision problems.

Keywords: multicriteria decision making, SAATY method, WSA method, MAPPAC method, TOPSIS method, ELECTRE method, health insurance

1. Introduction

Multicriteria decision making (MCDM) has seen a staggering measure of utilization. Its part in various application territories has expanded fundamentally, particularly as new techniques are created and as old strategies make strides. This chapter breaks down a few basic MCDM strategies and decides their relevance by assessing their relative points of interest and burdens. A survey of the utilization of these strategies and an examination of the development of their utilization after some time is then performed. The objective of this chapter is through setting up a situation as case study which uses the MCDM methods (WSA, MAPPAC, TOPSIS, and ELECTRE) to choose the best and most appropriate health insurance (UNIQA, SLAVIA, MAXIMA, and VZP) for international policyholder visiting the Czech Republic.

2. Applications of MCDM approaches in decision problems

MCDM is a branch of a general class of operations research (OR) models which deals with decision problems under the presence of a number of decision criteria. In light of the distinctive purposes and diverse information sorts, MCDM is isolated into multiobjective decision making (MODM) and multiattribute decision making (MADM). Within the field of OR, the development of MCDM is based on the simple finding in terms of environment: criteria, goals, attributes, objectives, and decision matrix.

The MCDM field is given to the advancement of suitable procedures that can be utilized in circumstances where different clashing choice elements must be considered all the while.

Customary enhancement, measurable, and econometric investigation approaches utilized inside the money-related building connections regularly taking into account the presumption that the considered issue is all around postured, very much defined with respect to the truth included, and they as a rule consider the presence of a solitary target, assessment standard, or perspective that underlies the led examination. In such a case, the arrangement of monetary issues is anything but difficult to acquire. In any case, in reality, the demonstrating of monetary issues depends on an alternate sort of rationale mulling over the accompanying components:

- The presence of various criteria
- The clashing circumstance between the criteria
- The mind boggling, subjective, and not well-organized nature of the assessment process
- The presentation of money-related chiefs in the assessment process

Financial related and operation specialists have as of late embraced this creative, thorough, and reasonable point of view, with results. On the premise of the diverse creators' view that it is conceivable to recognize primary reasons which have persuaded a change of perspective in the displaying of the money-related issues:

- Formulating the issue as far as looking for the ideal, objective get included in an exceptionally limit dangerous, regularly unessential to the genuine choice issue.
- The diverse monetary choices are taken by the people and not by the models; the leaders get increasingly profoundly included in the basic leadership process. With a specific end goal to take care of issues, it gets to be important to think about their inclinations, their encounters, and learning.
- For monetary choice issues, for example, the decision of venture undertakings, the portfolio choice, and the assessment of business disappointment hazard.

MCDM methodologies are appropriate for the investigation of a few money-related basic leadership issues. The broadened way of the components that influence monetary choices, the many-sided quality of the money related, business and financial situations, the subjective way of numerous budgetary choices are just a percentage of the elements of money-related choices which are as per the MCDM demonstrating system. **Table 1** outlines the utilizations of MCDM techniques.

http://dx.doi.org/10.5772/66790

Method	Advantages	Disadvantages	Areas of application
MAUT	Takes uncertainty into account; can incorporate preferences	Needs a lot of input; preferences need to be precise	Economics, finance, actuarial, water management, energy management, agriculture
AHP	Easy to use; scalable; hierarchy structure can easily adjust to fit many sized problems; not data intensive	Problems due to interdependence between criteria and alternatives; can lead to inconsistencies between judgment and ranking criteria; rank reversal	Performance-type problems, resource management, corporate policy and strategy, public policy, political strategy, and planning
CBR	Not data intensive; requires little maintenance; can improve over time; can adapt to changes in environment	Sensitive to inconsistent data; requires many cases	Businesses, vehicle insurance, medicine, and engineering design
DEA	Capable of handling multiple inputs and outputs; efficiency can be analyzed and quantified	Does not deal with imprecise data; assumes that all input and output are exactly known	Economics, medicine, utilities, road safety, agriculture, retail, and business problems
Fuzzy set theory	Allows for imprecise input; takes into account insufficient information	Difficult to develop; can require numerous simulations before use	Engineering, economics, environmental, social, medical, and management
SMART	Simple; allows for any type of weight assignment technique; less effort by decision makers	Procedure may not be convenient considering the framework	Environmental, construction, transportation and logistics, military, manufacturing and assembly problems
GP	Capable of handling large-scale problems; can produce infinite alternatives	It's ability to weight coefficients; typically needs to be used in combination with other MCDM methods to weight coefficients	Production planning, scheduling, health care, portfolio selection, distribution systems, energy planning, water reservoir management, scheduling, wildlife management
ELECTRE	Takes uncertainty and vagueness into account	Its process and outcome can be difficult to explain in layman's terms; outranking causes the strengths and weaknesses of the alternatives to not be directly identified	Energy, economics, environmental, water management, and transportation problems
PROMETHEE	Easy to use; does not require assumption that criteria are proportionate	Does not provide a clear method by which to assign weights	Environmental, hydrology, water management, business and finance, chemistry, logistics and transportation, manufacturing and assembly, energy, agriculture
SAW	Ability to compensate among criteria; intuitive to decision makers; calculation is simple does not require complex computer programs	Estimates revealed do not always reflect the real situation; result obtained may not be logical	Water management, business, and financial management
TOPSIS	Has a simple process; easy to use and program; the number of steps remains the same regardless of the number of attributes	Its use of Euclidean distance does not consider the correlation of attributes; difficult to weight and keep consistency of judgment	Supply chain management and logistics, engineering, manufacturing systems, business and marketing, environmental, human resources, and water resources management

Source: See Mark and Patrick (2013).

Table 1. Summary of MCDM methods.

3. Case study: MCDM of health insurance products in the Czech Republic

Everybody who visits the Czech Republic needs sufficient proof of health insurance. If you are a nonEU national and do not work for a Czech employer, you need to get travel health insurance before coming to the Czech Republic. Based on this background, the subject of the case study is an international tourist who wants to visit the Czech Republic and hence needs to make a decision for health insurance.

3.1. Health insurance for foreigners in the Czech Republic

Foreign nationals in the Czech Republic are required to have valid health insurance. There are two types of health insurance that are described below:

1. **Public health insurance**

   The following people have a legal right to public health insurance:

   - Anyone with permanent residency status in the Czech Republic
   - Employees whose employer is based in the Czech Republic

2. **Commercial (private) health insurance**

   There are two varieties of commercial health insurance:

   - Comprehensive medical insurance: it is suitable for foreigners who intend to stay for 90 days or longer and require long-term visa or long-term stay, or request an extension of a visa or residence permit. This health insurance is similar to public health insurance.
   - Basic medical insurance: it covers necessary treatment and hospitalization which cannot be postponed at all health care facilities. This insurance is recommended for individuals who do not fall under the public health system and plan only short-term stay. The insurance covers costs incurred as a result of an accident or sudden illness during the stay, including any costs related to repatriation to the country that issued the travel document or to the country where the foreigner has legal residence. Minimal coverage must be EUR 60,000 excluding any financial contribution to the aforesaid costs on the part of the insured person. See euraxess.cz.

3.2. Input data interpretation – weight calculation criteria (SAATY method)

Usually, before selecting the best and most appropriate health insurance, policyholder needs to consider the items of premium, claims, and minimum coverage maturity as criteria. Also, there are four insurance companies who provide health insurance for foreigner as alternatives, which are Pojišťovna VZP, a.s., UNIQA pojišťovna, a.s., MAXIMA pojišťovna, a.s., and SLAIA pojišťovna (see **Table 2**).

http://dx.doi.org/10.5772/66790

		C1	C2	C3
		Premium/CZK per month	**Claims/CZK per month**	**Min. coverage maturity/month**
A1	UNIQA	550	2,026,852	3
A2	SLAVIA	433	1,000,000	3
A3	MAXIMA	635	1,621,482	6
A4	VZP	1235	3,000,000	4
		MIN	MAX	MIN

Source: pojistenicizincu.cz.

Table 2. Input data.

In the process of making multiattribute decision, it should set weight for different criteria by using direct and indirect method, and in this case study will use the SAATY pairwise comparison method (Thomas, 2004, 2006, 2008),which is a kind of indirect method created by Thomas L. Saaty. **Table 3** presents a typical criteria matrix C.

	C1	C2	C3	C4
C1	C11	C12	C13	C14
C2	C21	C22	C23	C24
C3	C31	C32	C33	C34
C4	C41	C42	C43	C44

Table 3. Typical criteria matrix C.

$C_{i,j}$ in **Table 3** presents the preference on criteria i to criteria j. Which is also called the ratio w_i/w_j, and the preference can be judged from 1 to 9 in fundamental scale of absolute number which will be shown in **Table 4**, $C_{i,j} \in [\frac{1}{9}, 9]$ and $C_{i,j} \cdot C_{j,i} = 1$. Hence, if preferred i to j, then $C_{i,j} > 1$; if preferred j to i , then $C_{i,j} < 1$; if preferred the same preference on criteria i to j, then $C_{i,j} = 1$.

Table 5 provides the weights calculated by the SAATY method.

Intensity of importance	Definition	Explanation
1	Equal importance	Two activities contribute equally to the objective
2	Week or slight	
3	Moderate importance	Experience and judgment slightly favor one activity over another
4	Moderate plus	
5	Strong importance	Experience and judgment strongly favor one activity over another
6	Strong plus	
7	Very strong or demonstrate importance	An activity is favored very strongly over another; its dominance demonstrated in practice
8	Very, very strong	
9	Extreme importance	The evidence favoring one activity over another is of the highest possible order of affirmation
1.1–1.9	When activities are very close a decimal is added to 1 to show their difference as appropriate	A better alternative way to assigning the small decimals is to compare two close activities with other widely contrasting ones, favoring the larger one a little over the smaller one when using the 1–9 values
Reciprocals of above	If activity *i* has one of the above nonzero numbers assigned to it when compared with activity *j*, then *j* has the reciprocal value when compared with *i*	A logical assumption
Measurements from ratio scales		When it is desired to use such numbers in physical applications. Alternatively, often one estimates the ratios of such magnitudes by using judgment

Source: See Thomas (2008).

Table 4. Fundamental scale of absolute numbers.

	C1	C2	C3	v_i	w_j
C1	1	7	4	3.036589	**0.706365**
C2	0.142857	1	5	0.893904	**0.207938**
C3	0.25	0.2	1	0.368403	**0.085697**

Table 5. Weights calculate by SAATY method.

4. Results due to select methods

In this section, the procedure for four MCDM methods, which are WSA, MAPPAC, TOPSIS, and ELECTRE III, will be demonstrated.

4.1. WSA method

Weighted sum analysis method is based on the linear utility function construction at the scale 0–1. The worst variant based on the given criteria will have utility 0; the best variant will have utility 1 and other variants will have utility between both extreme values. WSA derives from the principle of utility maximization; however, the method presumes only linear function. For the maximization case, the best alternative is the one that yields the maximum total performance value.

First, the normalized criteria matrix will be created $R = (r_{ij})$, whose elements are derived from criteria matrix $Y = (y_{ij})$, based on

$$r_{ij} = \frac{Y_{ij} - D_j}{H_j - D_j}, \tag{1}$$

where r_{ij} is variant's utility of X_i when evaluated based on criteria Y_j, r_{ij} represents corresponding values from initial criteria matrix, D_j is the lowest criteria value of Y_j, and H_j is the highest criteria value of Y_j. This matrix represents matrix of utility values from ith variant based on jth criteria. Criteria values are linearly transformed that $r_{ij} \in \langle 0, 1 \rangle$. D_j corresponds to minimal criteria value of column j and H_j corresponds to maximum criteria value in column j. In case of minimization criteria normalization of column in matrix can be executed as

$$r_{ij} = \frac{H_j - Y_{ij}}{H_j - D_j}. \tag{2}$$

If it is necessary that all criteria in the matrix must be maximized then before executing standardization/normalization of matrix, it is necessary to recount elements in the column as follows

$$Y_{ij\,-\,\max} = H_{j\,-\,\min} - Y_{ij\,-\,\min},\ i = 1, 2, \ldots, p. \tag{3}$$

Meaning, deduct from the current highest element maximum $H_{j-\min}$ in the given column progressively with all other elements and by this the column with minimization criteria will be transformed to maximization. When using additive multicriterial utility function the variance utility a_i is then equal to

$$u(a_i) = \sum_{j=1}^{k} v_j \cdot r_{ij}. \tag{4}$$

Variant that reaches the maximum utility value is selected as the best, alternatively it is possible to rank the variants based on descending utility values, see Iveta and Jana (2015). The calculation of WSA method is done by the following procedures.

Obtaining the utilities and the preferred order of alternatives can be expressed as: A1 > A2 > A3 > A4. Hence, on the basis of the results above, the optimal choice is: UNIQA > SLAVIA > MAXIMA > VZP.

4.2. MAPPAC method

Multicriterion analysis of preferences by means of pairwise actions and criterion comparisons (MAPPAC) method, first introduced by Matarazzo (1986), is based on the comparison of pairs of feasible actions taking into account all possible pairs of criteria. The proposed method, known as MAPPAC, is based on a pairwise comparison of alternatives relative to each pair of criteria, defining the two relations *P* (preference) and *I* (indifference), which constitute a complete preorder. Moreover, by aggregating these preferences, it is possible to obtain a variety of relations on a set of feasible actions (Paruccini and Matarazzo, 1994). See Salt (2011).

The MAPPAC method has three assumptions (Matarazzo, 1990):

- For each K_i a quantitative, V_{ij} can be assigned to each alternative, a_j representing the performance of a_j with respect to K_i;
- A quantitative value $V_{ij}c$;
- An be assigned to each alternative, a_j on the basis of each criterion K_i;
- The value $V(V_{ij})$ of each V_{ij} can be quantified on the interval [0,1]; and
- The criteria are mutually difference and independent.

For each K_i a value V_{ij} is assigned to each a_j representing the performance of a_j on the basis of K_i. A numerical weight w_i is assigned to each K_i representing the importance of K_i with $\sum_{n=1}^{n} w_i = 1$. For each K_i representing the importance of $v(v_{ij})$ to each V_{ij} with $0 \le v(v_{ij}) \le 1$. see Hassan (2013). The calculation of MAPPAC method is done by the following procedures. The process of modified input data is the same in **Table 6**.

	MAX	MAX	MAX
	C1	C2	C3
A1	685	2,026,852	3
A2	802	1,000,000	3
A3	600	1,621,482	0
A4	0	3,000,000	2
Weights	0.70636	0.20794	0.08570

Table 6. WSA modified input data.

Obtaining the utilities and the preferred order of alternatives can be expressed as: A2 > A1 > A3 > A4. Hence, on the basis of the results above, the optimal choice is: SLAVIA > UNIQA > MAXIMA > VZP (**Tables 7–9**).

http://dx.doi.org/10.5772/66790

	MAX	MAX	MAX	
	C1	**C2**	**C3**	**U (alternatives)**
A1	0.85411	0.51343	1.00000	**0.79577**
A2	1.00000	0.00000	1.00000	**0.79206**
A3	0.74813	0.31074	0.00000	**0.59307**
A4	0.00000	1.00000	0.66667	**0.26507**
Weights	0.70636	0.20794	0.08570	

Table 7. WSA normalized criterion matrix R.

	C1	**C2**	**C3**
A1	0.85411	0.51343	1.00000
A2	1.00000	0.00000	1.00000
A3	0.74813	0.31074	0.00000
A4	0.00000	1.00000	0.66667
Weights	0.70636	0.20794	0.08570

Table 8. MAPPAC matrix C.

	A1	**A2**	**A3**	**A4**	**From above**	Total
A1	0.00000	0.37944	1.00000	0.81985	2	**2**
A2	0.62056	0.00000	0.81509	0.76695	1	**1**
A3	0.00000	0.18491	0.00000	0.71701	3	**3**
A4	0.18015	0.23305	0.28299	0.00000	4	**4**
From below	2	1	3	4		

Table 9. MAPPAC matrix P.

4.3. TOPSIS method

The technique for order of preference by similarity to ideal solution (TOPSIS) is based on the concept that the chosen alternative should have the shortest distance from the positive ideal solution and the longest distance from the negative ideal solution. It is a method of compensatory aggregation that compares a set of alternatives by identifying weights for each criterion, normalizing scores for each criterion and calculating the distance between each alternative and the ideal alternative, which is the best score in each criterion. The TOPSIS method is expressed in a succession of six steps as follows:

1. Calculate the normalized decision matrix. The normalized value r_{ij} is calculated by

$$r_{ij} = x_{ij}\sqrt{\sum_{i-1}^{m} x_{ij}^2}\ i = 1,2,...,m \text{ and } j = 1,2,...,n. \tag{5}$$

2. Calculate the weighted normalized decision matrix

$$v_{ij} = r_{ij}\, w_j\ i = 1,2,...,m \text{ and } j = 1,2,...,n. \tag{6}$$

where w_j is the weight of the jth criterion or attribute and $\sum_{j=1}^{n} w_j = 1$.

3. Determine the ideal (A^*) and negative ideal (A^-) solutions

$$A^* = \{(\max_i v_{ij}|j\in C_b), (\min_i v_{ij}|j\in C_c)\} = \{v_j^*|j = 1,2,...,m\} \tag{7}$$

$$A^- = \{(\min_i v_{ij}|j\in C_b), (\max_i v_{ij}|j\in C_c)\} = \{v_j^-|j = 1,2,...,m\} \tag{8}$$

4. Calculate the separation measures using the m-dimensional Euclidean distance. The separation measures of each alternative from the positive ideal solution and the negative ideal solution, respectively, are as follows

$$S_i^* = \sqrt{\sum_{j=1}^{m} (v_{ij} - v_j^*)^2}, j = 1,2,...,m \tag{9}$$

$$S_i^- = \sqrt{\sum_{j=1}^{m} (v_{ij} - v_j^-)^2}, j = 1,2,...,m \tag{10}$$

5. Calculate the relative closeness to the ideal solution

$$RC_i^* = \frac{S_i^-}{S_i^* + S_i^-}, i = 1,2,...,m \tag{11}$$

6. Rank the preference order.

The calculation of TOPSIS method is done by the following procedures (**Tables 10** and **11**).

	MIN	MAX	MIN
	C1	**C2**	**C3**
A1	0.35367	0.49543	0.35857
A2	0.27843	0.24443	0.35857
A3	0.40833	0.39634	0.71714
A4	0.79414	0.73329	0.47809
Weights	0.70636	0.20794	0.08570

Table 10. TOPSIS normalized matrix R.

http://dx.doi.org/10.5772/66790

	MIN	MAX	MIN			
	C1	C2	C3	di+	di-	ci
A1	0.24982	0.10302	0.03073	0.07260	0.31698	**0.81365**
A2	0.19668	0.05083	0.03073	0.10165	0.36557	**0.78243**
A3	0.28843	0.08241	0.06146	0.11946	0.27435	**0.69665**
A4	0.56096	0.15248	0.04097	0.36442	0.10370	**0.22152**
Weights	0.70636	0.20794	0.08570			
Ideal	0.19668	0.15248	0.03073			
Basal	0.56096	0.05083	0.06146			

Table 11. TOPSIS weighted criterion matrix W.

Obtaining the utilities and the preferred order of alternatives can be expressed as: A1 > A2 > A3 > A4. Hence, on the basis of the results above, the optimal choice is: UNIQA > SLAVIA > MAXIMA > VZP.

4.4. ELECTRE III method

The ELECTRE (for elimination and choice translating reality; English translation from the French original) method was first introduced in Benayoun et al. (1966). The basic concept of the ELECTRE method is to deal with "outranking relations" by using pairwise comparisons among alternatives under each one of the criteria separately. The outranking relationship of the two alternatives A_i and A_j denoted describes that even when the ith alternative does not dominate the jth alternative quantitatively, then the decision maker may still take the risk of regarding A_i as almost surely better than A_j in Roy (1973). Alternatives are said to be dominated, if there is another alternative which excels them in one or more criteria and equals in the remaining criteria.

The ELECTRE method begins with pairwise comparisons of alternatives under each criterion. Using physical or monetary values, denoted as $g_i(A_j)$ and $g_i(A_k)$ of the alternatives A_j and A_k, respectively, and by introducing threshold levels for the difference $g_i(A_j)-g_i(A_k)$, the decision maker may declare that he/she is indifferent between the alternatives under consideration, that he/she has a weak or a strict preference for one of the two, or that he/she is unable to express any of these preference relations. Therefore, a set of binary relations of alternatives, the so-called outranking relations, may be complete or incomplete. Next, the decision maker is requested to assign weights or importance factors to the criteria in order to express their relative importance.

Through the consecutive assessments of the outranking relations of the alternatives, the ELECTRE method elicits the so-called concordance index, defined as the amount of evidence to support the conclusion that alternative A_j outranks, or dominates, alternative A_k, as well as the discordance index, the counterpart of the concordance index. Finally, the ELECTRE method yields a system of binary outranking relations between the alternatives (**Tables 12** and **13**).

	A1	A2	A3	A4
A1	0.00000	0.20794	1.00000	0.79206
A2	0.70636	0.00000	0.79206	0.79206
A3	0.00000	0.20794	0.00000	0.70636
A4	0.20794	0.20794	0.29364	0.00000

Table 12. ELECTRE III matrix S.

Indifference classes	Alternatives
1.	UNIQA
2.	SLAVIA
3.	MAXIMA
4.	VZP

Table 13. ELECTRE III indifference classes.

The calculation of ELECTRE method is done by the following procedures. The first process of modified input data is the same in **Table 6**.

Obtaining the utilities and the preferred order of alternatives can be expressed as: A1 > A2 > A3 > A4. Hence, on the basis of the results above, the optimal choice is: UNIQA > SLAVIA > MAXIMA > VZP.

5. Discussion and summary

Using the Borda method to rank the results obtains the most appropriate insurance for policyholder. The Borda method is an election method in which the voters rank options or candidates in order of preference: the highest Borda count wins.

Table 14 presents the Borda method result for optimal health insurance.

Based on the Borda method result, UNIQA wins with highest Borda count, hence the optimal health insurance for policyholder is UNIQA.

	WSA	MAPPAC	TOPSIS	ELECTRE	SUM	Ranking
UNIQA	**1**	**2**	**1**	**1**	**5**	**1**
SLAVIA	2	1	2	2	7	2
MAXIMA	3	3	3	3	12	3
VZP	4	4	4	4	16	4

Table 14. Borda method for optimal health insurance.

http://dx.doi.org/10.5772/66790

6. Conclusion

Various MCDM techniques have been developed and used in the course of recent years. As of late, on account of easement cause by driving development, consolidating diverse techniques has ended up ordinary in MCDM. The blend of numerous techniques addresses gaps that might be found in specific strategies. These strategies, alongside the techniques in their unique structures, can be to a great degree fruitful in their applications, just if their qualities and shortcomings are appropriately surveyed. This chapter illustrates the case study of MCDM methods and evaluation for solving the problem which select the optimal choice of health insurance for foreigners who are willing to visit the Czech Republic.

Acknowledgements

The research was supported by the SGS project of VSB-TU Ostrava under No. SP2016/11.

JEL classification: C44, D81, C22

Author details

Haochen Guo

Address all correspondence to: haochen.guo@vsb.cz

Department of Finance, Faculty of Economics, VŠB-Technical University of Ostrava, Ostrava, Czech Republic

References

Benayoun, R., Roy, B., & Sussman, B. (1966). ELECTRE: Une méthode pour guider le choix en présence de points de vue multiples. Note de travail, 49.

Benedetto Matarazzo (1986). Multicriterion analysis of preferences by means of pairwise actions and criterion comparisons (MAPPACC). Applied Mathematics and computation. Volume 18. Issue 2. P119–141.

Benedetto Matarazzo (1990). A pairwise criterion comparison approach: the mappac and pragma methods. Springer Berlin Heidellberg. P253-273. ISBN 978-3-642-75935-2.

Dincer Erdal Salt. (2011). The structural analysis of key indicators of Turkish manufacturing industry: ORESTE and MAPPAC applications. European Journal of Scientific Research. Vol.60. No.1. 6-18. ISSN 1450-216X.

Dockalikova Iveta, Klozikova Jana. (2015). MCDM methods in practice: localization of suitable places for company utilization AHP and WSA, TOPISIS method. Proceedings of the 11th European Conference on Management Leadership and Governance. Portugal. ISBN 978-1-910810-77-4.

Jafari Hassan. (2013). Presenting an integrative approach of MAPPAC and FANP and balanced scorecard for performance measurements of container terminals. International Journal of Basic Sciences & Applied Research. Vol.2(4), 388–398.

Paruccini and Matarrazzo (1994). The use of MAPPAC and PRAGMA methods to support decision on industrial waste management. P95–110. ISBN 0792329228.

Roy.B (1973). Critères multiples et modélisation des préférences: l'apport des relations de surclassement. Université Paris IX-Dauphine.

Saaty L. Thomas. (2004). Decision making – the analytic hierarchy and network processes (AHP/ANP). Jounal of Systems Science and Systems Engineering. Vol.13, No.1, 1–35.

Saaty L. Thomas. (2006). Fundamentals of decision making and priority theory with the analytic hierarchy process. Vol. VI of the AHP series. RWS Publications. ISBN 978-0-9620317-6-2.

Saaty L. Thomas. (2008). Decision making with the analytic hierarchy process. Int. J. Services Sciences, Vol.1, No.1.

Velasquez Mark, Hester T. Patrick. (2013). An analysis of multi-criteria decision making methods. International Journal of Operations Research Vol.10, No.2, 55–66.

Chapter 14

China's "New Normal" and Its Quality of Development

Jin Han, Haochen Guo and Mengnan Zhang

Additional information is available at the end of the chapter

http://dx.doi.org/10.5772/66791

Abstract

China's new normal means a new higher stage of development, when an alternative is to improve the quality of economic development instead of accelerating growth rate by expansion policies. And the quality of development is the quality of living of most people. This study is to examine the current situations of China's quality of development by comparing China's human development index, inequality indices (Gini, quintile, and Palma), and development potential (human capital index) with the developed countries in Europe, North America, and Oceania, as well as countries with typical traits, such as the Latin American countries, Japan, and Czech Republic; further to put forward China's policy focuses in the new normal stage according to the concluded research results.

Keywords: China's new normal, human development index (HDI), inequality, human capital index (HCI), quality of life

1. China's new normal–a new higher stage

China's new normal is original from the slowdown of the GDP growth rate in recent years. **Graph 1** shows three obvious slowdowns since 1979. The three slowdowns are all accompanying with economic upheavals and big inflations, only the last and current one induces a new concept, "New Normal."

In May 2014, President Xi Jinping put forward the "new normal of China's economy," and described a series of new performances of China's economy. On December 5, 2014, the Politburo meeting of the Communist Party of China formally advocated to "take the initiative to adapt to the economic development of the new normal." Since then, the Chinese economy has entered a "new normal" stage.

Graph 1. Per capita GDP growth China 1979–2014 (1978 constant). Data source: Chinese statistics yearbook 2015: 3-1, 3-5.

Generally, the "new normal" has two characteristics: the first is about the slowdown from high-speed growth to high-middle-speed growth; the second is about the transformation of growth pattern from scale extensive growth to quality and intensive growth [1]. For the future strategy of Chinese government, there seem also two main streams: one is focusing on the growth speed, while thinking the transformation of growth pattern is given, for they think China need to sustain a growth speed to cross the middle-income trap that is the first priority of China [2–5]; another is to focus on the transformation of growth pattern and growth quality, while keeping the high-middle-speed growth even middle-speed growth [1, 6, 7]. We stand for the second view.

The speed slowdown of China's economic growth is not a bad thing. First, the growth rate from the high-speed down to high-middle-speed is suitable for China. China's GDP growth rate of 6.9% and per capita GDP growth rate of 6.3% in 2015, are still high enough in the context of the world (the world average of GDP growth rate is 2.5%, 2015). Second, the slowdown is beneficial from the consideration of the limit of natural resources and serious environmental problems of China, as the environment could no longer sustain the long lasting high-speed growth, even if it is further lower; after all, the ecological environment is the precondition of a country's sustainable development. Third, as a common sense, high-speed growth is apt to bring economic upheaval, and destroy the stability of development. Hence, in long run, keeping a high-middle-speed is better than high-speed for the sake of stable sustainable development.

Moreover, the speed slowdown is a good signal that indicates China has been entering a new stage of development, when an alternative is to improve the quality of economic development instead of accelerating growth rate by expansion policies. And the quality of development is the quality of living of most people; i.e., we can pay more attentions to most people's quality of life, as like a developed country's performances.

In brief, China's new normal means a new higher stage of development with the pursuit of a developed country. This study is to examine the current situations of China's quality of development by comparing China's human development index, inequality indices (Gini, quintile, and Palma), and development potential (human capital index) with the developed countries in Europe, North America, and Oceania, as well as countries with typical traits, such as the Latin American countries, Japan and Czech Republic; further to put forward China's policy focuses in the new normal stage, so to catch up with the developed countries in quality of development.

http://dx.doi.org/10.5772/66791

2. Material and methods

For comparing the quality of development, we arrange here with representative countries, comparable indicators and methodologies.

2.1. Countries considered

China is a large developing country with the largest population and large land mass in the world, and with socialist nature as its Constitution expressed. The countries as comparing counterparts, we choose mainly concerning: (1) well developed (at least its HDI higher than China's); (2) relative competent size of territory and population; and (3) representative in different regions and social models. By data testing, 14 countries have been selected as reference countries as follows.

The four countries, Norway, Denmark, Sweden, and Finland, are all Nordic countries, well developed with long-term stable sustainable qualified development, as generally accepted model of ideal society on the globe currently, the "Nordic model," which have more socialist component, such as generous social welfare and equal opportunity for public services to each family and individual all over the country.

These two countries, Germany and Switzerland, are high developed market economies with more socialist-natures in the "Rhine model," as major roles in mainland Europe with long-term stable qualified development and good performance in equality aspect.

The two countries, USA and UK, are well-developed market economies, natured as typical capitalist market in the "Anglo-Saxon model," and once the super powers in different ages.

The country of Australia is on the Oceania, tightly related with China in commercial intercourse; well-developed capitalist economy with sound social welfare as well.

The country of Japan is the next neighbor of China, the first and most developed economy in Asia, and has good performance generally but in depression for a long time in recent years.

The country of Czech Republic is a former socialist country located in central-eastern Europe, with the history of a member of former Soviet Union alliance, and keeps the most equal society record; not well developed but with very high value of human development index (Rank 28 in 2014 in nearly 200 countries).

The three countries, Argentina, Mexico, and Brazil, are also developing countries but capitalist natured in Latin America, ranking forefront of the world in inequality.

2.2. Indicators and methods

The chapter is to examine China's "new normal" state by comparing related indicators with 14 other countries typically scattered in the world (except Africa). Considering the paper's international angle, we make comparability and internationalism as the prime principles when selecting indicators utilized. Therefore, all indicators and data as follows are from UNDP, (http://hdr.undp.org) [8], the exception sources will be marked in addition at the right point.

2.2.1. Human development index (HDI)

The HDI represents a broader definition of well-being and provides a composite measure of three basic dimensions of human development: health (a long and healthy life), education (knowledge), and income (a decent standard of living) [9]. HDI is the most comparable and available indicator for measuring quality of life among countries.

2.2.2. Inequality indices (Gini, quintile, and Palma)

The World bank emphasizes, "To begin to understand what life is like in a country–to know, for example, how many of its inhabitants are poor–it is not enough to know that country's per capita income. The number of poor people in a country and the average quality of life also depend on how equally–or unequally–income is distributed" [10]. The **Gini Coefficient** is the most frequently used inequality index as "the mean difference from all observed quantities" [11]. However, the Gini does not capture where in the distribution the inequality occurs. For this reason, other two indicators, quintile ratio, and Palma ratio, are also chosen in the paper, which are more clearly reflect the high income and low income gap, successfully excluding the influence of middle income people.

The **quintile ratio** (20:20 or 20/20 ratio) compares how much richer the top 20% of populations are to the bottom 20% of a given population, which is actually a part of the Gini Coefficient that prevents the middle 60% statistically obscuring inequality, meanwhile highlighting the difference between two poles.

The **Palma Ratio,** meaning the ratio of the top 10% of population's share of gross national income (GNI), divided by the poorest 40% of the population's share of GNI–could provide a more policy-relevant indicator of the extent of inequality in each country, and may be particularly relevant to poverty reduction policy. It is based on the work of Chilean economist Jose Gabriel Palma who found that the "middle classes" tend to capture around 50% of national income, while the other half is split between the richest 10% and poorest 40% [12].

2.2.3. Human capital index (HCI)

"A nation's human capital endowment–the skills and capacities that reside in people and that are put to productive use–can be a more important determinant of its long-term economic success than virtually any other resource. This resource must be invested in and leveraged efficiently in order for it to generate returns–for the individuals involved as well as an economy as a whole" [13].

Graph 2 is drawn to show the relations among human development index and its three components, human, capital, and equality. Here, we emphasize that the HDI includes HCI, which account for two-thirds of HDI, even though education and health are not the whole HCI, but at least the major aspects; education and health are both capabilities residing in people, which is directly related to a person's income and in social level to both quantity and quality of economic development; Equalization and justice are important complement of HDI, which also have promoting effects on people's education and health by its benefiting mostly to the general public. That is, HDI, HCI, and equality are interrelated and tend to promote along the arrow directions, which constitute and cooperate the quality of development/quality of life.

http://dx.doi.org/10.5772/66791

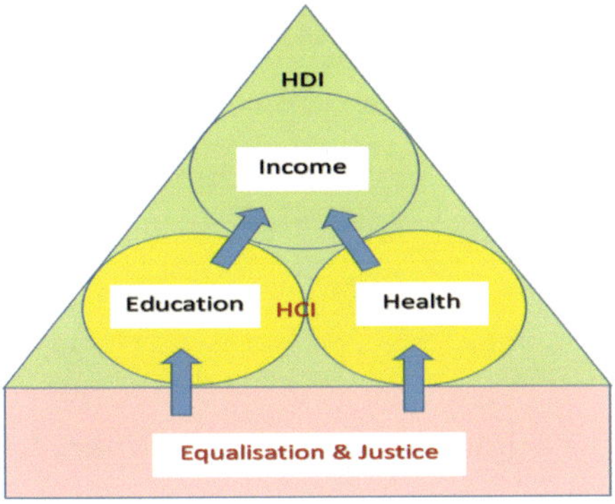

Graph 2. The promoting relations of equality, human capital, and human development.

All data used are registered in official sources. The international data for comparing among countries are from international organizations, UNDP. The method used in the chapter is mostly comparative analysis approaches with statistical graphs and tables.

3. Experimental

Here, we examine for comparing China's quality of development with the representative countries by using the three serials indicators; and conduct comprehensive comparative analysis and evaluation.

3.1. Human development and living quality

3.1.1. HDI overall status

Graph 3 shows the level of human development index of the 15 countries selected with various colors, which implies the overall quality of development and quality of life of different country groups. China is at the bottom of the row, ranked 90th in the world, and approximately accounts for 77% of the highest valued country, Norway; 79% of the United States, the typical capitalist country; and 82% of Japan, Asia's most developed country. That means we have a long distance to go in quality of life.

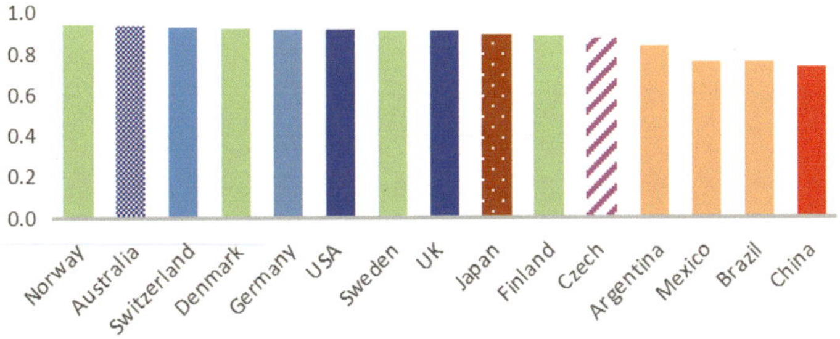

Graph 3. HDI in world context 2014.

Table 1 shows the overall level of HDI of four level groups, and the world and the developing countries. China, the second biggest economy in the world, is nearly 20% less than the level of the first 50 countries, and just at the average level of the world in quality of life.

Groups	HDI	China %
Very high human development	0.896	81.1
High human development	0.744	97.7
Medium human development	0.630	115.4
Low human development	0.505	144.0
World	0.711	102.3
Developing countries	0.660	110.2

Table 1. Overall level of human development in different groups 2014.

3.1.2. HDI components

In Annex **Table 1**, we make HDI and its component indicators in order respectively and make a sum rank in order to see the influence of each component. From Annex **Table 1** and **Graph 4**, we notice first that the general pattern does not change: (1) the upper ranked 8 countries are still upper but with changed ranks; (2) the lower seven countries are lower by the same rank with HDI order; (3) China retains at its bottom position by reordering, including total rank and almost all component cases (life expectancy of China is the only factor that does not row at the extreme bottom, which might somehow show off the medical condition or Chinese traditional medicine).

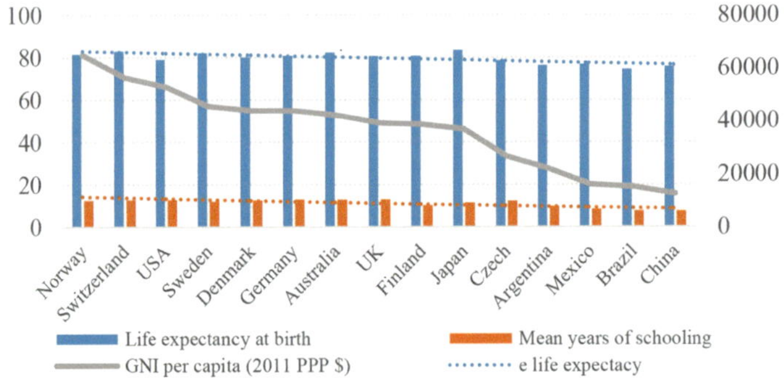

Graph 4. Components of HDI by GNI order 2014.

Moreover, we find some prominent features in Annex **Table 1** and **Graph 4**: (1) Both Germany and UK's re-ranks are upper by the same factor, "mean years of schooling" showing social sustainability, which imply the labor force and the civilized residents endowed by education; UK in Anglo-Saxon model with capitalist nature, has the similar pattern (8:1:8) with Germany (6:1:6) in "Rhine model," but far from the pattern of USA (10:4:3); Czech Republic (with similar

http://dx.doi.org/10.5772/66791

pattern 11:8:11) rows upper also by its "mean years of schooling," which means education gains much attention in Czech as well. (2) Australia (3:3:7) has almost the opposite pattern with USA, but with better momentum of development in practical economy than USA. (3) The life expectancy order of Japan is at the first, which might reflect Japanese life style is very healthy.

3.2. Inequality

Equalization and justice are important complement of HDI, so we here analyze income inequality standing for measuring social equality and justice, although which is far from comprehensive but essential and quantitative. According to the data of the National Bureau of Statistics, China's Gini coefficient has ever peaked to 49.1 in 2008, began to decline since 2010, to 46.9 in 2014, along with policy's functioning.

Graph 5 shows that, in the Gini coefficient case, China (2014) performs better than the three Latin countries and the two typical capitalist countries, USA and UK. However, the quintile ratio that shows the polarization in income distribution by the top 20% to the bottom 20%, has different performance: China' s value of quintile ratio is only better than that of the three Latin countries but worse than USA and UK, and far worse than other countries included; The Palma ratio, the richest 10% of population's share of gross national income divided by the poorest 40%'s share, provides support to the quintile's case.

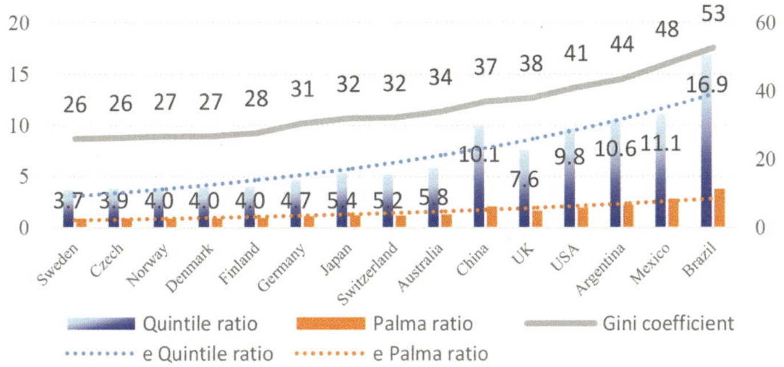

Graph 5. Income inequalities by Gini order 2014.

From the computing results in **Table 2**, we can see more clearly that China's polarization in income distribution, i.e., the highest income group to the lowest, excluding the influence of middle income people is conspicuous worse than the Gini performance with the influence of middle income populations included, by observing the deviations from the average of the 15 countries considered.

Of course, the income inequality in three Latin countries show much worse cases than in China; and their polarization is even much worse than their Gini case as well. That is probably the reason why the Latin countries could not performance better with so much endowment of natural resources. Therefore, equality and social justice in China as institutional environment given by the government should improve continuously for the sake of promoting the living quality of the people.

HDI rank total	Country	Indicators of income inequality		
		Quintile ratio	Palma ratio	Gini coefficient
14	Sweden	3.75	0.90	26.08
28	Czech	3.88	0.93	26.39
1	Norway	4.00	0.93	26.83
4	Denmark	3.96	0.94	26.88
24	Finland	4.04	0.98	27.79
6	Germany	4.72	1.14	30.63
20	Japan	5.39	1.22	32.11
3	Switzerland	5.23	1.21	32.35
2	Australia	5.85	1.32	34.01
90	China	10.08	2.08	37.01
14	UK	7.64	1.67	38.04
8	USA	9.79	1.96	41.12
40	Argentina	10.62	2.25	43.57
74	Mexico	11.13	2.84	48.07
75	Brazil	16.87	3.77	52.67
15 countries	Average	7.13	1.61	34.90
% deviation to average	China	41.41	29.29	6.04
	Argentina	49.00	40.03	24.83
	Mexico	56.11	76.27	37.72
	Brazil	136.59	134.55	50.90

Table 2. Fifteen countries' comparison of income inequality by Gini Order 2014.

In addition, China is a socialist country as its Constitution expressed, and in case any adverse effect happens, it is very necessary for China to have higher pursuit in equality and social justice, e.g., reach to 35/7/1.5 (Gini/quintile/Palma), equivalently the average level of listed 15 countries, close to the level of UK (38/7.6/1.7) or Australia (34/5.9/1.3), as the minimum pursuits in 5–10 year, from 37/10/2, the currently level of China by the inequality index.

3.3. Human capital

Generally observing the history and experiences of all developed countries, it is common nature that every country pays enough attention to two factors: labor force and ecological environment, which are two bases of a human society. We here focus on labor force only for which is the most active factor for social economic development, though ecological environment is a big problem in China.

A group of American economists, such as Gary S. Becker, T. W. Schultz, George J. Stigler, Milton Friedman, etc., advocate the concept "human capital" to describe the quality of labor force [14]. Now, that the concept of human capital has been widely spread and accepted, and for the sake of comparing the quality of labor force internationally, we take the advantage of data availability to use it, even though we are a bit shy to treat labors as capital.

3.3.1. *Human capital index and its aging structure*

From **Graph 6**, we can see that China's human capital level rows at the lowest position in the other 14 countries, and upper than Brazil. In aging structure, it seems a common problem currently for all other 14 countries but China. In fact, the aging issue in China is becoming a problem because of China's one-child policy which lasted 35 years. So, it becomes urgent to promote the quality of labors, if given the labor force participation and employment rate.

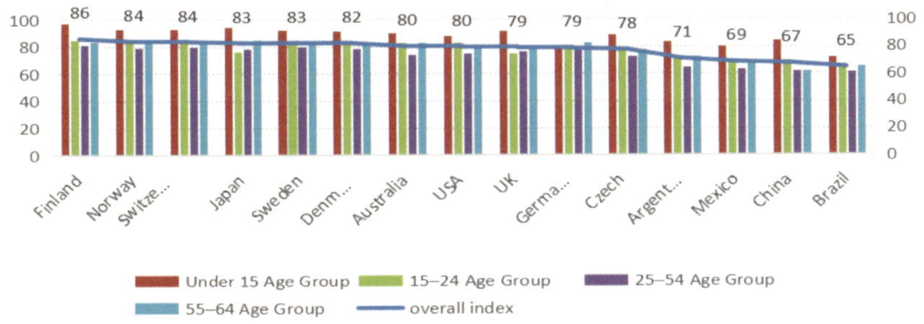

Graph 6. Human capital index and its structure by overall order 2015.

3.3.2. *Labor force participation and employment*

China has no doubt the best performance both in labor force participation and employment (**Graph 7**). Then, we see the quality of labor, for "education and training are the most important investments in human capital" [14].

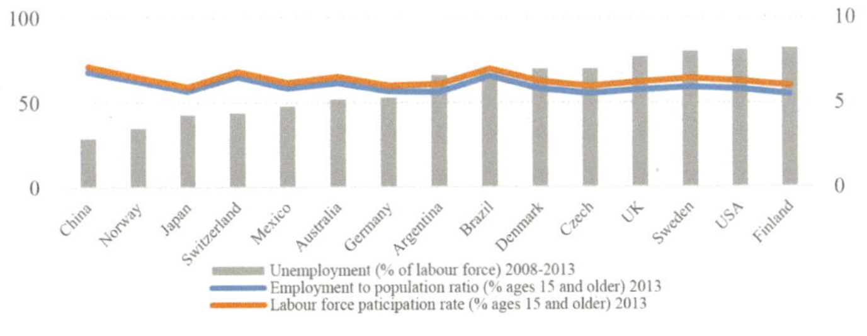

Graph 7. Employment and labour force paticipationparticipation by unemployment order

3.3.3. *Education efficiency*

From 15-year-old students' performance in 2012, we find that the quality of labor force in China is worth optimistic for the future. But on second thought, Chinese is so diligent and smart that China should have the highest quality of development, but China's HDI is at the 90th position, just at the middle level of the world. Why? There might be many reasons involved, may we have another paper to discuss the issue for the limit of article length.

4. Results and conclusions

From what has been discussed above, we conclude the following results:

(1) Equalization and justice are important complement of HDI; The HDI includes HCI; The two major parts of HCI, education and health, are both capabilities residing in people, which directly related to a person's income and in social level to both quantity and quality of economic development, and directly benefited from equalization and justice; Hence, HDI, HCI, and equality are inter relatedly constitute and cooperate the quality of development/quality of life. (**Graph 2**) The economy (income) is the business of market, while the education and health of labors and the income distribution should be supervised and guaranteed by the government; that is to say that the quality of life should be achieved by the combination of government and market.

(2) The overall level of HDI in China is nearly 20% less than the level of the first 50 countries, and just at the average level of the world in quality of life. Among the selected 15 countries, China is at bottom of the row, ranked 90th in the world, and approximately accounts for 77% of the highest valued country, Norway; 79% of the United States, the typical capitalist country; and 82% of Japan, the Asian most developed country. That means we have a long way to go in quality of life (**Table 1**, **Graph 3**).

(3) Both Germany and UK have best performance in "Mean years of schooling," which implying the labor force and the civilized residents endowed by education; UK in Anglo-Saxon model with capitalist nature, has the similar pattern (8:1:8, means rank of health/education/economy) with Germany (6:1:6) in "Rhine model," but far from the pattern of USA (10:4:3); Czech Republic (with similar pattern 11:8:11) rows upper also by its "Mean years of schooling," which means education gains much attention in Czech as well. Australia (3:3:7) has almost the opposite pattern with USA, but with better momentum of development in practical economy than USA. China should not take the model of USA, but learn more from Germany, UK and Australia, and Czech, that is, pay more attention to education for a civilized society in the future (Annex **Table 1**).

(4) In the Gini coefficient case, China (2014) performs better than the three Latin countries and the two typical capitalist countries, USA and UK; China' s quintile ratio is only better than that of the three Latin countries but worse than USA and UK; The Palma ratio provides support to the quintile's case. That is, China's polarization in income distribution is conspicuous worse than the Gini performance with the influence of middle income populations included. Hence, we should concern more of the low income groups (**Graph 5**, **Table 1**).

(5) The income inequality of three Latin countries shows much worse cases than in China, and their polarization is even much worse than their Gini case as well. Serious inequality cannot bring a developed economy from the lesson of Latin countries. Therefore, equality

and social justice in China as institutional environment given by the government should improve continuously for the sake of promoting the living quality of the people (**Table 1**).

(6) China is a socialist country as its constitution expressed, and in case any adverse effect happens, it is very necessary for China to have higher pursuit in equality and social justice, e.g., reach to 35/7/1.5 (Gini/quintile/Palma), equivalently the average level of listed 15 countries, close to the level of UK (38/7.6/1.7) or Australia (34/5.9/1.3), as the minimum pursuits in 5–10 years, from 37/10/2, the currently level of China by the inequality index (**Table 1**).

(7) China's human capital Index row at the lowest position among the countries, only better than Brazil's (**Graph 6**). But as the positive factor of HCI, China has the best performance in all 15 countries both in labor force participation and employment (**Graph 7**). From 15-year-old students' performance in education efficiency in 2012, the quality of labor force in China is worth optimistic for the future (**Graph 8**). Therefore, China has its advantages in human capital, and furtherly in the potential of development.

(8) It is possible to achieve better growth speed while we are focusing on the quality of development.

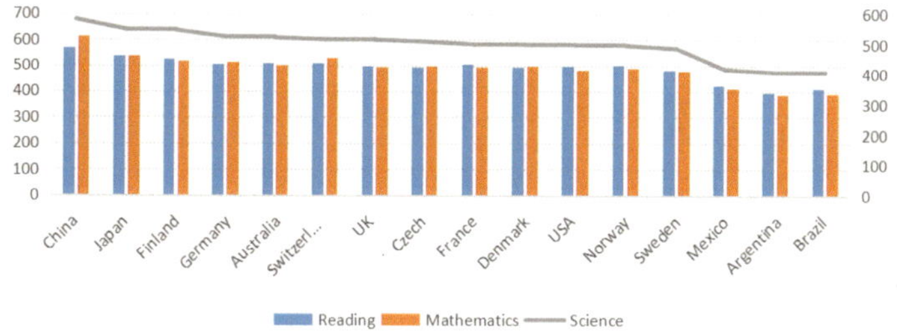

Graph 8. Education quality by order of science 2012 (Pperformance of 15-year-old student).

Acknowledgements

The authors would like to thank Dr. Tomáš Wroblowský, VSB, Czech Republic, for his feedback and suggestions regarding data and the quantitative methodologies used in the chapter.

We would also like to thank anonymous referees for their valuable comments and corrections to our English writing.

We would like to express our gratitude to both Social Science Foundation (Serial No: HB15LJ002), funded by Hebei Programming Office for Philosophy and Social Science, China, and Soft Science Foundation (Serial No: 16457699D), funded by Hebei Bureau of Science and Technology, China, for providing us with research funds.

The research is supported by the SGS project of VŠB-TU Ostrava Czech Republic under No. SP2016/11.

JEL classification: E6, F5, F6, O15, O5

Annex

HDI order origin	Country	HDI	HDI order	Life expectancy at birth (year)	LE order	Mean years of schooling (year)	SY order	GNI per capita (2011 PPP $)	GNI order	Total score	Score order	Country
1	Norway	0.94	1	81.6	5	12.6	7	64992.34	1	13	2	Norway
2	Australia	0.93	2	82.4	3	13	3	42260.61	7	13	2	Australia
3	Switzerland	0.93	3	83	2	12.8	5	56431.07	2	9	1	Switzerland
4	Denmark	0.92	4	80.2	9	12.7	6	44025.48	5	20	8	Denmark
6	Germany	0.92	5	80.9	6	13.1	1	43918.54	6	13	2	Germany
8	USA	0.91	6	79.1	10	12.9	4	52946.51	3	17	5	USA
14	Sweden	0.91	7	82.2	4	12.1	9	45635.5	4	17	5	Sweden
14	UK	0.91	8	80.7	8	13.1	1	39267.19	8	17	5	UK
20	Japan	0.89	9	83.5	1	11.5	10	36926.92	10	21	9	Japan
24	Finland	0.88	10	80.8	7	10.3	11	38694.77	9	27	10	Finland
28	Czech	0.87	11	78.6	11	12.3	8	26660.28	11	30	11	Czech
40	Argentina	0.84	12	76.3	13	9.8	12	22049.59	12	37	12	Argentina
74	Mexico	0.76	13	76.8	12	8.5	13	16055.97	13	38	13	Mexico
75	Brazil	0.76	14	74.5	15	7.7	14	15174.97	14	43	14	Brazil
90	China	0.73	15	75.8	14	7.5	15	12547.03	15	44	15	China

Annex 1. Component comparison of Human human development development 15 countries 2014.

http://dx.doi.org/10.5772/66791

Author details

Jin Han[1]*, Haochen Guo[2] and Mengnan Zhang[3]

*Address all correspondence to: hanjin99@126.com

1 Faculty of Economics, Hebei GEO University, Shijiazhuang, Hebei, China

2 Department of Finance, Faculty of Economics, VŠB-Technical University of Ostrava, Ostrava, Czech Republic

3 Graduate Department, Hebei GEO University, Shijiazhuang, Hebei, China

References

[1] Becker, Gary S. (1993 third edition). Human Capital. The National Bureau of Economic Research. The University of Chicago Press. Chicago and London. ISBN 0-226-04120-4.

[2] Cai, Fang. (2016). Let the world know China's New Normal of economic development. http://paper.people.com.cn/rmrb/html/2016-06/16/nw.D110000renmrb_20160616_2-07.html.

[3] Ceriani, Lidia. Verme, Paolo. (2012). The origins of the Gini index: extracts from Variabilità e Mutabilità (1912) by Corrado Gini. J Econ Inequal. (2012) 10: 424–443.

[4] Cobham, Alex. Sumner, Andy. (2013). Putting the Gini back in the bottle "The Palma" as a policy-relevant measure of inequality. Mimeograph. London: King's Collage London. http://www.socialprotectionet.org/sites/default/files/cobham-sumner-15march2013.pdf.

[5] Jahan, Selim, et al. (2015). Human development report 2015. Published for the United Nations Development Programme (UNDP). New York. eISBN 978-92-1-057615-4. http://hdr.undp.org/sites/default/files/2015_human_development_report.pdf.

[6] Jin, Bei. (2015). On the New Normal of China's economic development. China Industrial Economics. No.1.

[7] Leopold, Till, et al. (2015). The human capital report 2015. World Economic Forum. Geneva. ISBN 92-95044-49-5. http://www3.weforum.org/docs/WEF_Human_Capital_Report_2015.pdf.

[8] Li, Yang. (2015). The New Normal of China's economy is different to the New Normal of globle economy. http://soci.cssn.cn/ddzg/ddzg_ldjs/ddzg_jj/201503/t20150312_1542335.shtml.

[9] Lin, Yifu. (2016). Why China can increase as planned. http://beltandroad.zaobao.com/beltandroad/analysis/story20160407-602338.

[10] Liu, Wei. (2015). The New Normal of China's economy and its new strategy of economic development. China Academic Journal Electronic Publishing House.

[11] Tatyana P. Soubbotina. (2004). Beyond economic growth an introduction to sustainable development. World Bank Publications. ISBN 9780821359334. http://www.worldbank.org/depweb/english/beyond/global/chapter5.html

[12] Trends in the human development index, 1990–2014. http://hdr.undp.org/en/composite/Trends. Human Development Report 2015: Work for Human Development. United Nations Development Programme. New York. 2015. eISBN 978-92-1-057615-4.

[13] Wu, Jinglian. (2015). Accurately grasp the two characteristics of the New Normal. http://theory.people.com.cn/n/2015/0504/c49154-26942603.html.

[14] Yao, Yang. (2016). Across the Middle-Income Trap: advantages and challenges in China. Vol.028, National School of Development, Peking University.

Chapter 15

Intangible Influences Affecting the Value of Estate

Vladimír Kulil

Additional information is available at the end of the chapter

http://dx.doi.org/10.5772/66792

Abstract

The process of valuation of intangible influences was surveyed in China, Hong Kong, the USA, Canada, Japan, Germany, the UK, Poland, Russia, and Western Europe. Situation in mentioned locations is similar; valuation of intangible influences has not been determined by a concrete list of items and there has not been established, concrete clear process. This chapter proposes a method of valuation of goodwill (GW)-special effects that will impact assets' prices. It deals with proposed procedures for valuation of intangible assets and definitions of such property. Special effects are in particular name, historical value, design, quality of layout, security aspects, accessibility, conflict groups of inhabitants in or near the property, location, provenience, and other. The value of goodwill can be calculated as the difference between the market value and the material value. Part of the methodology is a general proposal for a method how to divide the assets into tangible and intangible part and author's software VALUE-RATUS 2015.

Keywords: market value, goodwill, bad will, price of real estate, tangible assets, intangible assets, coefficient of marketability

1. Introduction

Valuation of intangible assets includes certain specifics compared to cost assets. The specifics should be considered in the methodology and in final price. There exists a basic consensus in the way of tangible assets evaluation, in the case of intangible assets there is not. The aim is to introduce the scientific public with a different view on the essence of valuation. In scientific literature focused on real estate valuation there nearly does not exist a phrase intangible assets and real estate. Many appraisers consider buildings that are fixed with a land as a purely tangible thing. So it is not. Regarding the price, a property consists of intangible parts that have significant impact on valuation and this perspective is emphasized in this chapter.

Goodwill (GW) is an economic term denoting the difference between the market value of the company and substance price, less any liabilities. Indicates a value, intangible assets such as customer relationships, reputation, reflects market position, quality, and especially tradition. Goodwill can distinguish two ways: on the goodwill of the original and secondary. Initial goodwill is to create their own company's business activities, but not in the accounts of a company recognized because it is not reliably measurable. Secondary acquires goodwill on the acquisition of another company. Badwill (BW) is negative goodwill.

2. Overview of a current state of knowledge

Division of assets into tangible and intangible part has not been solved in the available literature and methodologies woxrldwide. Experts have no definite foothold for valuation. This division is very important and necessary for valuation of property as a whole; it brings a clearer overview of a quality and inner essence of a corporate assets, tangible subjects, and real estate. For example, it can represent an important measurement for investment decisions of investors. If the real estate as whole represents badwill , then it would be preferable to allocate capital to a location and environment with goodwill.

Usually, value of goodwill is determined by estimation with using complementary methods, there has not been reached a general consensus not in real concept goodwill meaning neither in its calculation. None of the definitions is fully accepted in general catholicity. For practical valuation, it is necessary to define the way of valuation of special influences–goodwill and badwill clearly.

In an international environment aside the Czech Republic, a survey focused on definitions and valuation of intangible assets was realized. In Germany, the final list of special intangible influences depends on an expert, who defines and evaluates them. In Great Britain so-called Red Book (RICS) is the basic document for valuation, the process is similar. The process of valuation of intangible influences was also surveyed in Poland, Russia, Hong Kong, the USA, Canada, Japan, China, and Western Europe.[1] Situation in mentioned locations is similar, valuation of intangible influences has not been determined by a concrete list of items and there has not been an established concrete clear process. In Slovakia, intangible influences are the part of price regulation[2] especially in the field of methods of location differentiation. Up to 21 factors of intangible effects for constructions and up to 22 factors for lands have been defined in Slovakia. The most detailed processes and unified definitions for intangible assets valuation abroad are in an international valuation standard IVS, where the issue is solved only at the general level without any concrete list of intangible special influences.[3] European valuation

[1]Seabrooke W., Kent P., Hwee Hong How H. International Real Estate an Institutional Approach. UK, USA, Australia: Blackwell Publishing Ltd., 2004, pp. 130–361.

[2]Decree No. 492/2004 Coll. of the Ministry of Justice of the Slovak Republic on the Determination of General Value of Real Estate.

[3]International Valuation Standards Committee: International Valuation Standards, 7. edition 2005, IVSC, London 2005, Change Proposal June 2010, Decree no. 1, no. 4, no. 6.

http://dx.doi.org/10.5772/66792

standards TEGoVA[4] have similar conception and they are formed in order to be conform to standards IVS and also in order to reach worldwide consensus in best practices in the valuation process (see **Table 1**).

Nr.	*Intangible pricing influences by modifying K_p per item with detailed list itemization*	*Recommended range %*
1.	Location favorableness in the locality compared to the average of the locality	From −10% to +10%
2.	Real estate name, prestigiousness, dominance	From −10% to +10%
3.	Historical value, cost of preservation of monuments	From −10% to +10%
4.	Architectonic rendering, the quality of disposition, view	From −10% to +10%
5.	Safety, users' privacy, conflict inhabitants in the surroundings	(From −70% to +10%)
6.	Danger of floods, landslides, damage from transport, bad smells, air pollution	From −10% to +10%
7.	Dangerous disposition and harmful material, radon	From −10% to 0%
8.	Transport accessibility with respect to the average standard in the surroundings	From −10% to +10%
9.	Impact of terraced housing, a building inside a row or at its end	From −10% to 0%
10.	Pricing perspective of real estate and other influences	From −70% to +10%

Table 1. Intangible pricing influences.

A good-quality system for intangible influences valuation according to price regulation[5] exists in the Czech Republic. Since 1997 a system has been developing and finishing with the aim to approximate maximally administrative prices to market prices. In the area of special influences, methodology introduced in Decree with appendixes is valid. For cost valuation under the price regulation from 1997 to 2013 marketability coefficients (K_p) were valid. The coefficients represent relationship between real estate price agreed according to buying contract and their prices determined on the basis of price regulation transferred to the unified price level (appendix no. 39 of Decree). While valuating by comparative way in decree appendixes up to 35 items of intangible special effects were determined, but in practice up to 100 current effects can be determined.

3. New findings and special influences valuation

The following procedure for valuation of goodwill and badwill types of assets of enterprises resulting from the mentioned model approaches appears as the most objective. Enterprise assets will be evaluated by a comparative, yield and cost method. The price will be adjusted in each used method according to an influence of special effects, which means good or bad

[4]EVS–European Valuation Standards, 5. edition 2003.

[5]Czech Act on Property Valuation No. 151/1997 Coll. with implementing decrees.

reputation and according to other special effects which influence usual market price. Other evaluated intangible assets of an enterprise (except goodwill) are included in the price if they really exist. On the basis of these data, market price is appraised. Similarly, valuation of goodwill and badwill by using the comparative, yield and cost method shall be worked out for real estate. A price in each method will be adjusted by direct special influences, it means good reputation, bad reputation, and other similar influences and it will influence the market price level. On the basis of this background, market price will be appraised. The amount of goodwill (GW) or badwill (BW) is the difference market price (CO) and cost price (CC) as:

$$GW(BW) = CO–CC. \tag{1}$$

The market value CO is determined by multiplying the cost value CC (replacement cost less depreciation, or material value) by marketability coefficient (*KP*) according to the relationship

$$CO = CC \times KP, \tag{2}$$

it follows that

$$KP = CO/CC. \tag{3}$$

The marketability coefficient is defined as the ratio between the average actual sales values achieved and the average cost prices of a comparable type of things at the particular time and location.

In context with intangible part of property, the valuer's approach is significant. An approach should be not only technical, but except experiences valuer should have also an expert feeling for fair and objective assessment of the matter of the circumstances, which have an impact on a property price. So, ethically right valuer processes are very important and it should be better integrated into the Czech law in the field of forensic experts. A judge is obliged by legal promise that he will decide conscientiously and fairly and since judge requires an expert opinion, an expert has to seek the market price which is fair and according to the best conscience.

Ownership of created movable or construction means a possibility of any manipulation in conformity to the laws. Even if the construction is defined as immovable we can move it to another land or a construction can be duplicated, certain depreciation exists here, lifespan is maximally hundreds or several thousand years. This property can be destroyed, concretely its material and also nonmaterial part. However, land as a part of the planet's surface cannot be created, transferred, or destroyed. The main component of land price is represented by nonmaterial part of a land. The tangible part of a land price has minimal or zero price as it is more described in the following text.

Permanent vegetation is a part of the plot of land and has a material substance which can be determined by using a cost, yield, or comparative method especially in relation to economic

http://dx.doi.org/10.5772/66792

benefit. Permanent vegetation also has an intangible component that is valuable as in the case of plots of land and buildings. It implies to special influences of actual demand and usability for the owner or potential buyer. An intangible component of permanent vegetation is represented by the landscape and aesthetic function, ensuring privacy and recreation, security, defense, windbreak function, protection against noise, odors, dust, pollutants, against inclement weather and climate, providing reinforcement of subsoil slope, and land. The following factors belong also to this group: erosion as influences, hydrological function, oxygen production, the possibility of the existence of fauna, flora, production of fragrances, cultural and historical features, for example, with protected trees, etc.

4. Methodology proposal for special influences valuation in the field of real estate

In the field of real estate, marketability coefficients (K_p) have been worked out for valuation with administrative price. Marketability coefficients K_p take into consideration location of structures and plots of land on the basis of statistical assessment of all realized sales in the Czech Republic. The effect of location on the price is very important. Administrative cost prices with K_p in some categories of property are multiplies of the cost price determined without K_p.

A utility value is clearly defined in a German literature: the utility value of real estate consists of a quality and location parts. A quality part of real estate refers to a technical quality and to the architectural design and equipment. Location part respects a structure of build-up area, traffic availability, availability of connection to local infrastructure, influence of noise, industrial emissions, influence of historical development of a town and so-called very valuable addresses. The utility value of real estate is therefore a critical item in relation to a price. According to Anglo-Saxon literature, a goodwill has to be considered as a documentary only if a long-term income in the context with goodwill can be expected. It is assumed that for this property, a buyer paid an extra charge, which is as an intangible asset supported by the utility value.[6] Just so defined extra charge can be considered as special valuation of surveyed special intangible influences that help to create a value of movables, real estate, and enterprise. Intangible assets of a goodwill type are the subject of financial reporting and accounting depreciation in enterprises, but in an international trade accounting rules are inconsistent. In the acquisition of enterprises, a very significant sum for the goodwill can be reported. Acquisition also contains immovable assets, which are part of the intangible assets of firms. For example, when buying the American company Gerber Product Co. by Swiss company Sandoz Ltd. goodwill amounted to 3.2 bn. USD in transactions totaling 3.7 billion. USD, it is about 86% of the purchase price.[7]

[6]Horne Van J. Financial Management and Policy. New Jersey, USA: Englewood Cliffs, 1989, s. 647.

[7]Shetty A. a kol. Finance and Integrated Global Approach, USA: Austen Press, Homewood, 1995, s. 577, s. 600.

5. List of the groups of special influences

Marketability coefficient (*KP*) is a product of marketability coefficient (K_p) determined by Czech price regulation and index of additional special influences if they exist and have an impact on a price. If K_p is not determined or does not correspond to an average market price, then it can be determined by an expert appraisal, for example, with the help of statistical office data.

$$KP = K_p \times \left(1 + \sum_{i=1}^{10} KPi\,\% \times 0.01\right) \qquad (4)$$

The percentage range is recommended and was determined by an expert estimation, according to a general specialized literature for valuation and also with regard to determined margins in previous regulations valid since 1977. A correction scale is proposed as 50% of the recommended range in the case of advantage or disadvantage due to special influence and at the same time, the scale is proposed as 100% of the recommended range due to significant advantage or significant disadvantage. The ranges were proposed as a price adjustment compared to the average standard of real estate in a given location. Based on the calculations of special influences in the case of real estate, it is clear that the rate must be applied sensitively with the principle of prudence, depending on actual market demand.

Only influences with reasonably justifiable impact on property price can be calculated. Rates of surcharges and reductions percentage for those hundred items of proposed special influences need to be considered both in terms of cost, and particularly in terms of the market value–the usual market price, thus impact on the marketability of a specific property in real time and place. Recommended rates and limits are valid jointly for the whole collection of structures, plots of land, and permanent vegetation. Final valuation of special influences will be realized by the sum of increases and reductions with conclusive justification.

6. Intangible assets ownership and valuation

6.1. Ownership

Who owns the permanent vegetation, buildings, or landscaping (even future) on the plot of land, usually owns the goodwill or badwill relating to this land.[8] The price of land is also a reflection of the external construction work on the adjacent land, or even in distant surrounding, for example in terms of access to the land or flood prevention measures, to build public facilities. Goodwill and badwill in this case automatically become the property

[8]Kulil, V. (2015).Goodwill and Valuation. Saarbrücken Germany: OmniScriptum GmbH & Co. KG. [monografie]

http://dx.doi.org/10.5772/66792

of the owner. Mentioned external investments have an impact on the owners' property. It is a free acquisition of intangible assets. Only some cases of badwill can be compensated by an investor, usually only in the case of exceeding the health, sanitary, and technical standards (e.g., dust, odors, and noise from the road). The land without the possibility of any building, construction, and technical adjustments without a permanent vegetation, which has not any use for a human even prospectively, has not any value for a human and any intangible part of the price.

6.2. Yield value VH

For yield valuation assessment of goodwill (badwill) will be realized by a reasonable adjustment of the capitalization rate through adequate reduction or surcharges compared with an average standard of the real estate quality. An identical correction of the sum of the influences by the cost method is proposed as a standard level of capitalization, which is tied to the material component of the property, should be modified only by aspects of intangible assets, considering assets risk in the future. Capitalization rate (*P*) used for calculating the yield value will be adjusted by the identical percentage according to price influences (adjustment of risk surcharge) compared to the standard character of the property, with an average capitalization rate (*k*) and the average amount or profitability risk. Generally while using the calculation according to the formula for perpetual annuity, the yield value is the ratio of net annual return (CV) and (*P*) capitalization rate as a percentage:

$$\mathrm{VH} = \mathrm{CV}/P\,\%, \tag{5}$$

$$P\,\% = k\,\% + 0.01 \times \left(\sum_{i=1}^{10} KPi\%\right). \tag{6}$$

6.3. Comparative value PH

In this case valuation will be realized as comparison toward standard etalon in relation to goodwill (badwill). A comparative value (*PH*) will be adjusted by the rate $\pm\sum KPi$ compared to an average comparative value of real estate (*Ph*) without special influences. Using the indexes should be reasonably justified. Identical influences and rates of individual groups of items from no. 1 to no. 10 for comparison will be used. It is also possible to use a direct comparison with other property regarding existing special influences if there are resources for comparing the appropriate quality. However, it is necessary to take into account the need of comparing a large number of qualitative characteristics, which may generate significant mistakes with regard to the limited information on the compared properties which are available. An aforementioned method of comparison with average real estate etalon therefore appears as a more accurate method and in the latest phase it is recommended to adjust the results by extraordinary special influences:

$$PH = Ph \times \left(1 + \sum_{i=1}^{10} KPi\% \times 0.01\right). \tag{7}$$

6.4. Market value CO

According to cost, yield, and comparative valuation mentioned in previous sections, there will be realized an appraisal of market value. The amount of price of special influences–goodwill and badwill–will be the difference between market value of property and cost price without *KP* (cost price CC). The amount of harm in connection with the easement will be counted as a standard yield method and subtracted from the market value of the property. The maximum discount is not determined.

6.5. Coefficient of an intangible asset

From the concept of marketability coefficient *KP* or K_p (in German-speaking countries a similar term market hopefulness is used) its fundamental as an index for determining the degree of special influences–intangible assets (NM) in a positive or negative amount toward the current price (CC) and usual market price–value of assets (CO) as a whole is not obvious.

$$\mathrm{CO} = \mathrm{CC} + \mathrm{NM}. \tag{8}$$

Coefficient of an intangible asset (K_{NM}) appears to be more accurate term. An intangible character of valued property results from the mentioned term. And it shall not be determined as the estimated generally not well-understood constant, which an expert established. This coefficient can be expressed by the following formula:

For real estate

$$K_{\mathrm{NM}} = (\mathrm{CC} + \mathrm{NM})/\mathrm{CC}. \tag{9}$$

For movables

$$K_{\mathrm{NM}} = (\mathrm{CC} + \mathrm{NM})/\mathrm{CC}. \tag{10}$$

For enterprises from the material value (*S*)

$$K_{\mathrm{NM}} = (S + \mathrm{NM})/S. \tag{11}$$

7. New methodology

The separate system of valuation tangible and intangible assets was worked out.[9] For the field of special influences, there were proposed and defined apposite terms goodwill and badwill for valuation analogically according to the terms used by economists and appraisal experts while appraising enterprises. The character and fundamental of marketability coefficients *KP*

[9]Kulil, V. (2014). Goodwill and Valuation. Brno: Akademickénakladatelství CERM s.r.o. [monografie]

are clarified from the point of view of their relationship to tangible and intangible assets. Ten main areas and hundred items of intangible influences affecting real estate price are complexly defined. It is a modular method, influences are evaluated in percentage. A higher number of influences more than 100 are possible, for practical use, however, it can be misleading.

The author created a software NEMO-RATUS 2015[10] for practical use. The proposed procedures in the whole extend of valuation including table analysis of proposed hundred special influences for real estate with logarithmic regression and including intangible assets into cost yield, comparative, and market price are applied in the computer system. The market price of property is automatically divided into tangible and intangible part. For more details see http://www.ekf.vsb.cz/k166/cs/. The proposed procedures and detailed listing of special influences represent a comprehensive, practical, and unequivocal support for the valuation practice of experts.

Real estates have only two parts of market price. It is cost-quantifiable tangible cost price, which is adjusted by intangible goodwill (GW) or badwill (BW), its price can only be appraised. The intangible and relative character of special influences results from their external character impact. The proposed methodology enables to divide each movable and immovable property into the tangible part (cost price) and the intangible part (GW, BW) with sufficiently estimated accuracy for practical use. Then also enterprise assets can be newly divided into tangible and intangible parts. Goodwill or badwill as a summary of specific intangible impacts on the market price is calculated as the difference between the market value of the property and its cost price. This rule applies generally to movable property, immovable property, and enterprises.

Plots of land price are represented by all rights related to human activities on a land including construction and construction rights. The plot of land is to full extend intangible asset only of the goodwill type. The land price is not determined randomly, but is a reflection ("shadow") of values of specific structures located there or planned or it is a reflection of future use. Owner of the intangible part of the price of real estate is an investor, that plan and finance modifications of the property. It may not always be real estate owner. Goodwill or badwill caused by investments or investment plans in the area of valuated real estate passes automatically and free of charge to the ownership of the property owner.

8. Summary

The aim of the chapter was to work out a proposal for valuation of special influences that have an impact on real estate price . Controllable procedures for the valuation of intangible assets were proposed. A system of valuation with direct implementation in cost, yield and comparative methods from which we can estimate the market value was proposed. In the case of real estate special influences are defined mostly as good or bad name of locality real estate, historical value, design, quality of layout, safety aspects, transport accessibility, conflict inhabitants in the surroundings, influence of terraced house, other influences, and price perspective.

[10]Kulil V. (2015). *Software for goodwill valuation VALUE-RATUS*. http://www.ekf.vsb.cz/k166/cs/.

Terms goodwill (GW) in the case of positive impact and badwill (BW) in the case of negative impact were defined for each surveyed special influence.

JEL classification: M21, M31

Additional sources

Czech Act *on Property Valuation No. 151/1997 Coll. with implementing decrees.*

ČSÚ Praha. *(1993–2015). Statistickéúdaje.* https://www.czso.cz/csu/czso/domov.

EUROSTAT Brusel. (1998–2015). *Statistickéúdaje EU.* http://ec.europa.eu/eurostat.

EVS–*European Valuation Standards,* 5th edition, 2003.

IVS–International Valuation Standards Committee: *International Valuation Standards,* 7th edition 2005, IVSC, London 2005, Change Proposal June 2010, Decree no. 1, no. 4, no. 6.

Author details

Vladimír Kulil

Address all correspondence to: vladimir.kulil@vsb.cz

Faculty of Economics, VŠB-Technical University of Ostrava, Ostrava, Czech Republic

References

Brachmann, R. (1993). *Construction Costs of Industrial Buildings, Commercial Factory Price of Real Estate, Insurance Rates.* Praha, Czech Republic: CONSULTINVEST. [monografie]

Bradáč, A. a kol. (2009). *Theory of Real Estate Evaluation, VIII. edition.* Brno, Czech Republic: AN CERM s.r.o. [monografie]

Horne, V. J. (1989). *Financial Management and Policy.* New Jersey, USA: Englewood Cliffs. [monografie]

Kulil, V. (2014). *Goodwill and Valuation.* Brno, Czech Republic: AN CERM s.r.o. [monografie]

Kulil, V. (2015). *Goodwill and Valuation.* Saarbrücken Germany: OmniScriptum GmbH & Co. KG. [monografie]

Kulil, V. (2015). *Software for goodwill valuation VALUE-RATUS.* Ostrava, Czech Republic: http://www.ekf.vsb.cz/k166/cs/.

http://dx.doi.org/10.5772/66792

Ross, F., Brachmann, R., Holzner, P. (1993). *Detection of Construction of the Buildings and Commercial Real Estate Values*. Praha, Czech Republic: CONSULTINVEST. [monografie]

Seabrooke, W., Kent, P. A., Hwee, H. (2004). *International Real Estate an Institutional Approach*. UK, USA, Australia: Blackwell Publishing Ltd. [monografie]

Shetty, A., McGrath, F. J., Hammerbacher, I. M (1995). *Finance and Integrated Global Approach*. USA: Austen Press, Homewood. [monografie]

Telec, I. (2007). *Overview of Intellectual Property Rights I, Human Rights Foundation, License Agreement*. Brno, Czech Republic: Doplněk. [monografie]

Chapter 16

Capital Adequacy Ratio, Bank Credit Channel and Monetary Policy Effect

Li Qiong

Additional information is available at the end of the chapter

http://dx.doi.org/10.5772/64477

Abstract

This chapter, based on the theory model derived from bank credit transmission channel and the balanced panel data of 18 Chinese commercial banks from 2008 to 2012, conducts empirical test and carries out classified study of China's bank credit transmission channel and the effect of capital adequacy ratio on credit channel. Results show that the bank characteristics have, on a micro level, heterogeneity effects on credit transmission channel of monetary policy. Banks of higher capital adequacy ratio and smaller asset size are more vulnerable to the impacts of monetary policy. Therefore, the author proposes policy suggestions from the perspectives of optimizing the structure of financial markets and improving the monetary policy effect.

Keywords: monetary policy, bank credit channel, bank characteristics, capital adequacy ratio

1. Introduction

The credit channel is the main transmission mechanism by which bank capital adequacy ratio affects monetary policy effect. The credit channel emphasizes the important role of bank assets and liabilities in the transmission mechanism of monetary policy. The bank loan channel theory claims that monetary policy will affect the real economy by changing the bank loan supply behaviors. After the 2008 financial crisis, the Basel Agreement III rushed out. International agencies generally raised the bar of commercial bank capital regulation, which stipulated the minimum capital ratio and improved the capital adequacy ratio at the same time. Tougher capital constraints will reduce the interest rate elasticity of the loan supply, change the loan supply behaviors of commercial banks, and influence the monetary policy effect. Banks of different capital levels have difference in characteristics, the effect of loan supply to interest rate elasticity and influence on monetary policy.

Since the 1980s, the idea of credit channel of monetary policy transmission began to be widely accepted. Credit channel claims that because of incompleteness of financial markets, information asymmetry, contract cost, and other problems existing in economy operation process, banks play a special and important role in spreading risks, reducing transaction costs, and alleviating the adverse selection and moral risks caused by incomplete information in the credit market, etc. It is because of the special role of banks that a specific borrower can only borrow the needed funds from the banks. So the credit constitutes a significant channel of monetary policy transmission [1–4]. A number of scholars at home and abroad verify the existence of bank credit channel through empirical analysis.Taking the bank capital supervision system, the goal conflicts of central bank, and banking regulator authorities into consideration, new exogenous factors will make the traditional monetary policy credit transmission mechanism different. Scholars at home and abroad have conducted fruitful researches on this issue at the beginning of the 1990s. Kopecky-VanHoose [5] constructed a microeconomic model of a single banking sector, which proved that capital adequacy regulation will change the short-term and long-term credit behaviors of banks, and affect the credit transmission mechanism of monetary policy. Chun and Xiamin [6] examined banks' behaviors by establishing the model of individual bank, and eventually concluded that the bank credit market is a supply-dominated market, in which bank credit rationing behaviors will weaken the effectiveness of monetary policy, but capital regulation can improve the effect of monetary policy transmission. Liu Bin [7] studied the effect of capital adequacy supervision on the transmission mechanism of monetary policy and bank credit behavior by using the banking sector model and general equilibrium model, and concluded that the impacts of monetary policy to credit behavior vary with different capital scales. Wang Tao and Jiang Zaiwen's [8] research confirmed that tightening monetary policies can effectively control the bank credit supply, and is more effective to the smaller-scaled banks. Feng Ke and He Li's [9] research confirmed that the capital adequacy ratio and the bank credit scale are significantly positively correlated, but the transmission function of the bank credit channel to monetary policy is very limited. Li Tao and Liu Mingyu [10] claimed that the impact of bank credit channel on the banking sector varies with the different characteristics of banks represented by the capital adequacy ratio.

The existing literature has analyzed the relationship between the capital adequacy ratio and the bank credit transmission channel from various aspects, but there remain some shortcomings. First, most of the existing literature chooses reserve and disposable deposits as the operation target of monetary policy, while often ignores the role of interest rates[1] in monetary policy transmission. Second, the existing researches ignore the possible differences between the individual banks because of limited sample sizes.

---

[1]On July 19, 2013, the Central Bank of China announced completely unlock financial institutions loan rate control from July 20, 2013 on, which marked China's market-oriented interest rate reform entering a new stage. Interest rates unlocking was likely to be achieved in the last 1 or 2 years.

http://dx.doi.org/10.5772/64477

This chapter emphasizes the impacts of interest rate on bank credit behaviors and analyses the influences of the banks' characteristics represented by capital adequacy ratio on the credit transmission channel of monetary policy when constructing the theoretical model, and uses 18 banks' panel data to study the effectiveness of the monetary policy credit channel under capital supervision.

2. Theoretical model

In order to study the effect of capital adequacy to bank credit behavior, the author, refers to Feng Ke and He Li's [9] and Jiang Chun and Yu Xiamin's [6] theoretical models on the microscopic characteristics of banks and makes certain simplification to their models.

Take a simplified bank balance sheet as an example. Bank asset is made up of deposit reserve (R), government bonds (SEC), and loans (L), which are composed of total deposit (D) and equity capital (*K*):

$$R + L + \mathrm{SEC} = D + K \tag{1}$$

According to the requirement of statutory deposit reserve:

$$R = \rho D \tag{2}$$

where ρ is the statutory reserve ratio.

According to the requirement of capital adequacy ratio:

$$K = \theta L \tag{3}$$

where θ is the capital adequacy ratio.

L is the function of τ_L,

$$L = L(r_L) \tag{4}$$

where τ_L is the bank loan rate.

The bank's function is:

$$\pi = r_L L + r_{SEC} SEC – r_D D – r_K K \tag{5}$$

where τ_D is the deposit interest rate, τ_{SEC} is the interest rate of government bonds, and τ_K is the return on equity capital requirements.

According to Eq. (2)–Eq. (5) we can further get:

$$\pi = r_L L(r_L) + r_{SEC} SEC - r_D D - r_K \theta L(r_L) \tag{6}$$

In order to maximize the profit, the bank needs to meet the first-order necessary conditions:

$$\pi = r_L L'(r_L) + L(r_L) + r_K \theta L'(r_L) = 0 \tag{7}$$

$$L'(r_L) = \frac{L}{\theta r_K - r_L} \tag{8}$$

So, according to Eqs. (1)–(8), we can clearly see, when $\theta r_K > r_L \theta_L$, $L'(r_L) < 0$, the number of bank loans is a decreasing function of the loan interest rate; when $\theta r_K < r_L$, $L'(r_L) > 0$, the number of bank loans is an increasing function of the loan interest rate.

This demonstrates when loan supply dominates the credit rationing, banks will weigh the cost of capital rate (θr_k) and the size of loans (r_L) to determine the increase or decrease in the loan supply. If the capital cost rate is higher than the loan interest rate, the number of bank loans will decrease as the interest rates rises. If the capital cost rate is lower than the loan interest rate, even though the loan interest rate rises, banks will increase the number of loans for more profit.

With conditions Eq. (1)–Eq. (8), the partial derivative of capital adequacy ratio is:

$$\frac{\partial(L'(r_L))}{\partial\theta} = \frac{L'(\theta r_K - r_L) - r_K L}{(\theta r_K - r_L)^2} = \frac{L - r_K L}{(\theta r_K - r_L)^2} = \frac{L(1 - r_K)}{(\theta r_K - r_L)^2} > 0 \tag{9}$$

Results show that when there is a capital constraint, the larger the θ is, the more sensitive the bank loans will be to the monetary policy. When there is a stringent capital constraint, tightening monetary policy raises interest rates, reduces the profitability of banks, and the size of the endogenous capital accumulation, thus banks can only reduce loans to meet the minimum capital standards.

According to the analysis above, the author gets the following hypothesis:

Hypothesis 1: For banks of high capital cost rate, the number of loans changes in reverse direction of interest rates. And for banks of low capital cost rate, the number of loans changes in the same direction of interest rates.

Hypothesis 2: The higher the capital adequacy ratio is, the greater the magnitude of the number of loans will be as the interest rates change.

3. Empirical test

3.1. Construction and data processing of empirical model

Kashyp and Stein's [3] two-stage regression model with panel data bank microscopic analysis the effect of the capital adequacy ratio standards to bank credit channel. The first stage is the regression of the credit scale and capital adequacy ratio, and the second stage is the regression of the proxy variable of currency policy and capital adequacy ratio. On this basis, a cross-term panel regression model of the proxy variables of monetary policy capital adequacy ratio is introduced to conduct regression analysis of the bank characteristics and the cross-term coefficient with the credit scale to study the effect of bank characteristics to the credit transmission channel of monetary policy.

To better reflect the characteristics of banks, the author uses bank total assets (TA, Total Asset) and bank assets adequacy ratio (CAR, Capital Adequacy Ratio) to represent the sizes of banks and capital levels. Since the central bank continues to take the money supply as an intermediate target, and regulates mainly through the statutory deposit reserve ratio, loans and deposit benchmark interest rates, the author takes the one-year benchmark loan rate r, the statutory deposit reserve ratio R, and the narrow money supply M1 (because the central bank has a stronger control of M1 than M2) as the proxy variables of monetary policy.

3.2. The construction of empirical model

Based on the variable selection above, the author constructs the empirical model in this study as follows:

$$\Delta In(L_{i,t}) = \alpha_1 + \alpha_2 MP_{i,t} + \alpha_3 \Delta In(TA_{i,t}) + \alpha_4 CAR + \alpha_5 \Delta In(TA_{i,t}) MP_{i,t} + \alpha_6 CARMP_{i,t} + \varepsilon_{i,t} \tag{10}$$

Among which $MP_{\mathrm{i,t}}$, is the proxy variable of monetary policy, which includes one-year benchmark $\mathbf{r}_{Lt}$ loan rate, the legal deposit rate R_{t} and money supply amount $M1_{\mathrm{t}}$. We use $\partial L/\partial \mathrm{r}$ and $\partial L/\partial M1$ to show the effect of monetary policy changes on bank credit, and cross-term coefficient represents bank characteristics' role on that effect. $\partial^2 L/\partial \mathrm{r} \partial \Delta \ln(TA)$ and $\partial^2 L/\partial \mathrm{r} \partial CAR$ indicate the response to currency policy of bank credit of different asset scale and level of capital. The author focuses on the cross-term coefficient α_6.

This author selects five big state-owned commercial banks and 13 small and medium-sized joint-stock banks' 2008–2012 annual data,[2] all data are derived from the 2008 to 2012 annual financial statements of these banks' official websites. Proxy variables of monetary policy are based on data from China Statistical Yearbook website. The reserve ratio and the one-year loan rates are calculated based on the adjusted weight according to the time. In order to eliminate the heteroscedasticity, the author gets natural logarithm and takes the growth rates of all scale variables.

3.3. Descriptive statistics

Table 1 shows the descriptive statistics regression variables in this chapter.

Table 1 shows that from 2008 to 2012, the samples' average commercial banks' loan is 1.8292 trillion Yuan, the highest being Industrial and Commercial Bank of China in 2012, reaching

[2]The 18 commercial banks include the Industrial and Commercial Bank of China, Agricultural Bank of China, Bank of China, China Construction Bank, Bank of Communications, China CITIC bank, China Merchants Bank, Hua Xia Bank, China Minsheng Bank, Ping An Bank, China Everbright bank, Shanghai Pudong Development Bank, Industrial Bank, Nanjing Bank, Bank of Ningbo, Anhui Merchants Bank, Guangdong Development Bank, and Bank of Shanghai.

Variable	Maximum	Minimum	Median	Average	Standard deviation
Loan L (hundred million Yuan)	88,037	402	8787	18,292	21,974
Total assets TA (hundred million Yuan)	1,81,300	919	16,146	36,852	45,768
Capital adequacy ratio CAR	24.12%	8.58%	11.52%	11.98%	2.11

Table 1. Descriptive statistics results.

8.8037 trillion Yuan. From the viewpoint of the total assets, sample banks' average asset is 3.6852 trillion Yuan, the highest being the Bank of China in 2011, reaching 18.13 trillion Yuan. From the viewpoint of the banks' capital strength, sample banks' capital average adequacy ratio is 11.98%. And in 2008, Nanjing Bank reached as high as 24.12% and Pingan Bank being 8.58%.

3.4. Regression results

Due to the different natures of property rights, China's banks' structure presents an echelon distribution, mainly made up of state-owned commercial banks and joint-stock commercial banks. People's Bank of China also classifies them as large financial institutions and small- and medium-sized financial institutions. The author uses Eviews software to all banks, large state-owned banks, and joint-stock small- and medium-sized banks respectively to conduct mixed-panel and fixed-effect-panel regression analysis.

Table 2 shows the regression results of the one-year loan interest rate to the bank credit effect. As a result, it can be seen that all banks and joint-stock small- and medium-sized banks have significant regression coefficients, while the regression results of the large state-owned banks are not significant, which means large banks' credit scale is not sensitive to interest rate changes.

	All banks		Large state-owned banks		Joint-stock small and medium-sized banks	
	Mixed panel regression	Fixed panel regression	Mixed panel regression	Fixed panel regression	Mixed panel regression	Fixed panel regression
C	1.5570**	1.2107*	2.6421	2.2802	2.0561***	1.8251**
$\Delta lnTA$	3.4865***	2.9308***	4.0940	3.8629	4.2240***	3.45191**
CAR	−0.1521***	−0.1251**	−0.2212	−0.1988	−0.2270***	−0.1973**
$\Delta lnTA^{*}$ rl	−51.3821***	−43.0599***	−62.6583	−59.2758	−61.3452***	−49.0242**
CAR^{*} rl	2.4588***	1.5638	3.5830	2.9542	3.6939***	2.9117**
rl	−23.5738**	−12.1744	−41.2525	−31.8385	−32.3879***	−24.9495*

Table 2. Regression results of one-year loan rate to banks' credit effect.

Table 3 shows the regression result of deposit reserve rate (RRR) to bank credit effect. As a result, it can be seen that small- and medium-sized joint-stock banks' regression coefficients are significant, while in the regression results of all banks and large state-owned bank, only the banks' balance increment and its cross-term with the RRR are significant, which means bank assets will affect the sensitivity of the credit scale to the changes of RRR.

	All banks		Large state-owned banks		Joint-stock small and medium-sized banks	
	Mixed panel regression	Fixed panel regression	Mixed panel regression	Fixed panel regression	Mixed panel regression	Fixed panel regression
C	0.9796	1.1547	1.6342	1.6364	2.9754***	2.5665**
$\Delta lnTA$	2.3094***	1.8839***	5.5258**	4.7805	5.0446***	4.3115**
CAR	−0.0857	−0.0910	−0.1665	−0.1642	−0.3124***	−0.2708**
$\Delta lnTA^{*} R$	−0.1091***	−0.0877***	−0.2642**	−0.2287	−0.2733***	−0.2302**
$CAR * R$	0.0047	0.0035	0.0083	0.0074	0.0188***	0.0153*
R	−0.0486	−0.0399	−0.0767	−0.0664	−0.1757***	−0.1391*

Note: *, **, and *** represent significant under the significance level of 10, 5, and 1%, respectively.

Table 3. Regression results of reserve requirement ratio (RRR) to bank credit effect.

Table 4 shows the regression results of M1 to bank credit effect. It shows only the asset scale of all banks and small- and medium-sized joint-stock banks have significant regression relationship with their cross-terms with M1, which means bank asset scales can affect the sensitivity of credit scales to the changes of M1.

	All banks		Large state-owned banks		Joint-stock small and medium-sized banks	
	Mixed panel regression	Fixed panel regression	Mixed panel regression	Fixed panel regression	Mixed panel regression	Fixed panel regression
C	−2.2316	−6.0126	12.2154	8.8906	0.0645	−2.7256
$\Delta lnTA$	27.6874***	25.2591***	11.6582	17.3209	28.8552***	28.6159***
CAR	0.1749	0.4983	−0.6223	−0.4625	−0.1535	0.0587
$\Delta lnTA^{*} lnM1$	−2.1892***	−1.9984***	−0.9203	−1.3714	−2.2766***	−2.2596***
$\Delta CAR^{*} lnM1$	−0.0138	−0.0406	0.0495	0.0366	0.0124	−0.0050
$lnM1$	0.1859	0.5003	−0.9628	−0.6948	0.0009	0.2324

Table 4. Regression results of narrow money supply growth rate to bank credit effect.

4. Empirical results and policy implications

4.1. Empirical result analysis

First, through the analysis of proxy variables of monetary policy, it can be seen that the regression coefficient of the one-year loan rate is significantly negative, indicating that tight monetary policy will lead to incremental reduction of bank loans, which are mainly manifested as the incremental reduction of small- and medium-sized banks' loans and large banks' loan scales unaffected by the rate. Statutory RRR increase will result dramatic fall of the incremental loans of small- and medium-sized banks and the scale of banks' loans have no response to the narrow currency amount changes, which is consistent with Hypothesis 1. Bank capital absorbs risk and compensates for the loss, therefore, on average, medium- and small-sized banks' capital ratios are higher than that of the large banks [11]. And their core capital ratios are significantly higher than large banks. According to pecking order theory, equity financing cost is the highest among all manners [12]. So, because of a higher capital cost ratio, the scales of the loan of medium and small-sized banks will change in the opposite direction as the rate varies. That is to say, small- and medium-sized banks are more sensitive to interest rate changes. China's monetary policy is mainly transmitted through the small- and medium-sized banks. Large state-owned banks, due to government funding, are barely restrained to capital cost, so large banks are not sensitive to monetary policy changes and show stronger credit rationing ability.

The reasons are as follows: first, on the one hand, China's large state-owned banks occupy more credit resources, large banks are able to choose the loan customers and decide the pricing; On the other hand, state-owned banks' main loan customers are state-owned enterprises, while small- and medium-sized banks loan customers are mainly small- and medium-sized enterprises. State-owned enterprises, because of its own particularity, their fund and financing are more policy-oriented, so state-owned banks' loan increments are not sensitive to the market rate, but small- and medium-sized banks are mainly dominated by the marker rate.

Second, through the analysis of the proxy variables of bank characteristics, it can be seen that the regression coefficients of all banks' and small banks' assets increment is significantly positive, the incremental loans of large banks are not affected by the scale of bank assets, indicating the size of bank loans will increase with the expansion of assets, among which small banks have greater loan expansion impulse. Because compared to other banks, the higher the capital assets scale are, the safer they are. So, when the total assets scale expands, the public's confidence increases. And at the same level of interest rates, the public will be more inclined to put money into larger-scaled banks. This demonstrates that, in China, the size of the banks will produce a heterogeneous impact on loan business, namely to say, small- and medium-sized banks' loan business development space is limited, while large banks will get a broader space for development.

The regression coefficients of the capital adequacy ratio of small- and medium-sized banks are significantly negative and the incremental loans of large banks are not affected by the capital adequacy ratio, indicating that the banks' loan increment will decrease as the capital adequacy ratio increases. Take the Bank of Nanjing as an example, the capital adequacy rate Bank of

Nanjing in 2012 reached 14.98%, the highest among 18 sample banks, while its loan increment is 2.2464 trillion Yuan, the lowest among 18 sample banks. This is because as Chinese government's emphasis on capital supervision increases and supervision ability improves, banks must incorporate the regulatory requirements into operation development strategies. But the capital adequacy ratio, as an important indicator of bank balance sheets, its increase will make the banks maintain more capital to meet regulatory requirements, and the loan expansion scales will be reduced accordingly.

Third, through the analysis of cross-term variables of banks' characteristics and the proxy variables of monetary policy, it can be seen that the cross term variables of the proxy variables of monetary policy and capital adequacy ratio of small and medium-sized banks are significantly positive, which is consistent with the theoretical model of formula (9) and proves Hypothesis 2. As the banks' capital assets ratio increase, their sensitivity of the credit scale to the monetary policy will gradually increase, and the sensitivity of large banks' credit increment to monetary policy is not affected by the capital adequacy ratio. The currency policy will bring greater impact to banks of higher capital adequacy ratio. One possible explanation is that tight monetary policy will firstly influence the loanable funds and as banks' loan supply and interest margin decreases, the banks' profit are compressed, capital value declines, and equity capital financing becomes more difficult. To maintain a high capital adequacy ratio, banks need to reduce the size of the loan more dramatically, thus the amplification effect of capital constraints on bank credit channel appears. Expansionary monetary policy increases the value of the banks' capital and under the original capital constraints; more capital can promote larger scale of loans.

4.2. Policy implications

Based on the above empirical results, the author gets the following policy implications.

First, the existence of China's banks' credit channel is consistent with many domestic scholars'[3] study results [13]. The proportion of indirect financing in China is more than 70%, bank loans being the most important external funding source in nonfinancial sectors. The banks' credit market controls the capital availability by credit rationing, which is very important in the transmission of monetary policy. Therefore, in order to improve the effect of monetary policy, the central bank should attach importance to the credit scale control and incorporate the total credit amount into the current monetary policy framework.

Second, banks' characteristics will have influence on the heterogeneity of credit transmission channel of monetary policy at the micro level, and make the monetary policy show asymmetric effects. Banks of high capital adequacy ratio and small asset sizes are more vulnerable to the impact of the monetary policy. The Central Bank and China Banking Regulatory Commission (CBRC) should adopt differentiated treatment, set different standards for small- and medium-sized banks, and large state-owned banks, provide appropriately relaxed policy environment

[3]Sheng Zhaohui (2007) finds credit channels are still China's main channel of the monetary policy transmission by using data since the 1990s.

to small- and medium-sized joint-stock banks. And for large state-owned banks, because its capital adequacy ratio is much higher than the basic standard and capital constraints have little impact on them, the government should appropriately raise the capital adequacy ratio and other standards to strengthen the credit transmission channel, promote fair competition in the financial market, so as to achieve better monetary policy effect.

Third, credit channel of monetary policy in China is mainly transmitted through the small- and medium-sized joint-stock banks, so to further optimize the structure of the banking market is an effective way to improve monetary policy effect. Since the 1980s, along with the restructuring of Bank of Communications, China Merchants Bank, Guangdong Development Bank, and other joint-stock commercial banks have been established, and a rudiment of joint-stock banks in China emerged. By the 1990s, the market system reformed, to relax the access conditions for the banking industry, a large number of joint-stock commercial banks came into being, accelerating the process of marketization of the banking system. With China's access to the WTO in 2001, small- and medium-sized joint-stock banks begun to attract investment and go into the market. The number of local banks and rural credit cooperatives is expanded continually. Although the development of small- and medium-sized joint-stock banks and promoting full competition in the financial markets has become the general trend, it is undeniable that the state-owned commercial banks is still in the monopoly position in the credit market. At the end of the December of 2013, the loans balance of financial institutions was 71.9 trillion Yuan, among which, the balance of loans of Chinese large national banks[4] was 38.2 trillion Yuan, accounting for 53% of the total RMB loans. It is seen that to enhance competition in China's credit market, the government also need to vigorously develop small- and medium-sized banks and further optimize the market structure of banks.

At the same time, small- and medium-sized enterprises, as an important part of China's economy, its external funding sources rely mainly on small- and medium-sized joint-stock banks, the small- and medium-sized joint-stock banks, therefore, undertake the important responsibility to activate the economy market and promote employment. To improve the monetary policy formulation and supervision, the government should provide some favorable policies to small- and medium-sized joint-stock banks to make them better develop, which will help to clear the financing channel of small- and medium-sized enterprises, tackle the problems caused by the shortage of funds. Compared with the large state-owned banks, small- and medium-sized joint-stock banks are more vulnerable to policy impacts because of the small asset scales and low quality. Therefore, small- and medium-sized joint-stock banks should improve their asset quality and vigorously develop intermediary business, expand funding channel, reduce the pressure of capital occupancy, and form diversified income and profit structure to improve the their own competitiveness.

---

[4]Chinese national large Banks refers to the banks of more than 2 trillion Yuan total quantity of domestic and foreign currency (take the total amount of domestic and foreign currency of every financial institutions by the end of 2008 as reference standard), including ICBC, China Construction Bank, Agricultural Bank, Bank of China, China Development Bank, Bank of Communications, and Postal Savings Bank of China.

http://dx.doi.org/10.5772/64477

Acknowledgements

This article is funded by the National Social Science Fund Project (15BJY143), the social science research project from the Education Department of Hubei Province (2012G064), and Dr. Start-up research project from Hubei University of Technology (BSQD12086).

Author details

Li Qiong

Address all correspondence to: Liqiong@mail.hbut.edu.cn

Economy and Management School, Hubei University of Technology, Hubei, Wuhan, China

References

[1] Bernanke, Ben S., Alan S. Blinder. Credit, Money, and Aggregate Demand. The American Economic Review, 1988, 78(2): 435–439.

[2] Bernanke, Ben S., Alan S. Blinder. The Federal Funds Rate and the Channels of Monetary Transmission. American Economic Review, 1992, (82)4: 901–921.

[3] Kashyap, A., Stein, J., Wilcox, D. Monetary Policy and Credit Conditions: Evidence from the Composition of External Finance. The American Economic Review, 1993(83): 78–98.

[4] Li Qiong. China's Monetary Policy "Credit-Cost" Channel Research. Huazhong University of Science and Technology, 2011.

[5] K Kopecky, D VanHoose. Bank Capital Requirements and the Monetary Transmission Mechanism. Journal of Macroeconomics, 2004, 26: 443–464.

[6] Jiang Chun, Yu Xiamin. Capital Accord and Monetary Policy Transmission: An Analysis Based on Bank Loan Behavior. Jinan Finance, 2006, 07: 16–19.

[7] Liu Bin. Development of DSGE Model in China and Its Application in Monetary Policy Analysis. Financial Research, 2008, 10: 5–25.

[8] Wang Tao, Jiang Zaiwen. An Analysis on Monetary Policy Credit Transmission Mechanism and Effect. Reform of Economic System, 2011, 01: 97–100.

[9] Feng Ke, He Li. Study on the Effect of Credit Expansion on Monetary Policy Transmission Mechanism in China's Bank Listing and Financing. Economic Research, 2011, 12: 52–63.

[10] Li Tao, Liu Mingyu. Empirical Study on the Asymmetry of Bank Capital and Monetary Policy. Economic Review, 2013, 01: 79–86.

[11] Li Haihong. China's City Commercial Bank Capital Structure Analysis of Accounting Research. 2012, 5: 47–49.

[12] Li Zhihui. Business Operation and Management of Commercial Banks. China Financial Publishing House, 2004: 121.

[13] Sheng Zhaohui. An Analysis on effect of Monetary Policy. Transmission Mechanism in China, 1994–2004.

Chapter 17

Innovation Measurement in the Czech Republic and People's Republic of China

Jindra Peterková and Zuzana Wozniaková

Additional information is available at the end of the chapter

http://dx.doi.org/10.5772/66794

Abstract

For companies innovations are vital to ensure their continued growth and ability to survive in a highly competitive business environment. Realization of successful innovations has positive impact on countries and their economies. At the same time strong economy in a country is an assumption of strong economy of its regions. Interested party receives an information feedback about innovation performance and character of innovation environment and enables to implement measures to eliminate any shortcomings. Innovations can be measured at enterprise, state and global levels. Measurement of innovations in connected with following questions: What is emphasized in measuring innovation at appropriate level? What structure do indicators that are used when measuring have? What types of indicators are used when measuring? What are the differences in innovation measurement at the level of the Czech Republic and People's Republic of China? These questions are answered in this chapter. Aim of the chapter is to monitor possible ways of innovation measurement at the enterprise, state and global levels and at the same time to compare differences in innovation measurement in the Czech Republic and People's Republic of China. For that purpose, analysis, synthesis, description and comparison were used.

Keywords: innovations, technical innovations, measurement

1. Introduction

Not only advanced economies but also developing nations recognize that innovation is one of the main drivers of economic growth, leading to emergence of new industrial enterprises and branches, develops manufacturing, increases production level while inputs remain unchanged and lead to increasing revenues. Several authors (Brynjolfsson and McAfee,

2015; Zelený, 2011) point out that innovations are fundamental great inventions, or recombination of things that have already existed. Contemporary innovations take the form of digital technologies, based on hardware, software, and network. At the same time, digitalization enables the use of a huge amount of data that can be reproduced again. Innovation measurement represents necessary assumption for the right innovation management. Qualitative and quantitative methods are used for innovation measurement in business practice, while qualitative methods alone do not enable to quantify relationship among given values. We can divide them into nominal values (in the case of two values it can be determined whether they are the same or different) and ordinal values (in the case of two values their order can be determined). On the other hand, quantitative values can quantify the relationship between two values and can be divided into interval values (in the case of two values their difference can be determined) and indicators for shares (we can determine how many times the values vary).

Innovations are measured at the enterprise, state, and global levels. Enterprise innovation activities are measured at the enterprise level. At the state level, state innovation activities are assessed while at the global level innovation is measured by the capacity of a given economy in a particular territorial unit (state unit).

The aim of the chapter is to monitor the way of innovation measurement at the enterprise, state, and global levels. The chapter also involves the comparison of innovation measurement at the level of two states, concretely the People's Republic of China and the Czech Republic. Data from the Czech Statistical Office, EUROSTAT, and other sources are used in this chapter.

We address the question of how enterprise innovations have been measured under the conditions of two very different states. For that purpose, analysis, synthesis, description, and comparison are used. First, the Czech system of innovation measurement at the enterprise level is analyzed and structured into blocks, dimensions, and indicators which are described in more detail. Concrete data from the Czech Statistical Office focused on innovative enterprises are analyzed. Second, description of innovation measurement at the state level in the Czech Republic including actual results of more and less innovative states in EU follow. We also introduce a third level—a global level of innovation measurement represented by Global Innovation Index. In Section 4, the China's national innovation system is described. Comparison and synthesis are needed for finding the differences between innovation measurement in the Czech Republic and People's Republic of China and for final conclusion.

2. Theoretical background

Considerable variety exists in the definition and measurement of concepts related to what can be broadly termed "innovation". A range of labels such as radical, discontinuous, breakthrough, and new is given to phenomena touching upon different dimensions of inventive outcomes (Verhoeven et al., 2016). This section structures different meanings of innovation and provides an overview of different classification of innovations introduced by firms.

http://dx.doi.org/10.5772/66794

2.1. Characteristics of innovation

According to broad approach, innovation means any change in social life (Valenta, 2001). Innovation can be represented by a new way of working, which results in a positive change (Gallo, 2011). Therefore, innovation in business practice is a narrow segment. According to OECD innovation goes far beyond R&D. It goes far beyond the confines of research labs to users, suppliers, and consumers everywhere—in government, business, and nonprofit organizations, across borders, across sectors, and across institutions. Scholars define innovation as a creative process of devising a useful product, service, or mode of action from a pure concept located within a company (Bogdanienko et al., 2004; Amabile et al., 1996). Anything new may be perceived as innovation, if its qualities or attributes distinguish it from its existing counterparts (Burnett, 1953; Damanpour, 1991). Drucker (1993) claims that innovation is a specific tool of entrepreneurs, the means by which they exploit change as an opportunity for a different business or a different service. Entrepreneurs need to search purposefully for the sources of innovation, the changes, and their symptoms that indicate opportunities for successful innovation. The innovation equation model considers creativity as generating an idea and risk-taking as taking action on the idea, innovation = creativity + risk-taking (Pearl, 2011). Innovations are beneficial for enterprises as well as for customers in terms of value for customer.

2.2. Different classification models used for discussing innovation types

The Oslo Manual, developed jointly by Eurostat and the Organization for Economic Co-operation and Development (OECD), provides a framework to enable innovation measurement. The manual proposes innovation types of:

- product (good or service that is new or significantly improved; this includes significant improvements in technical specifications, components and materials, software in the product, user friendliness, or other functional characteristics);
- process (a new or significantly improved production or delivery method. This includes significant changes in techniques, equipment, and/or software);
- marketing methods (a new marketing method involving significant changes in product design or packaging, product placement, product promotion, or pricing);
- a new organizational method in business practices, workplace organization, or external relations.

Innovations may also be classified according to "type." Schumpeter (1934) distinguished between five different types: new products, new methods of production, new sources of supply, the exploitation of new markets, and new ways to organize business. In economics, most of the focus has been on the first two types. The terms "product innovation"' and "process innovation" have been used to characterize the occurrence of new or improved goods and services and improvements in the ways to produce these good and services, respectively. However, the focus on product and process innovations, although useful for the analysis of some issues, should not lead us to ignore other important aspects of innovation.

Considering originality Kuratko (2009) distinguishes four types of innovations: invention (a totally new product, service or process), extension (new use of or different application of an already existing product, service, or process), duplication (creative replication of an existing concept), and synthesis (combination of existing concepts and factors into a new formulation or use).

Classification of the Slovak researcher Valenta (2001) introduced eight types of innovations from the zero level to the seventh level:

- innovations of the zero level: generation of initial properties,
- innovations of the first level: the simple target adaptation to quantitative requirements while preserving the functions of a business system or its parts,
- innovations of the second level: regrouping or organizational change,
- innovations of the third level: adaptation changes,
- innovations of the fourth level: the elementary qualitative change
- innovations of the fifth level: higher qualitative change of functional properties of system or its parts,
- innovations of the sixth level: qualitative change of functional properties of a business system or its part.
- innovations of the seventh level: the highest radical change of functional properties of a business system or its part changing its basic functional principle.

According to Albury (2005), successful innovation is the creation and implementation of new process, products, services, and methods of delivery which result in significant improvements in outcomes, efficiency, effectiveness, or quality. However, current experts suggest that in order to gain competitive success, business needs to be able to effectively implement, monitor, and measure the innovation process (Hassanien and Dale, 2013).

3. Innovation measurement

3.1. Innovation measurement at the enterprise level

While measuring innovation at the enterprise level it is necessary to distinguish two other levels. One measurement of business innovations is realized by the Czech Statistical Office and the second one is worked out by enterprises alone. Czech Statistical Office monitors innovations according to the Oslo Manual 2005. The manual was developed on the basis of OECD initiative. The same method of measurement has been used in all the EU member states. The main sense of using identical statistical data gathering lies in obtaining comparable data about innovation environment and innovation activities in businesses with the whole European Union. According to Oslo Manual 2005, innovations are divided into technical and nontechnical innovations—see according to actualized methodology of EUROSTAT, in 2010 enterprise that introduced product or process innovation or had continuing or had interrupted innovation activities (technical innovations), or introduced marketing, or organizational innovation (nontechnical innovation) is considered to be innovative enterprise. The Czech Statistical Office that realizes statistical gathering in two years' cycles found out that between 2004 and 2012 the share of innovative enterprises in the whole group of enterprises was around 50% which means that each second enterprise innovated, see **Figure 1** and **Table 1**.

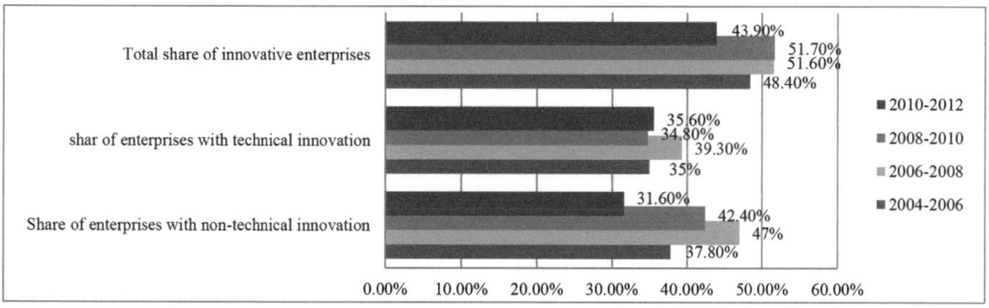

Figure 1. Number of innovative enterprises from the whole group of enterprises including classification of technical and nontechnical innovations.

Blocks	Dimensions	Indicators
Technical innovations	Product innovations	Introduction of a good or service that is new or significantly improved with respect to its characteristics or intended uses. This includes significant improvements in technical specifications, components and materials, incorporated software, user friendliness, or other functional characteristics. Product innovations can utilize new knowledge or technologies, or can be based on new uses or combinations of existing knowledge and technologies.
	Process innovation	Implementation of a new or significantly improved production or delivery method. This includes significant changes in techniques, equipment and/or software. Process innovations can be intended to decrease unit costs of production or delivery, to increase quality, or to produce or deliver new or significantly improved products.
Nontechnical innovation	Organizational innovation	Implementation of a new organizational method in the firm's business practices, workplace organization, or external relations.
	Marketing innovation	Implementation of a new marketing method involving significant changes in product design or packaging, product placement, product promotion, or pricing.

Source: Peterková and Ludvík (2015) and www.global-innovation.net

Table 1. Innovation measurement at the enterprise level.

The Czech Republic first worked out statistical innovation survey in 2002 in the framework of being a new member of European Union. In European Union, the first statistical survey focused on innovations was worked out in 1993. In the Czech Republic, seven surveys about innovations were carried out, while some changes in methodology of gathering the data appeared. Last survey covered period 2010–2012. The period of 2008–2010 seems to be favorable for innovations (51.7%), on the other hand the period of 2010–2012 seems to be less favorable, as the share of innovative enterprises is 43.9% from the whole number of economically active enterprises. In the period of 2004–2008, technical innovations dominated (35.6%) over non-technical innovations (31.6%). According to statistical survey in 2010–2012, the biggest number of innovative enterprises appears in information and communication technologies (64.8%), while technical innovations were introduced mostly (57%), and were followed by nontechnical innovations (45.7%). The second most innovative branch is finance and insurance (55.9%) followed by manufacturing (48.3%) with predominance of technical innovations over nontechnical ones. The last place belongs to mining and quarrying branch (23.2%) and enterprises in transportation and storage (10.8%). In all branches with the exception of wholesale, technical innovations predominated over nontechnical. Measurement of innovations at enterprise level belongs to managers' and owners' competencies. They use their own innovation techniques. Managers use hard metrics which are available without any additional costs and are transferable to financial expressions, or soft metrics which are used for evaluation of the rate of meeting internal targets in the area. One-third of all companies appeared at the list of top-ranked companies Fortune 1000 uses innovation metrics published by Innovation Point, see **Table 2.**

Innovation metrics
Annual R&D budget as a percentage of annual sales
Number of patents filed in the past year
Total R&D headcount or budget as a percentage of sales
Number of active projects
Number of ideas submitted by employees
Percentage of sales from products introduced in the past X year(s)
Source: http://www.innovation-point.com/innovationmetrics.htm

Table 2. Set of innovative metrics.

3.2. Innovation measurement at the state level in the Czech Republic

In order to become the most competitive and dynamic knowledge-based economy with sustainable growth, the European Union established European Innovation Scoreboard. The first scoreboard was proposed of 17 countries increased to about 30 and the number of indicators increased to 29. Scoreboard is divided into three parts: enablers, firm activities, and outputs. Enablers are the main drivers of innovations such as new doctorate graduates, finance support, or venture capital. Firm activities contain firm investments, collaborating enterprises, or intellectual assets. Outputs include innovators such as SMEs with product and process innovations or with marketing or organizational innovations and economic effects such as license and patent revenues from abroad (Gupta and Trusko, 2014).

This scoreboard was later renamed innovation union scorecard that helps to provide benchmarking among 27 member states in the sphere of innovation implementation. The aim of the benchmarking based on innovation union scorecard is to strengthen research and innovation. The structure of the scorecard has three blocks, eight dimensions, and 25 indicators, see **Table 3.**

http://dx.doi.org/10.5772/66794

Blocks	Dimensions	Indicators
Enablers	Human resources	New doctorate graduates Population aged 30–34 when tertiary educationYouth with at least upper secondary education
	Open, excellent, attractive research systems	International scientific co-publications top 10% most cited scientific co-publications Non-EU doctorate students
	Finance and support	R&D expenditures in the public sector Venture capital
Firm activities	Firm investments	R&D expenditures in the business sector Non-R&D innovation expenditure
	Linkages and entrepreneurship	SMEs innovating in-house Innovating SMEs collaborating with others Public-private co-publications
	Intellectual assets	PCT patent applications PCT patent applications in societal challenges Community trademarks Community designs
Outputs	Innovators	SMEs with product and process innovations SMEs with marketing or organizational innovations High-growth innovative firms
	Economic effects	Employment in knowledge-intensive activities Medium and high-tech product exports Knowledge-intensive services exports Sales and new-to-market and new-to-firm innovations License and patent revenues from abroad

Source: Gupta and Trusko (2014).

Table 3. European Union innovation scoreboard framework.

According to average innovation performance, the member states are grouped into four performance groups: innovation leaders, innovation followers, moderate innovators, and modest innovators. To be an innovation leader, the member state has to demonstrate a balanced innovation system.

In comparative assessment of the research and innovation performance of the EU member states was found following findings: Sweden has confirmed its innovation leadership. It is followed by Denmark, Finland, and Germany as European innovation leaders. Compared to 2014, innovation performance has increased in 15 EU countries, while it declined in 13 others. Latest results showed that (Innovation Union Scoreboard, 2015):

- Sweden, Denmark, Finland, and Germany are "innovation leaders" with innovation performance well above that of the EU average;
- Austria, Belgium, France, Ireland, Luxembourg, the Netherlands, Slovenia, and the United Kingdom are "innovation followers" with innovation performance above or close to that of the EU average;
- The performance of Croatia, Cyprus, Czech Republic, Estonia, Greece, Hungary, Italy, Lithuania, Malta, Poland, Portugal, Slovakia, and Spain is below that of the EU average. These countries are "moderate innovators";
- Bulgaria, Latvia, and Romania are "modest innovators" with innovation performance well below that of the EU average (Innovation Union Scoreboard, 2015).

Sweden's innovation system is once more in the first position in the EU with the overall ranking remaining relatively stable. The performance group memberships have remained relatively stable compared to the previous IUS edition, with Cyprus and Estonia being the only countries that changed group membership, in their case changing from the innovation followers to the moderate innovators. Within the moderate innovators, Estonia is the top performer followed by the Czech Republic that has overtaken Italy and Cyprus. The most innovative countries have balanced innovation systems with strengths in all dimensions, but some other countries reach top scores in individual dimensions. Sweden, Ireland, Finland, and the United Kingdom score the best in human resources; the Netherlands, Sweden, and Denmark reach top positions in open, excellent, and attractive research systems; Estonia, Denmark, Finland, and Sweden perform best in finance and support; Germany, Sweden, Estonia, and Finland are the best performers in firm investments; Belgium, the United Kingdom, and Denmark are top performers in linkages and entrepreneurship; Sweden, Denmark, Finland, and Germany reach the top positions in intellectual assets; Ireland, Luxembourg, and Germany are the best performers in the innovators dimension; and Ireland, Denmark, and Luxembourg reach the highest results in economic effects. Over a longer time period of 8 years, the EU has been improving its innovation performance, with Latvia, Bulgaria, and Malta being the innovation growth leaders but innovation growth differences exist also within the groups and the innovation gap between the member states closes slowly. However, compared to the last year, innovation has not been improving. A direct comparison with the results of last year's edition is not possible as there have been some changes in the measurement framework, but a comparison with innovation performance as it would have been last year using the same measurement framework shows that innovation performance has declined for 13 member states, in particular for Romania, Cyprus, Estonia, Greece, and Spain. For the EU at large innovation performance has not changed and for 15 member states it has improved, most notably for Malta, Latvia, and Bulgaria (Innovation Union Scoreboard, 2015).

At a wider European level, Switzerland confirms its top position outperforming all EU member state. Taking into account European countries outside the EU, also this year Switzerland confirms its position as the overall innovation leader by continuously outperforming all EU member states and by being the best performer in as many as six indicators. Internationally, South Korea and the US defend their positions as the top global innovators.

3.3. Innovation measurement at the global level

For the measurement of country's innovation extent and how is integrated into its political, business, and social aspects we can use the Global Innovation Index. For the first time, this index was published by the business school INSEAD, World Intellectual Property Organization, an entity of United Nations (WIPO). Global Innovation Index measures the capability of economy to innovate, and its innovation performance (Jewell, 2012). Global Innovation Index is based on two pillars: innovation input subindex and innovation output subindex. The area of institutions, human capital resources, infrastructure, market

http://dx.doi.org/10.5772/66794

sophistication, and business sophistication create innovation input subindex. Innovation output subindex results from knowledge and technology outputs and creative outputs, see **Table 4**.

Blocks	Dimensions	Indicators
Innovation input subindex	Institutions	Political environment Regulatory environment Business environment
	Human capital	Education
	Infrastructure	ICT General infrastructure Ecological sustainability
	Market sophistication	Credit Investment Trade and competition
	Business sophistication	Knowledge workers Innovation linkages Knowledge absorption
Innovation output subindex	Knowledge and technology outputs	Knowledge creation Knowledge impact Knowledge diffusion
	Creative outputs	Creative intangibles Creative goods and services Online creativity

Source: Gupta and Trusko (2014).

Table 4. Global innovation index framework.

Practical use of Global Innovation Index showed that average innovation ranking increases with the income level of a country. North America leads in innovation followed by Europe, Southeast Asia, Northern Africa, and Western Asia, Latin America and Caribbean, Central and Southern Asia, and sub-Sahara Africa (Gupta and Trusko, 2014).

Although there exist many global indexes or other measurement systems of innovation activities, each country can use its own measurement corresponding to the particular conditions. The following section of the chapter focuses on comparison of innovation measurement in China and in European Union.

4. Comparison of innovations measurement in the Czech Republic and People's Republic of China

4.1. China's national innovation system

Measurement innovation in China has been more focused on assessing intellectual capital in terms of patents, literature citations, and growth in research and development in terms of addition of R&D functions in corporations, R&D expenditures, and R&D personnel as a percentage of total employment (Gupta and Trusko, 2014). Literature citations are based on published total and joined Chinese science and technology papers. The measures are shown in **Table 5**.

Blocks	Dimensions	Indicators
Total and joined Chinese science and technology papers	Universities	Total university papers University papers as % total Joint papers with universities (%) Joint papers with R&D institutes (%) Joint papers with firms (%)
	R&D institutes	Total R&D institute papers R&D institute papers as % total Joint papers with universities (%) Joint papers with R&D institutes (%) Joint papers with firms (%)
	Firms	Total firm papers Firm papers as % total Joint papers with universities (%) Joint papers with R&D institutes (%) Joint papers with firms (%)
Technology-based spin-offs from universities and research institutes	Universities	Number Profit (RMB)
	Research institutes	Number Profit (RMB)
Patenting activity by organization type and patent type	Invention patents	R&D (% total) Universities (% total) Firms (% total)
	Utility patents	R&D (% total) Universities (% total) Firms (% total)
	Design patents	R&D (% total) Universities (% total) Firms (% total)
Sources of R&D funding	Government	Share of total (%) Increase (%) Amount (RMB billion)
	Enterprises	Share of total (%) Increase (%) Amount (RMB billion)
	Banks	Share of total (%) Increase (%) Amount (RMB billion)
	Other	Share of total (%) Increase (%) Amount (RMB billion)

Blocks	Dimensions	Indicators
Total and joint patenting activity	Universities	Total university patents University patents as % total Joint patents with universities (%) Joint patents with R&D institutes (%) Joint patents with firms (%)
	R&D institutes	Total R&D institute patents R&D institute patents as % total Joint patents with universities (%) Joint patents with R&D institutes (%) Joint patents with firms (%)
	Firms	Total firm patents Firm patents as % total Joint patents with universities (%) Joint patents with R&D institutes (%) Joint patents with firms (%)
Share of national R&D expenditure by sector	Research institutes	Share (%)
	Universities	Share (%)
	Enterprises	Share (%)
Regional variation in innovation inputs and outputs	Eastern Region	Invention patents (number and % China) Regional GDP (RMB billions and % China) RRD&E Personnel (thousands and % China) R&D expenditure (RMB millions and % China)
	Zhejiang	Invention patents (number and % China) Regional GDP (RMB billions and % China) RRD&E Personnel (thousands and % China) R&D expenditure (RMB millions and % China)
	Central region	Invention patents (number and % China) Regional GDP (RMB billions and % China) RRD&E Personnel (thousands and % China) R&D expenditure (RMB millions and % China)
	Western region	Invention patents (number and % China) Regional GDP (RMB billions and % China) RRD&E personnel (thousands and % China) R&D expenditure (RMB millions and % China)

Source: Liu and White (2001).

Table 5. Measurement of innovation activities in China.

Chinese system of innovation measurement at the state level points out that the most important institutions in the area of innovation are universities, firms, and research institutes. Significant attention is paid to design, utility, and invention patents. Sources of research and development funding are divided into three main groups: enterprises, banks, and others.

Funding of innovation activities is also monitored in connection with sectors or branches as source acceptors. Total and joined Chinese science and technology papers are also evaluated.

From historical point of view although the Chinese government has made dramatic progress toward a more effective and efficient national innovation system compared to its performance under central planning, a number of important issues remain such as an inadequate legal environment that cannot yet provide a reliable environment for inter-organizational relationships that are crucial in the innovation process. This issue is the biggest and growing discrepancy among regions in terms of innovative activity, which the Chinese government has recognized but has been largely ineffective in addressing (Liu and White, 2001).

Beijing has a strong science base, including the Chinese Academy of Sciences (CAS), and top universities; these are national R&D centers with global connections. Shanghai has a large-scale, R&D-intensive industry base. Guangdong province has a foreign (manufacturing) firm-based innovation system and accounts for more than half of China's PCT patent applications (almost two-thirds in ICT). In contrast, China's western regions lack the absorptive capacity needed to capture knowledge flows from coastal areas and abroad. Collaboration, as shown in patent data, is weak across regions.

4.2. Comparison of innovation measurement in the Czech Republic and People's Republic of China

The state level of innovation measurement provides data about innovation investments and innovation performance of particular state, which refers to system ex post. In the Czech Republic, innovations are measured by using innovation scoreboard which is focused on innovation conditions (human resources, research systems, finance, and support), enterprise activities (firm investments, partnership and enterprise, intellectual property), and innovation outcomes which include effects of enterprise innovation activities (economic effects). Innovation scoreboard takes into account character of realized innovations which means that technical (product, process) and nontechnical innovations (marketing, organizational) are evaluated.

In China, measurement of innovations at the state level is based on seven factors. Significant attention is paid to the evaluation of patents, scientific papers, and research and development. Simultaneously research and development expenditures as well as sources of financing are monitored. In China and European Union, we can find monitoring of spin-off firms founded by universities and also research organizations. The used measurement system shows that technical and also nontechnical innovations are monitored separately in China and also in the Czech Republic.

Both measurements at the state level point out an importance of cooperation among universities and business practice and commercialization of research findings. Both countries monitor a number of spin-offs and cooperation businesses with nonprofit organizations.

5. Conclusion

Innovations are an important assumption of economic growth of enterprises, states, and even global economies. Innovations represent quantitative or qualitative improvement of product, process, or business model. A process of innovation measurement depends on the innovation

http://dx.doi.org/10.5772/66794

type and institution approach to innovations success measurement. At the same time, each innovation has a different character and institutions in various countries have different priorities, which lead to the fact that particular method frameworks and approaches differ. Mainly combinations of quantitative as well as qualitative indicators have been used. Differences are obvious at all levels: enterprise, national, and global. For enterprise level of innovation measurement, own business innovation metrics are used. Czech Statistical Office monitors technical and nontechnical innovations separately.

In the Czech Republic, for the state level innovation scoreboard is used. This innovation measurement is focused on enablers, firm activities, and outcomes as effects from carried out innovation activities by firms. The innovation scoreboard includes 25 indicators.

At global level innovation measurement is focused on innovation performance and innovation environment, wider regions such as European Union, Central and Southern Asia.

Comparison of innovation assessment in two different countries the Czech Republic and People's Republic of China showed that both states emphasize commercialization of university outputs in business practice and at the same time in both states, research expenditures are reported. Provided comparison did not show whether character of innovation (technical or nontechnical) is recorded.

Systems for innovation monitoring and later evaluation of these two countries come out from different business environment that is influenced by various political, legal, and cultural values. This implies that the united system of innovations management and measurement cannot be implemented globally.

Author details

Jindra Peterková and Zuzana Wozniaková*

*Address all correspondence to: zuzana.wozniakova@vsb.cz

VSB-Technical University of Ostrava, Ostrava, Czech Republic

References

Albury, D., 2005. Fostering innovation in public services. Public Money and Management, 25, pp. 51–56.

Amabile, T.M., Conti, R., Coon, H., Lazenby, J., Herron, M., 1996. Assessing the work environment for creativity. Academy of Management Review, 39 (5), pp. 1154–1184.

Bogdanienko, J., Haffer, M., Popławski, W., 2004. Enterprise innovation. UMK, Toruń.

Burnett, H.G., 1953. Innovation: The basis of cultural change. McGraw-Hill, New York.

Brynjolfsson, E., Mcafee, A., 2015. Druhý v?k strojů. Práce, pokrok a prosperita v éře špičkových technologií. Jan Melvil Publishing, Brno, Czech Republic.

Damanpour, F., 1991. Organizational innovation: A meta-analysis of effects of determinations and moderators. Academy of Management Journal, 34 (3), pp. 555–590.

Hassanien, A., Dale, C., 2013. Facilities management and development for tourism, hospitality and events. CABI, Boston, MA.

Kuratko, D., 2009. Entrepreneurship: theory, process, practice. South-Western Cengage Learning, Mason, OH.

Pearl, M., 2011. Grow globally: opportunities for your middle-market company around the world. John Wiley & Sons, Hoboken, NJ.

Drucker, P. F., 1993. Management: tasks, responsibilities, practices. Harper Business, New York.

Gallo, C., 2011. Tajemství inovací Steva Jobse. Computer Press, Brno, Czech Republic.

Gupta, P., Trusko, E. B., 2014. Global innovation science handbook. McGraw-Hill Education, New York.

Innovation Union Scoreboard, 2015. Luxembourg: Office for Official Publications of the European Communities, Brussels.

Jewell, C., 2012. Global innovation index. WIPO Magazine.

Liu, X., White, S., 2001. Comparing innovation systems: A framework and application to China's transitional context. Research Policy [online], 30 (7), pp. 1091–1114. http://linkinghub.elsevier.com/retrieve/pii/S0048733300001323/ (accessed 2016-04-22).

Valenta, F., 2001. Inovace v manažerské praxi. Velryba, Praha, Czech Republic.

Verhoeven, D., Bakker, J., Veugelers, R., 2016. Measuring technological novelty with patent-based indicators. Research Policy, 45 (3), pp. 707–723. http://linkinghub.elsevier.com/retrieve/pii/S0048733315001857/ (accessed 2016-05-30).

Peterková, J., Ludvík, L., 2015. Řízení inovací v průmyslovém podniku. SAEI, vol. 42. VŠB-TU Ostrava, Ostrava.

Schumpeter, J., 1934. The theory of economic development. Harvard University Press, Cambridge, MA.

Zelený, M., 2011. Hledání vlastní cesty. Computer Press, Brno, Czech Republic.

Section 3

Materials – Technologies – Environment

Comparison of Human Resource Management Practices in Czech and Chinese Metallurgical Companies

Martin Čech, Andrea Samolejová, Jun Li,
Wenlong Yao and Pavel Wicher

Additional information is available at the end of the chapter

http://dx.doi.org/10.5772/66795

Abstract

The objective of this chapter is to analyze and compare various aspects of human resource management (HRM) practices in Chinese and Czech metallurgical companies. A questionnaire consisting of 58 questions devoted to specific aspects of HRM such as recruitment, performance evaluation and remuneration, and training and development was designed to acquire necessary data. Data acquired from 42 Chinese and 36 Czech companies were analyzed in order to yield the most beneficial outcomes. This chapter focuses on recruitment and selection of employees, evaluation, remuneration, and motivation of employees and career management. Results show significant differences in various aspects of HRM between both countries. Differences and some similarities are discussed and managerial implications are presented in the chapter.

Keywords: human resource management, metallurgical companies, China, Czech Republic, recruitment, selection, remuneration, motivation, career, benefits

1. Introduction

This chapter presents results of the project Support of VŠB-TUO activities with China with financial support from the Moravian-Silesian Region. The field of HRM practices in manufacturing companies in both countries was researched by Li et al. [1], who presented comparison of practices in various aspects of HRM based on data acquired since 2011. Some preliminary results of the project have been published earlier, focusing on different HRM aspects, such as setting of HR department, number of its employees, HR planning, and training and development in Chinese manufacturing companies [2]. This chapter focuses on situation in metallurgical companies of both countries, which were selected from the total sample of surveyed companies.

As human resource management (HRM) plays an irreplaceable role, HR managers must meet the demands of dynamically changing environment and maintain and motivate human resources in order to increase competitive advantage of organizations [3]. Also in this field of management, planning plays an essential role in overall performance and success as is discussed by Samolejova et al. [4]. In the area of recruitment or selection, it was found so far that particular selection methods are used more or less frequently in different areas in relation to the cultural values of that country, e.g., high uncertainty avoidance culture used more test types, more interviews [5]. Different cultures emphasize different attributes on the selection of employee, people in achievement-oriented country consider skills, knowledge, and talent, in ascription-oriented culture, age, gender, and personal relationship are important [6]. This finding is proved again and is discussed in this chapter. Performance management has developed over the past two decades as a strategic, integrated process which incorporates goal-setting, performance appraisal, and development into a unified and coherent framework with the specific aim of aligning individual performance goals with the organization's wider objectives [7].

In this chapter, the differences originated in above-mentioned culture specifics are discussed. It was suggested that collectivist societies are more likely to use informal, subjective appraisal. The concept of performance appraisal sits uncomfortably with character assessment. Cultural variations in the area encompass both how people should appraised and by whom [8]. As discussed below, the historical background of Confucius philosophy of China is very relevant in determining the nature of HRM. Our research showed significant differences in HRM practices between both countries, considering the emphasis on different characteristics such as seniority and group achievement in China and personal performance and experience in the Czech Republic.

2. Sources and methods

Data for this study was collected using a questionnaire designed by Czech and Chinese co-researchers. The questionnaire has a total number of 58 questions divided into several sections referring to various aspects of HR management. It was designed to reflect both the Chinese and Czech industrial environment that would provide a relevant material to compare both environments in future research. Fifty local metallurgical companies of Hubei province and 41 metallurgical companies in Czech Republic were invited to participate in the research.

The questionnaire was sent via email together with an introduction letter. We received 43 completed questionnaires from China, one per company, which makes a good return rate of 84% and 36 questionnaires from Czech Republic, with a return rate of 87%. The size distribution of the companies is described in **Table 1**. For the purpose of this chapter, we used three size categories of companies. Not all the data acquired were analyzed for the purpose of this chapter, which is focused on analysis and comparison of questions dealing with recruitment, selection, evaluation, remuneration, motivation, and career management among Czech and Chinese metallurgical companies.

http://dx.doi.org/10.5772/66795

Company size	Medium	Big	Very big	
No. of employees	100–250	250–750	more than 750	Total
Czech Republic	21	8	7	36
Czech Republic [%]	58.30%	22.20%	19.50%	x
China	11	16	16	43
China [%]	25.58%	37.21%	37.21%	x

Table 1. Size distribution of the companies according to number of employees.

3. Results and discussion

3.1. Recruitment

Recruitment is the process of generating a pool of candidates from which the appropriate person is selected to fill a job vacancy. It is composed of recruitment, selection, and employment. Preliminary results of Chinese part of our study were published in our previous article. This chapter presents comparison of practices in Chinese and Czech metallurgical companies in that field.

In this survey, 80% of firms have prepared descriptions of job positions. When asked about the source of recruitment, 66% of Chinese and 100% of Czech companies give priority to internal sources and turn to external sources only when necessary. The situation is reversed when important positions need to be filled–47% of Chinese firms fill important positions exclusively from internal sources, however only 33% of Czech metallurgical companies do the same. This fact also illustrates the general cultural difference and emphasis on loyalty on Chinese side more than on the Czech side, which does not necessarily say which approach is better taking into account the best achievable performance for the company. As regards to the sources of recruitment, a similar portion of about 60% of companies in both countries choose to cooperate with universities and vocational schools. 75% of surveyed Chinese companies cooperated with labor markets and government employment organizations. In the Czech Republic this source of recruitment is slightly more common and was used in 85% of companies surveyed. The situation is very different in the field of use of employment agencies which temporarily lease the workforce. This phenomenon is very common and 81% of Czech companies use it regularly, but only 33% of Chinese companies do so. A similar ratio can be observed in the use of recruitment agencies, only 44% of Chinese companies use them while it is very common in the Czech Republic where nearly 70% use this method of recruitment. Further research in this field should be focused on the ratio of in-house and outsourced recruitment in companies as only yes/no answers were possible in our research so no information on the rate of utilization of this method is available.

Based on the above-mentioned results several recommendations could be considered. Chinese companies could motivate their employees by using them more often as the first choice for positions (and not only the managerial ones) staffing. Czech companies should seek for managers in internal sources more often because a career growth is one of the most powerful tools of qualified and ambitious employees' motivation. As for the most valued personal characteristics, it is difficult to recommend anything to the firms as the differences obviously reflect experiences in both countries and, unfortunately, Czech companies are experienced that skills and knowledge stated in candidate CVs vary from the real ones and so rely more on recommendations.

3.2. Selection of employees

Concerning the question on selection, the surveyed manager was required to rank certain characteristics of candidate, based on those significant to the employer, such as educational and professional knowledge and skills, work experience, social abilities and personality, social status, references of previous employers, and recommendation from known persons or existing staff and others. Very interesting differences were showed by results of the above-mentioned question on importance of candidate characteristics. Results again corresponded with the basic values of the societies. In China, the three most valued personal characteristics were professional knowledge and skills, work experience, and social and personal characteristics. Contrary to that, in the Czech Republic suitable candidates to fill the job positions are mostly selected according to recommendations of known person or current staff or reference of the previous employer, while the work experience was at the last position as the position of recommendation of known person or staff and previous employer reference in Chinese companies.

Each company has its own values on account of the corporate culture, the nature of business, size, and other factors, but most frequently used selection methods can be identified. There are many methods to select employees and their use varies between both countries. In the Czech Republic, resume analysis is the most used and adopted by 100% of metallurgical companies. In China, the most used method is interview, adopted by almost 99% companies in the study. Both methods are combined in both countries. As we distinguish between structured and informal interview, we can say that the rate of use of structured interview is the same in both countries, about 65% of companies use it. The situation is different with the use of informal interview. Only 35% of Chinese companies use this form of interview, while more than 76% of Czech companies do the same. Psychological tests, professional tests and test of language skills are performed most frequently in Chinese companies but assessment center is used twice more, in nearly 29% of Czech companies compared to China. In both countries HR managers pay attention to reference, and check and verify those recommendations.

Final decision in employee selection is usually made by head of department, where the requirement originated, this situation is similar in both countries. In China, it is very rare to let this decision to be made by HR department even when it is established, in Czech it is more common and the situation is the same with use of committee appointed by management. The last mentioned methods are supposed to be used more frequently when important position is to be filled that for blue-collar positions.

3.3. Evaluation of employees

To manage the company performance it is necessary to have information on how efficiently human capital is working. To provide a feedback to employees the evaluation of their performance in fulfilling management expectations takes place. Companies were asked about specifics of their evaluation system in order to acquire relevant perspective on how employees are evaluated among surveyed metallurgical companies in both countries. As managers realize the importance of evaluation, an active evaluation system able to fairly diversify the employee performance is somehow implemented in 86% of Czech and 78% of Chinese companies. The period of evaluation varies a lot among the samples in both countries, but there is no significant difference in the period between the Czech Republic and China. In both countries more than 60% of companies evaluate employees once a year, as one year is considered optimal time to check and manage company performance, let employees to progress and observe a trend in performance but of course also the allocated resources are taken into account as the evaluation is not simple and easy and it takes significant amount of time, energy, and other resources. Considering the above mentioned, only 10% of Czech and 20% of Chinese metallurgical companies perform evaluation twice a year. A significant portion of 28% of Czech companies–while only half of that in China–perform evaluation also in different time periods, usually every 3 months. In this case, evaluation is not always the same, according to different fields of evaluation, such as personal development, work performance, and others. The basic characteristics of the evaluation in both countries are similar–mostly the direct supervisor performs the evaluation and feedback of employee is allowed and taken into account. What is quite common in China, where 24% of companies perform evaluation done by HR department, is very rare in the Czech Republic where less than 5% of companies perform employee evaluation using this method.

Top three most commonly used evaluation methods are same for both countries with almost the same proportion. The most frequently used method is comparison with stated objectives in more than 60% of companies, followed by forced distribution into performance groups in about 37% companies, and comparison with other employees in about 25% of companies in both countries.

A significant difference can be seen in the following methods. Comparison with corporate standards is used more than two times in China, compared to only 19% of Czech metallurgical companies. Even more different is the approach to evaluation using critical cases – markedly good and markedly bad performance during the evaluated period. This method is commonly used in more than 44% of Chinese companies, while only 14% of Czech companies perform this kind of evaluation. Considering other researched evaluation methods, evaluation interview is more commonly used in the Czech Republic, where it is performed in more than 47% of cases compared to only 13% penetration of the method in China. The situation is opposite with the method of 360" evaluation which was implemented in more than 30% of Chinese companies but the penetration of the method in the Czech Republic is still less than 20%.

Employee appraisal is an essential part of performance management and a necessary part of efficient employee-motivation systems. Any company should implement employee appraisal and its results should be reflected in wages, usually in the variable component, and should be used when deciding on future education and career growth of employees. It is remarkable that 14% of Czech and 22% of Chinese metallurgical companies do not use this HR tool. This activity should be done primarily by the closest superior and can be only supported by HR department which is not a general rule in Chinese companies. Also we recommend to implement evaluation interview in all Chinese companies. Without a feedback from the assessed person, the appraisal process cannot be effective. During the interview, the assessed can respond immediately to find out problems, he/she can communicate the causes of the identified deficiencies, which may not always be on his/her side. Moreover his/her personal participation in the formation of the objectives for the next evaluation period is likely to have him align internally with these and fulfill. As for other evaluation methods used, again, they strongly reflect the employee approach and habits in both countries that vary and we take them as given. For example, we see the 20% use of the 360" evaluation method in Czech companies as sufficient as this method is suitable more for managerial and key positions than for blue-collar positions and there is a not higher share of such positions in the Czech metallurgical industry.

3.4. Remuneration and motivation

The main objective of this part of the questionnaire and research was to determine the most frequent factors that influence remuneration of employees of metallurgical companies and differences between both countries. Obviously, the most commonly mentioned factors were the job requirements and the actual work performance. In both countries, more than 80% of companies had taken into account this factor when remunerating its employees. As we could assume employee behavior could be another important remuneration factor. Interesting is that, this factor was mentioned in only slightly more than 52% of Czech companies and only 36% of Chinese companies. The size of the company is one of the influencing factors as almost none of the medium companies were emphasizing the importance of employee behavior. We anticipated that all companies would consider the current state of the organization and the ambient conditions when remunerating, but the research has not proved it. In both countries less than 47% of companies did not mention this factor. We assume that the reason is all companies take the factors of labor costs, conditions in labor market, level of competition, and the financial situation into account. However, these factors are incorporated into basic principles of financial management of the company, which means people do not consider it on the lower management level of remunerating although it is definitely one of the determining factors.

Another important issue on remuneration is a method of determining wages. As the time salary is very common in Czech Republic–more than 90% of companies perform this type of wage determination, situation is slightly different in China, where only 48% of companies do the same.

Almost 15% of Chinese companies implement piecework wage, which is used in only 10% of metallurgical companies that participated in research. However, this question should be the subject to further examination as the influence of job description was not taken into account. The method of pay for the expected results has proved to be used more in China–31% of companies to 14% of Czech companies. In both countries about 9% of companies stated that they use some other methods of wage determination such as wage linked to the KPIs and others.

Besides the basic wages, an additional wage occurs in some form in all companies. The situation is different in Czech Republic and China. The most frequently used additional wage methods in Czech Republic are rewards, used in 100% of surveyed metallurgical companies,

http://dx.doi.org/10.5772/66795

followed by supplements and premiums, used in 90 and 85% of companies. In China, there is no method used in all companies and generally the use of additional remuneration is not as common as in the Czech Republic. The most common methods are rewards in 78% of surveyed Chinese companies followed by monthly personal evaluation in 70% of companies and supplements in 63%. Premiums are used in 44% of companies. The size of the company has a significant role in the number of additional wage methods used in both countries–more methods were used in big end very big companies. This field proved to be very interesting and could be assumed to have a significant influence on employee motivation. Further research studies in this field should be focused on the differences in the ratio of basic and supplementary wages in both countries.

Survey has also showed an important finding about rationalization proposals. As this additional remuneration is very common in the Czech Republic where nearly 72% of companies use it, in China it is used in less than 30% of metallurgical companies. The significant difference was also proved in use of shares on profit of company. The method is used in more than 50% of Czech companies, but in China there still remains to be a great potential of this motivating factor as it is used in only 23% of surveyed metallurgical companies. The less used method of remuneration was employee shares, which remains the domain of big and very big joint-stock companies. The situation in the Czech Republic and China is very similar as employee shares were used in 14% of total number of surveyed companies in each country, which is 33% of big and very big companies together in Czech Republic and 18% of them in China.

3.5. Employee benefits

As employee benefits are considered as important motivational factors and competitive advantage, we focused on determination of differences in the set of benefits in both countries. The results are not very surprising in the context of cultural and historical difference. The interesting finding is that Chinese companies are progressive in some aspects of HRM and in some of them they are not and that differs a lot in comparison with the Czech Republic. The situation could be described with an example of employee benefits. In the Czech Republic, almost every employee or potential employee expects a certain set of employee benefits, such as mobile phone, drinking regime on the site, additional education or language courses, some extra days of vacation, company car, and some others. Our survey proved the above benefits are the most commonly used in Czech Companies to fulfill employee expectations. All of the benefits above-mentioned are used in about 80% of companies, no matter what the size is. In China the situation is different. Benefits most commonly provided in China are extra-vacation days, allowance for travelling to work, medical examinations, and contribution to culture followed by sick days, which almost all, except medical examinations and extra vacation days, are the less used benefits in the Czech Republic. Chinese companies rarely provide contribution to vaccination, vouchers, and soft loans but they provide some benefits not so common in the Czech Republic such as contribution to accommodation and transport or gifts for birthday and other anniversaries.

Differences mentioned above could be explained by different perception of life values in China and the Czech Republic. The commonly known difference in time perception –monochronic and polychronic systems –can be used to describe this phenomenon. Western cultures, Czech

Republic included, consciously or unconsciously, emphasize the social status, welfare of the individual, and career. It is quite bold simplification, but it suits the main idea of the difference in employee benefits. Compared to that, eastern and southern cultures, in this case China, are more based on group achievement, seniority, relationship, family, health, and that kind of core values. Companies follow this assumption and provide people benefits they value. This could be very inspirational for the Czech Republic metallurgical companies, especially when the current benefits are getting standardized and people consider them more as a basic part of the wage than a real benefit, based on employee performance and motivation. Further research in this field should be focused on the financial resources allocated to various benefits and benefits at all as this was not covered by this survey and it could provide the different perspective on employee benefits in metallurgical companies.

In the areas of remuneration and benefits we do not dear to recommend changes to any country. This part of HR we consider as very paternalistic and specific for each region and citizens' mentality, something that develops ages to change. We only recommend to learn from findings of research, and take them into account for the possible cooperation between companies from both countries.

3.6. Career management

The field of career and its management was covered by few questions of the questionnaire and has showed some interesting differences in perception of career in both countries. Most of the Czech companies–71%–stated that the length of the employment is not an essential factor in employee evaluation and career growth. It means that also young people, ambitious newcomers have an opportunity to be promoted according to their performance and overall benefit to the company. The fact that the experience and expertise go along with age or employment length is obvious, but it should not be defining precondition. In contrast to the conditions in the Czech Republic, almost the same portion of about 74% of Chinese metallurgical companies stated that they consider length of employment very important and an essential factor to consider promotion of employees and their evaluation. Both approaches offer pros and cons, as the ambitious and extraordinary performing newcomers have an opportunity of fast and steep career growth and as the other approach takes into account the indisputable influence of time–employment length–on trust, loyalty, and experience.

What might be surprising and proving that China as a fast-growing strong economy and progressive country is that more than 44% of surveyed metallurgical companies use career growth plan for their employees. Promoting and career growth is then based on fulfillment of worker's tasks in each successive stage. On comparison, only 19% of Czech companies in the survey and mostly big and very big has the same career management tool implemented. The above-mentioned explains the difference in approach to promotion described in the first paragraph of this chapter. In case all Czech companies had career plan and the promotion was based on fulfillment of successive stages it would be difficult to proceed very quickly, which somehow proves the influence of length of employment.

In this area, Czech companies may learn from the more adaptive Chinese companies. Career growth, as we already mentioned, is a very important part of employee motivation and the 19% share gives Czech managers a huge space for improvement.

4. Conclusions

As far as human resource management is recognized as one of the most important managerial aspect in the life of every company, this study contributed to the knowledge of HRM practices in Chinese and Czech metallurgical companies. This paper explored various aspects of HRM in both countries and provided HR managers from both countries some implications and points to get mutually inspired. The presented research proved some well-known and described many not so known differences in approach to various domains of management, HRM included, whose origin could be found in the historical development of both countries. But not only differences have been found. Many similarities of HRM practices were described in this paper, which proved that the world is now much smaller than before and people are willing to learn and experience different cultures, countries, and individuals. During work on the project very strong partnership between of the Hubei University of Technology and the VSB–Technical University of Ostrava has been established and proved to be very promising in various fields of research and authors of the chapter are very grateful to all participants which were willing to share their experiences.

Acknowledgements

This publication has been created within the project Support of VŠB-TUO activities with China with financial support from the Moravian-Silesian Region.

Author details

Martin Čech[1*], Andrea Samolejová[1], Jun Li[2], Wenlong Yao[2] and Pavel Wicher[1]

*Address all correspondence to: martin.cech@vsb.cz

1 VSB-Technical University of Ostrava, Ostrava-Poruba, Czech Republic

2 Hubei University of Technology, Nanhu, Wuhan, China

References

[1] Li, Y. 2016. Comparison of HRM practices between Chinese and Czech companies, Perspectives in Science, 7, 2–5.

[2] Cech, M. et al. 2016. Human resource management in Chinese manufacturing companies, Perspectives in Science, 7, 6–9.

[3] Koubek, J. 2009. Managing human resources in the Czech Republic. In: Morley, M. J., Heraty, N., Michailova, S. (Eds.), Managing Human Resources in Central and Eastern Europe. Routledge, London.

[4] Samolejova, A. et al. 2015. Factors of human resource planning in metallurgical company, Metalurgija, 54, 243–246.

[5] Cranet, 2011. Cranet survey on comparative human resource management. International Executive Report 2011. [online]. [cit. 2016-4-15]. Available at www: http://www.ef.uns.ac.rs/cranet/download/cranet_report_2012_280212.pdf

[6] Ryan A. K. 1999. An international look at selection practice: nation and culture as explanations for variability in practice, Personnel Psychology, 52, 23–33.

[7] Dessler, G. 2007. Human resource management, 10th ed. Pearson Edu. Prentice Hall, New Jersey.

[8] Stone, R. E. F., Stone, D. L. 2002. Cross-cultural differences in response to feedback: implications for individual, group, and organizational effectiveness. In: Ferris G. R. (Ed.), Research in Personnel and Human Resource Management, 21, 275–331.

Chapter 19

The Influence of Loading System Stiffness on Empirical Correlations for Determination of Tensile Characteristics from the Results of SP Tests

Karel Matocha, Ondřej Dorazil, Miroslav Filip,
Jinbin Zhu, Yuan Chen and Kaishu Guan

Additional information is available at the end of the chapter

http://dx.doi.org/10.5772/66796

Abstract

The present chapter describes the influence of the loading system stiffness on empirical correlations for determination of yield and tensile strengths at laboratory temperature from the results of small punch (SP) tests. The results obtained proved that measuring of test specimen deflection during SP test eliminates the significant effect of the loading system stiffness on the above-mentioned correlations.

Keywords: small punch test, load-displacement curve, empirical correlation, yield strength, tensile strength, specimen deflection

1. Introduction

The need for evaluating the actual mechanical properties of structural components by direct testing method has led to the development of innovative techniques based on miniaturized specimens. Among these, a technique called the small punch (SP) test has emerged as a promising candidate (Hurst and Matocha, 2010, Lucon, 2001, Lucas, 1990). It is a mechanical testing method used presently to obtain tensile, fracture, and creep data from very small quantities of experimental material. In 2007 CWA 15627 "Small Punch Test Method for Metallic Materials" (CWA 15627:2007 D/E/F, 2007) was issued by CEN (European Committee for Standardization).

The objective of the SP test is to produce a load-displacement (punch displacement, crosshead displacement, specimen deflection) record (see **Figure 1**), which contains information about the elastic-plastic deformation and strength properties of the material.

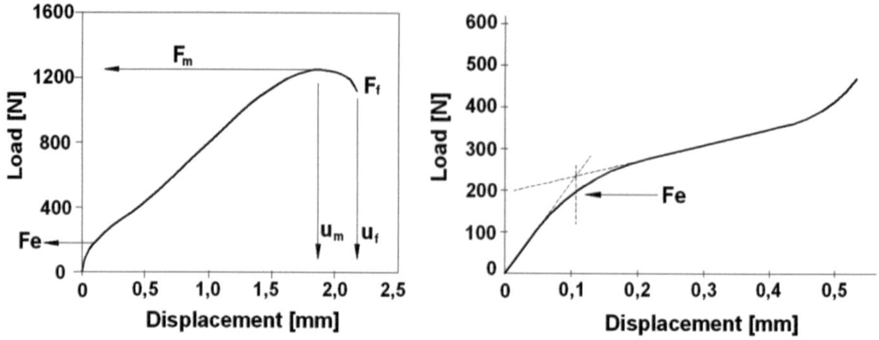

Figure 1. Load-displacement curve recorded during a time-independent small punch (SP) test.

The following parameters are determined from the load-displacement curve during the time-independent SP tests (CWA 15627:2007 D/E/F, 2007):

F_m [N]: maximum load recorded during SP test.

F_e [N]: load characterizing the transition from linearity to the stage associated with the spread of a yield zone through the specimen thickness. It is determined according to the Code by the two tangents method (see **Figure 1**).

u_m [mm]: displacement corresponding to the maximum load $F_{m.}$

u_f [mm]: displacement corresponding to 20% load drop.

E^{SP} [J]: SP fracture energy obtained from the area under the load-displacement curve up to the u_f.

The load-displacement curves obtained can be utilized to derive empirical correlations between SP and standardized test results (Mao and Takahashi, 1987, Hurst and Matocha, 2012, Rodriguez et al., 2012) or they can be analyzed in terms of elastic-plastic finite element methods (Nakata et al., 2010, Hůlka et al., 2012, Prakash and Ramesh, 2012, Madia et al., 2013).

Most of the empirical correlations for the determination of the yield strength from the results of penetration tests found in the literature are expressed as the dependence of the yield strength on the parameter F_e/h_0^2, where h_0 is the initial thickness of the disc, because it was proved that this parameter eliminates the effect of any differences in disc specimen thicknesses on load F_e (Dymáček and Ječmínka, 2014, Hurst and Matocha, 2012). Tensile strength is correlated either with the parameter F_m/h_0^2 or the parameter $F_e/(u_m \cdot h_0)$, because it was proved that this parameter eliminates the effect of any differences in disc specimen thicknesses on load F_m and u_m (Hurst and Matocha, 2012). There is, however, an important factor affecting the shape of the load-displacement record. This is a procedure used for the displacement monitoring. Only few authors have paid attention to the different possibilities for the displacement monitoring, i.e.,

http://dx.doi.org/10.5772/66796

punch displacement, bottom central point displacement (deflection), testing machine crosshead displacement, and loading system stiffness (Moreno et al., 2016, Matocha et al., 2014).

In the present chapter, the influence of the displacement monitoring method on empirical correlations for determination of yield and tensile strength for P92 steel was followed up. Both testing machine crosshead displacement and bottom central point displacement (deflection) were monitored during SP tests at laboratory temperature. It was proved that the monitoring of deflection eliminates the effect of loading system stiffness on the above-mentioned correlations.

2. Testing material

A steam pipe ø 219.1 × 22.2 mm made of P92 steel in as-received state was used as the testing material. Controlled chemical composition of the testing materials is shown in **Table 1**. The testing material was heat-treated to four significantly different strength levels (see **Table 2**). Tensile tests were carried out at room temperature on MTS 100 kN servohydraulic testing machine using round testing bars of 8 mm diameter.

C	S	Mn	Si	P	Cu	Ni	Cr
0.12	0.009	0.53	0.24	0.012	0.050	0.13	8.56
Mo	V	Ti	Nb	W	Co	N	$Al_{tot.}$
0.43	0.19	<0.005	0.062	1.63	0.007	0.045	0.009

Table 1. Controlled chemical composition of testing material [wt.%].

State	YS [MPa]	UTS [MPa]	A [%]	Z [%]
As received	679	808	18.0	61
HT1: tempering 800°C/2 hours/air	502	675	26.8	67
HT2: tempering 760°C/2 hours/air	541	707	24.4	65
HT3:tempering 750°C/2 hours/air	574	726	21.4	66
HT4:tempering 740°C/2 hours/air	584	733	21.6	63

Table 2. Tensile properties after selected heat treatments.

SP tests at laboratory temperature were carried out on the servomechanical testing machine LabTest 5.10ST under control at crosshead speed of 1.5 mm/min. Both the crosshead displacement and the specimen deflection were measured during the SP test using testing jig for monitoring test specimen deflection (bottom central point displacement) (see **Figure 2**).

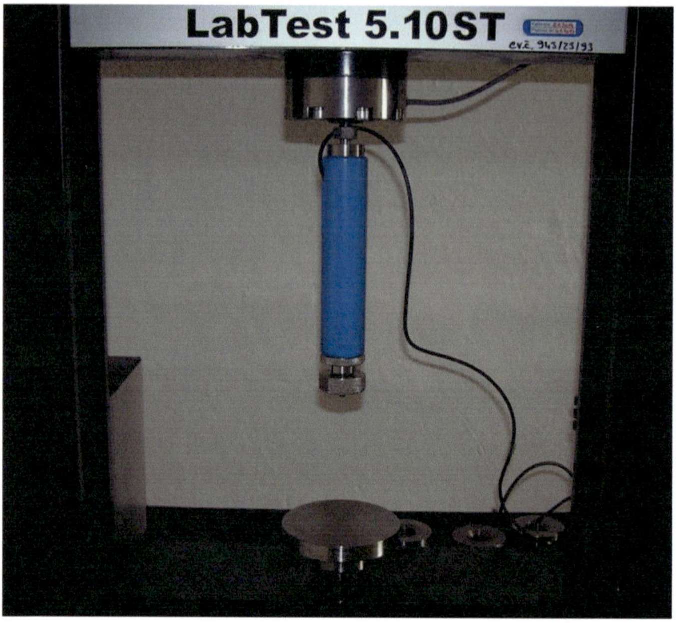

Figure 2. Testing facilities with the jig for monitoring test specimen deflection.

3. Results and discussion

Figure 3 shows the stiffness of the loading system used for SP tests at laboratory temperature.

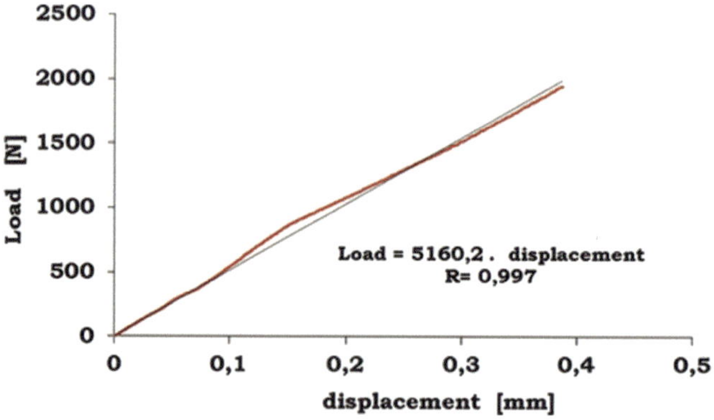

Figure 3. Stiffness of the loading system used for SP tests at laboratory temperature.

http://dx.doi.org/10.5772/66796

Figures 4 and **5** show the influence of displacement monitoring method (crosshead displacement, deflection) on empirical correlations for determination of yield and tensile strengths from the results of SP tests for P92 steel.

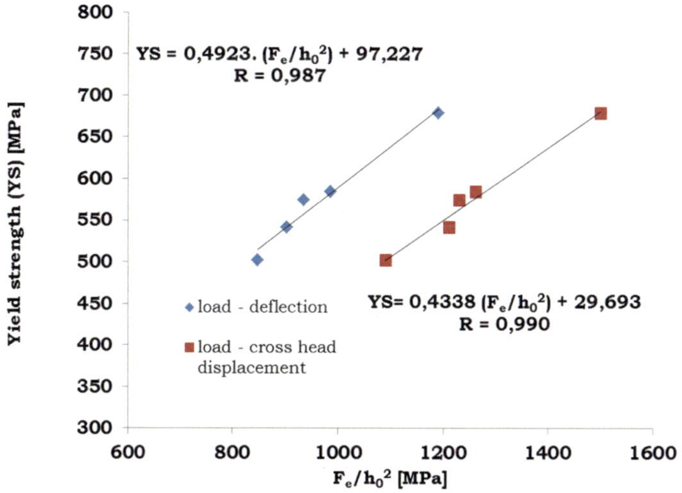

Figure 4. The influence of displacement monitoring on empirical correlation for determination of yield strength from the results of SP tests for P92 steel.

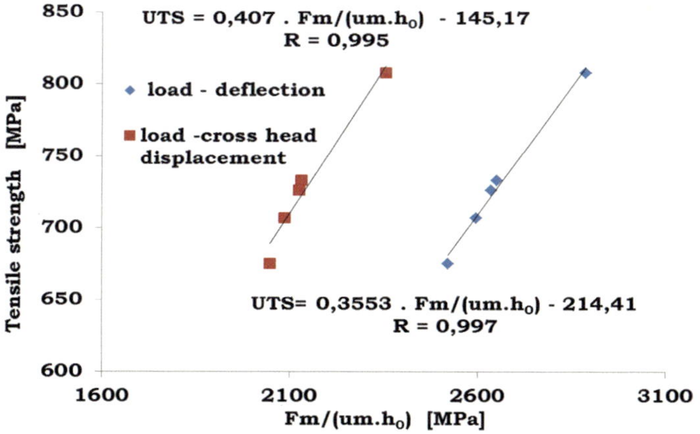

Figure 5. The influence of displacement monitoring on empirical correlation for determination of tensile strength from the results of SP tests for P92 steel.

The results obtained have shown significant influence of displacement monitoring method on both empirical correlations. To explain this difference, the stiffness of the loading system was deducted from load—crosshead displacement records (see **Figure 6**) of test specimens in as-received state and after heat treatment 1 and heat treatment 4.

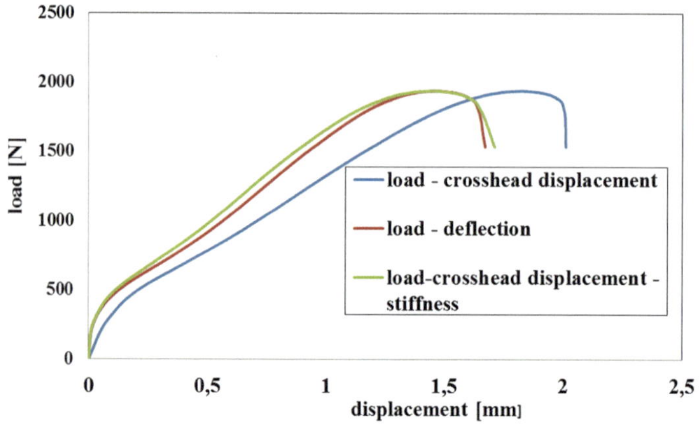

Figure 6. Load-displacement records of the test specimen after heat treatment 4 tested at laboratory temperature.

Figure 7 shows the empirical correlation for yield strength obtained for deflection of test disc monitoring together with parameters F_e/h_0^2 obtained after correction of load—crosshead displacement record for the loading system stiffness.

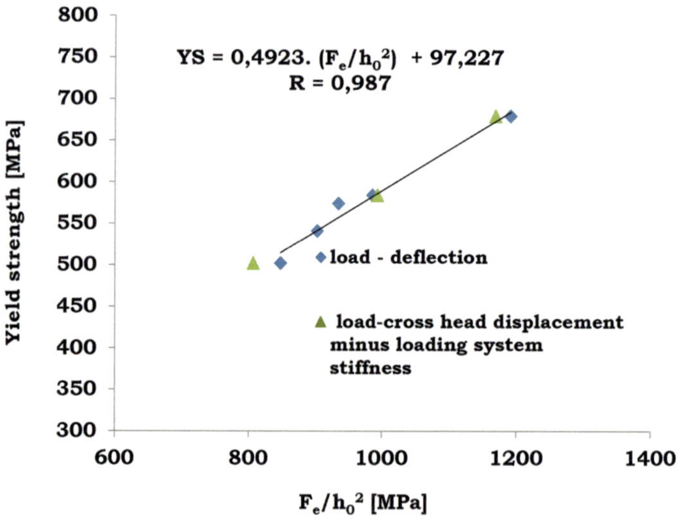

Figure 7. Empirical correlation for yield strength obtained for deflection of test disc monitoring together with parameters F_e/h_0^2 obtained after correction of load—crosshead displacement record for the loading system stiffness.

Figure 8 shows the empirical correlation for tensile strength obtained for deflection of test disc monitoring together with parameters $F_m/(h_0 \cdot u_m)$ obtained after correction of load—cross-head displacement record for the loading system stiffness.

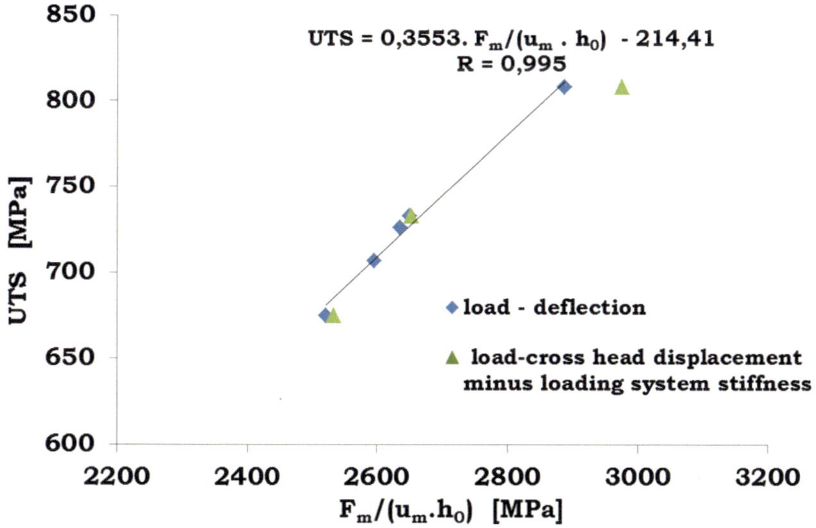

Figure 8. Empirical correlation for tensile strength obtained for deflection of test disc monitoring together with parameters $F_m/(u_m \cdot h_0)$ obtained after correction of load—crosshead displacement record for the loading system stiffness.

4. Conclusions

1. The method used for monitoring displacement during the SP test affects load-displacement record significantly and therefore empirical correlations for the determination of yield and tensile strengths from the results of SP tests.

2. Differences observed between the empirical correlations for determinations of yield and tensile strengths from the results of SP tests obtained on different testing machines can be attributed to the different stiffness of the loading system.

3. Only load- deflection record monitored during the SP test eliminates the influence of loading system stiffness.

4. This fact should be taken into account when revising the document CWA 15627.

Acknowledgements

This paper was created in the project *Support of VŠB – TUO activities with China* with the financial support from Moravian-Silesian Region and in the Project No. *LO1203 "Regional Materials*

Science and Technology Centre - Feasibility Program" funded by the Ministry of Education, Youth and Sports of the Czech Republic.

Author details

Karel Matocha[1,2*], Ondřej Dorazil[1], Miroslav Filip[1], Jinbin Zhu[3], Yuan Chen[3] and Kaishu Guan[3]
*Address all correspondence to: matocha.karel.mmvyzkum.cz

1 Material and Metallurgical Research, Ltd., Ostrava-Vítkovice, Czech Republic

2 Faculty of Metallurgy and Materials Engineering, VŠB-Technical University of Ostrava, Czech Republic

3 School of Mechanical Engineering, East China University of Science and Technology, Shanghai, China

References

[1] Hurst, R., Matocha, K., 2010. The European Code of p for small punch testing – where do we go from here? Proc. of 1st Int. Conf. "Determination of Mechanical Properties of Materials by Small Punch and Other Miniature Testing Techniques", August 31–September 2, Ostrava, Czech Rep, pp. 5–11. ISBN 978-80-254-7994-0.

[2] Lucon, E., 2001. Material damage evaluation and residual life assessment of primary power plant components for long-term operation using specimens of non-standard dimensions. La Revue de Métallurgie, December, 2001 p. 1079.

[3] Lucas, G.E., 1990. Review of small specimen test techniques for irradiation testing. Metallurgical Transactions A, Vol. 21A, May, pp. 1105–1119.

[4] CWA 15627:2007 D/E/F, 2007. CEN Workshop Agreement "Small Punch Test Method for Metallic Materials".

[5] Nakata, T. et al., 2010. Tensile property evaluation by stress and strain analyses of small punch test specimen using finite element method. Proc. of 1st Int. Conf. SSTT "Small Sample Test Techniques", Metallurgical Journal. Vol. 63, pp. 146–150.

[6] Hůlka, J. et al., 2012. FEM sensitivity analysis of small punch test. Proc. 2nd Int. Conf. SSTT, "Determination of Mechanical Properties by Small Punch and other Miniature Testing Techniques", Ostrava, Czech Rep., pp. 329–338.

[7] Prakash, R.V., Ramesh,T., 2012. Numerical simulation of shear punch and small punch tests using Gurson-Tvergaard-Needleman Damage Model. Proc. 2nd Int. Conf. SSTT, "Determination of Mechanical Properties by Small Punch and other Miniature Testing Techniques", October 2–4, Ostrava, Czech Rep., pp. 355–365. ISBN 978-80-260-0079-2.

[8] Madia, M. et al., 2013. On the applicability of the small punch test to the characterization of the 1CrMoV aged steel: Mechanical testing and numerical analysis. Engineering Failure Analysis, Vol. 34, December, pp. 189–203.

[9] Mao X., Takahashi H., 1987. Development of a further-miniaturized specimen of 3 mm diameter for TEM disk (Ø 3 mm) small punch tests. Journal of Nuclear Materials. Vol. 150, pp. 42–52.

[10] Hurst, R., Matocha, K., 2012. Where are we now with the European Code of practice for small punch testing? Proc. 2nd Int. Conf. SSTT, "Determination of Mechanical Properties of Materials by Small Punch and other Miniature Testing Techniques", October 2–4, Ostrava, Czech Rep., pp. 4–18. ISBN 978-80-260-0079-2.

[11] Rodriguez, C., 2012. The application of the small punch test to the mechanical characterization of different steel grades. Proc. 2nd Int. Conf. SSTT "Determination of Mechanical properties of Materials by Small Punch and other Miniature Testing Techniques", October 2–4, Ostrava, Czech Rep., pp. 188–195. ISBN 978-80-260-0079-2.

[12] Dymáček, P., Ječmínka, M., 2014. Applicability of small punch testing methods for material mechanical properties at room and high temperatures. Proc. 3rd Int. Conf. SSTT 2014 "Determination of Mechanical Properties of Materials by Small Punch and other Miniature Testing Techniques", September 23–25, Castle Seggau near Graz, Austria, pp. 65–74. ISBN 978-80-260-6722-2.

[13] Moreno, M.F., Bertolino, G., Yawny, A., 2016. The significance of specimen displacement definition on the mechanical properties derived from small punch test. Materials and Design, Vol. 95, pp. 623–631.

[14] Matocha, K. et al., 2014. The effect of stiffness of the loading system on determination of tensile characteristics from the results of small punch tests. Acta Metallurgica Slovaca, Vol. 20, No. 4, pp. 389–396.

Chapter 20

Synthesis and Characterization of Gadolinium Oxide Nanocrystallites

L. Kuzníková, K. Dědková, L. Pavelek, J. Kupková,
R. Váňa, M. H. Rümmeli and J. Kukutschová

Additional information is available at the end of the chapter

http://dx.doi.org/10.5772/66797

Abstract

Lanthanide oxide nanocrystallites have gained a lot of attention due to their diverse use for potential applications and for this reason it is very important to find a suitable preparation method that would be economically inexpensive and easy to implement. The chapter describes the preparation of gadolinium oxide nanocrystallites (nano Gd_2O_3) through thermal decomposition of a complex formed by $Gd(NO_3)_3 \cdot 6\ H_2O$ and glycine. Decomposition of the complex occurs at temperatures about (250 ± 10)°C. An ultrafine white powder of the gadolinium oxide nanocrystallites was obtained. The resulting nanocrystallites were characterized by X-ray powder diffraction analysis, which revealed the size of the gadolinium oxide nanocrystallites equal to 10 nm. The morphology of the gadolinium oxide nanocrystallites was examined by scanning electron microscopy. The elemental composition of the product was confirmed by EDS analysis.

Keywords: thermal decomposition, nanocrystallites, gadolinium oxide, XRD, EDS, SEM

1. Introduction

Nanomaterials are defined as materials with at least one direction usually in the range of 1–100 nm [1] and because these materials have different physical, chemical, and electrical properties in comparison with traditional bulk materials, they may be used for new products and applications and may also be incorporated into various industrial processes [2].

Lanthanide oxides have gained a lot of attention due to their diverse use for applications such as in the nuclear industry, electronics, lasers, and optical materials [3]. Gadolinium oxide

(Gd_2O_3) is the most researched of all the lanthanide oxide. A great deal of interest in gadolinium oxide exists because of its physicochemical properties, such as the crystallographic stability up to temperatures of 2325°C, high mechanical strength, excellent thermal conductivity, and a wide band optical gap [4]. Generally, nanoparticles of lanthanide oxides can be prepared using a variety of methods, such as homogeneous precipitation [5], thermal decomposition [6], combustion method [7], microemulsion techniques [8], hydrothermal crystallization [9], spray pyrolysis [10], sol-gel [11], sonochemical methods [12], and other methods [13]. Most often nanocrystallites of lanthanide oxides are prepared through calcination methods using a suitable precursor [14].

The aim of this work was the preparation of gadolinium oxide nanocrystallites through a thermal decomposition method and their subsequent characterization using a combination of techniques.

2. Experimental

2.1. Synthesis of gadolinium oxide nanocrystallites

Gadolinium oxide nanocrystallites (nano Gd_2O_3) were prepared by the thermal decomposition [15] of the complex formed by the salt $Gd(NO_3)_3{\cdot}6\ H_2O$ and glycine. An aqueous solution of $Gd(NO_3)_3{\cdot}6\ H_2O$ and glycine (NH_2CH_2COOH) with a concentration of 0.5 mol·dm^{-3} were mixed. The resulting complex was dried at 120°C and calcined at 600°C for 1 hour. Decomposition of the complex occurred at about of (250 ± 10)°C. Other components of the complex evaporated in the form of the following gases: N_2, CO_2, and H_2O. The following scheme illustrates synthesis of samarium oxide nanocrystallites:

$$Gd(NO_3)_3{\bullet}6H_2O + NH_2CH_2COOH \xrightarrow[N_2,\, CO_2,\, H_2O]{600^\circ C,\, 1hod} Gd_2O_3 \qquad (1)$$

2.2. Characterization of gadolinium oxide nanocrystallites

X-ray powder diffraction analysis was performed using the X-ray diffractometer Ultima IV Rigaku (Rigaku, Japan), operated at 40 kV and 40 mA with CuKα radiation (reflection mode, Bragg-Brentano arrangement, scintillation counter). The XRD patterns were recorded in the 10–70° 2θ range with a scanning rate of 2°·min^{-1}. The samples were placed in a ground glass depression in the sample holder and flattened with a glass slide. X-ray beam was demarcated by 2/3° divergence, 10 mm divergent height limiting, 2/3° scattering, and 0.6 mm receiving slits. Phase analysis was evaluated by database PDF-2 Release 2011. Graphics processing XRD patterns was made using OriginPro8. The Gd_2O_3 reflection of the (222) plane was used to determine crystallite size using the Scherrer formula [16]

$$L_c = \frac{K \bullet \lambda}{\beta \bullet \cos\theta}, \qquad (2)$$

http://dx.doi.org/10.5772/66797

where K is the factor of microstructure, λ is the wavelength of radiation, β is the full-width at half-maximum (FWHM), and θ is the diffraction angle.

Scanning electron microscope MAIA3 GMU (TESCAN)—ultra-high resolution SEM with Schottky field emission cathode—was used for electron micrographs. Images were taken using a combination of InBeam SE + Low-Energy BSE detector at 2.5 kV. Furthermore, the product morphology was also observed by scanning electron microscope Quanta FEG (FEI), and EDS analysis was performed using the APOLLO X analyzer (EDAX).

3. Results and discussion

The thermal decomposition of the gadolinium salt and glycine produced a white powder of gadolinium oxide nanocrystallites.

The obtained XRD pattern in **Figure 1** shows the single phase of Gd_2O_3 (data from JCPDS file No. 03-065-3181) with cubic crystal structure. Four major reflections of Gd_2O_3 were observed and correspond to the (222), (400), (440), and (622) crystal lattice planes. Other smaller reflections were assigned to the (211) and (431) planes, respectively. The crystallite size of reflection (222) was about 10 nm.

EDS spectrum (**Figure 2**) confirmed the presence of gadolinium and oxygen. Gold present in the EDS pattern is caused by that coated with Au thin layer.

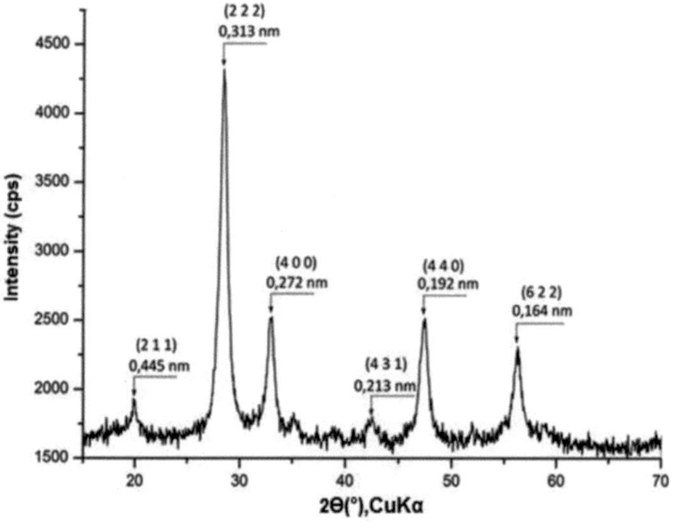

Figure 1. X-ray powder diffraction pattern of the prepared nanomaterial.

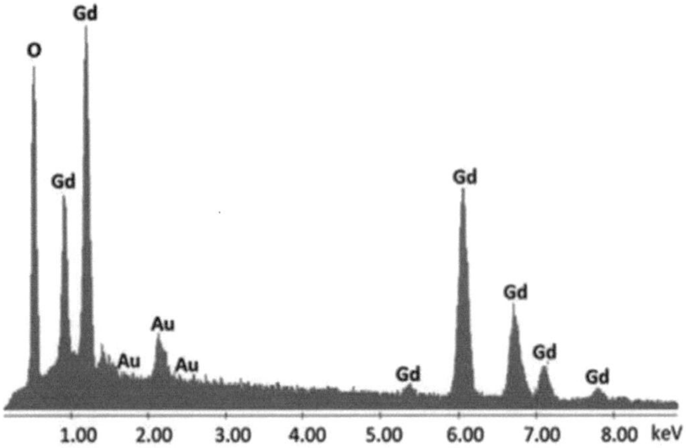

Figure 2. EDS spectrum of the sample.

The SEM images were taken in secondary electron modes (**Figure 3**). It can be seen that gadolinium oxide nanocrystals form aggregates which can be explained by the electrostatic forces. At lower magnifications, the network configuration of the material can be observed, and at the higher magnifications, it can be seen that the material appears as porous mousse with meso- and macropores.

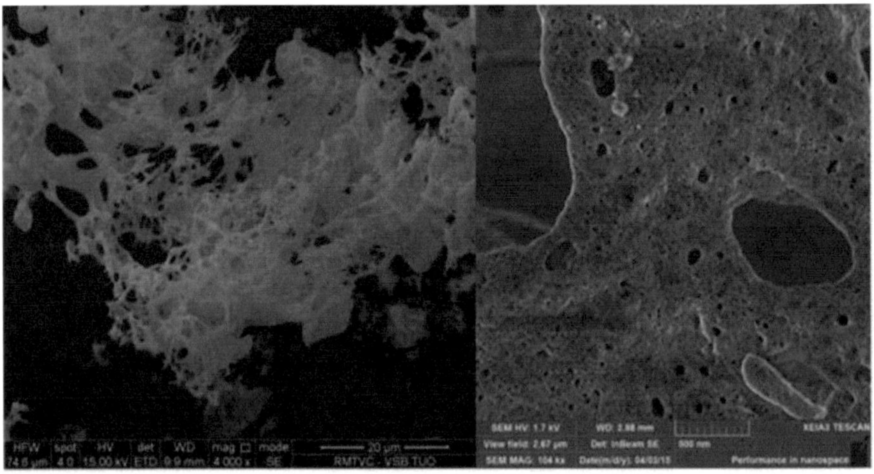

Figure 3. Examples of SEM images of the sample at different magnifications.

http://dx.doi.org/10.5772/66797

4. Conclusions

Gadolinium oxide nanocrystallites with crystallite size of 10 nm were prepared by the thermal decomposition of $Gd(NO_3)_3 \cdot 6\ H_2O$ and glycine. Currently, increasing production and usage of nanocrystallites for various industrial applications may raise questions and concerns about their impact on human health and environment. Therefore, potential toxic effects of gadolinium oxide nanocrystallites prepared by the thermal decomposition method should be evaluated in future as well.

Acknowledgements

This chapter was created on the Faculty of Metallurgy and Materials Engineering in the Project No. LO1203 "Regional Materials Science and Technology Centre - Feasibility Program" funded by Ministry of Education, Youth and Sports of the Czech Republic.

Author details

L. Kuzníková[1*], K. Dědková[1,2], L. Pavelek[3], J. Kupková[1,2], R. Váňa[4], M. H. Rümmeli[5,6,7] and J. Kukutschová[1,2]

*Address all correspondence to: lubomira.kuznikova.st@vsb.cz

1 Nanotechnology Centre, VŠB-Technical University of Ostrava, Ostrava–Poruba, Czech Republic

2 Regional Materials Science and Technology Centre, VŠB-Technical University of Ostrava, Ostrava–Poruba, Czech Republic

3 Department of Chemistry, Faculty of Metallurgy and Materials Engineering, VŠB-Technical University of Ostrava, Ostrava–Poruba, Czech Republic

4 TESCAN Brno, s.r.o., Brno, Czech Republic

5 College of Physics, Optoelectronics and Energy & Collaborative Innovation Center of Suzhou Nano Science and Technology, Soochow University, Suzhou, China

6 IFW Dresden, Dresden, Germany

7 Centre of Polymer and Carbon Materials, Polish Academy of Sciences, Zabrze, Poland

References

[1] Lövestam, G., H. Rauscher, G. Roebben, B. Sokull Klüttgen, N. Gibson, J. P. Putaud and H. STAMM. Considerations on a Definition of Nanomaterial for Regulatory Purposes. Luxembourg: Publications Office of the European Union, 2010. ISBN 978-92-79-16014-1.

[2] Gómez-Rivera, F., J. Field, D. Brown and R. Sierra-Alvarez. Fate of cerium dioxide (CeO_2) nanoparticles in municipal wastewater during activated sludge treatment. *Bioresource Technology*. 2012, **108**, 300–304.

[3] Tsuzuki, T., E. Pirault and P. Mccormick. Mechanochemical synthesis of gadolinium oxide nanoparticles. *Nanostructured Materials*. 1999, **11**(1), 125–131.

[4] Tamrakar, R., D. Bisen and N. Brahme. Comparison of photoluminescence properties of Gd_2O_3 phosphor synthesized by combustion and solid state reaction method. *Journal of Radiation Research and Applied Sciences*. 2014, **7**(4), 550-559.

[5] Muccillo, E., R. Rocha, S. Tadokoro, J. Rey, R. Muccillo and M. Steil. Electrical conductivity of CeO_2 prepared from nanosized powders. *Journal of Electroceramics*. 2004, **13**(1–3), 609–612.

[6] Kamruddin, M., P. Ajikumar, R. Nithya, A. Tyagi and B. Raj. Synthesis of nanocrystalline ceria by thermal decomposition and soft-chemistry methods. *Scripta Materialia*. 2004, **50** (4), 417–422.

[7] Purohit, R., B. Sharma, K. Pillai and A. Tyagi. Ultrafine ceria powders via glycine-nitrate combustion. *Materials Research Bulletin*. 2001, **36**(15), 2711–2721.

[8] Lee, J.-S., J.-S. Lee and S.-C. Choi. Synthesis of nano-sized ceria powders by two-emulsion method using sodium hydroxide. *Materials Letters*. 2005, **59**(2–3), 395–398.

[9] Masui, T., H. Hirai, N. Imanaka, G. Adachi, T. Sakata and H. Mori. Synthesis of cerium oxide nanoparticles by hydrothermal crystallization with citric acid. *Journal of Materials Science Letters*. 2002, **21**(6), 489–491.

[10] Xu, H., L. Gao, H. Gu, J. Guo and D. Yan. Synthesis of solid, spherical CeO_2 particles prepared by the spray hydrolysis reaction method. *Journal of the American Ceramic Society*. 2002, **85**(1), 139–144.

[11] Liu, Z., B. Guo, L. Hong and H. Jiang. Preparation and characterization of cerium oxide doped TiO_2 nanoparticles. *Journal of Physics and Chemistry of Solids*. 2005, **66**(1), 161–167.

[12] Yu, J., L. Zhang and J. Lin. Direct sonochemical preparation of high-surface-area nanoporous ceria and ceria–zirconia solid solutions. *Journal of Colloid and Interface Science*. 2003, **260**(1), 240–243.

[13] Zawadzki, M. Preparation and characterization of ceria nanoparticles by microwave-assisted solvothermal process. *Journal of Alloys and Compounds*. 2008, **454**(1–2), 347–351.

[14] Hu, J.-D., Y-X. Li, X.-Z. Zhou and M.-X. Cai. Preparation and characterization of ceria nanoparticles using crystalline hydrate cerium propionate as precursor. *Materials Letters*. 2007, **61**(28), 4989–4992.

[15] Kuzníková, L., K. Dědková, L. Pavelek, J. Kupková, R. Váňa and J. Kukutschová. Synthesis and Characterization of Samarium Oxide Nanocrystallites. *Journal of Nanoscience and Nanotechnology*. 2016, **16**(1–3), 7829–7831.

[16] DE Graef, Marc and Michael E. Mchenry. *Structure of materials: an introduction to crystallography, diffraction, and symmetry*. Second edition, fully revised and updated. New York: Cambridge University Press, 2012. ISBN 978-110-7005-877.

Chapter 21

Numerical Simulation on Seismic Behavior of UHTCC Beam-Column Joints

Wei Li

Additional information is available at the end of the chapter

http://dx.doi.org/10.5772/66798

Abstract

To study the effects of ultra-high toughness cementitious composite (UHTCC) on the seismic behaviors of local enhancement beam-column joints under reversed cyclic loading, three specimens were tested and performed with half-scale interior joints in the finite element software Open System for Earthquake Engineering Simulation (OpenSees) and ANSYS in this chapter. The element "Concrete02" was used to simulate the material properties of UHTCC. The comparison of simulated results and experimental results indicated that displacement-beam-column element could be efficiently used to simulate the hysteresis response and the characteristic of energy dissipation of joints. The cementitious composites with ultra-high toughness could significantly improve the seismic performance of core area and had better ductility. Compared with ANSYS, the OpenSees finite element model was proved that preferably reflected the UHTCC enhanced nonlinear characteristic of frame nodes and effectively analyzed beam-column joint bearing capacity and seismic behavior.

Keywords: beam-column joint, OpenSees, ANSYS, numerical simulation, UHTCC

1. Introduction

Reinforced concrete (RC) beam-column joints were the vital components of structure that transfer and distribute internal forces and maintain structural integrity to ensure the safety. Most previous earthquake damage and research study showed that beam-column joints were vulnerable in seismic-prone areas and difficult to be repaired after the destruction. Accordingly, extensive research about the beam-column joints for seismic performance was triggered. From these research communities in the last decades, several studies focused on the application of steel fiber to enhance the joint shear strength or deformability, while the softening properties of concrete still existed. Ultra-high toughness cementitious composite

(UHTCC) [1–3], as one kind of reinforcing concretes was composed by polyvinyl alcohol fiber and cement paste as the interior performance was determined by micromechanics. Under the bending and tensile load, the UHTCC material exhibits pseudostrain hardening, high toughness, and great strain capacity under and multiple fine cracking behavior with the average crack width of 1 mm. Due to its superior strain behavior, UHTCC was an ideal material to replace the concrete in the joint zone of interior beam-column joints to substantially improve the load capacity and energy absorption capacity.

In this research, three half-scale interior beam-column joint specimens were tested. The results of the analysis and comparison of the numerical simulation data by ANSYS and Open System for Earthquake Engineering Simulation (OpenSees) can provide reference for the design and research of this kind of beam-column joints.

2. Experimental investigation

2.1. Experimental overview

To verify the toughness effectiveness of the UHTCC, three half-scale interior beam-column joint specimens whose core area was replaced by this material, as the specimen geometry was summarized in **Figure 1**, were tested. All the specimens had the same cross-sectional dimension (150 × 250 mm), and the detailed material parameters of concrete and UHTCC and summary of test parameters of beam-column joint specimens were listed in **Tables 1** and **2**, respectively. Axial compression ratio and volume-stirrup ratio were the main parameters. All material parameters of steel were uniformly presented in **Table 3**.

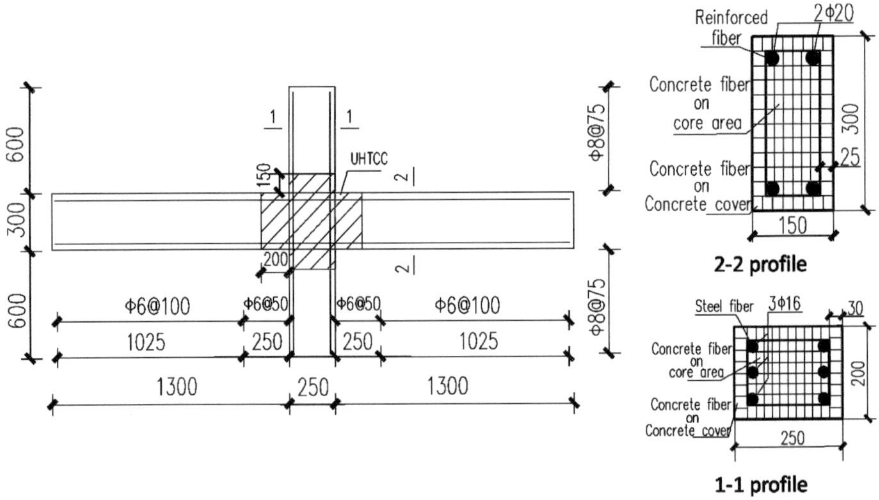

Figure 1. Specimen size and reinforcement figure.

http://dx.doi.org/10.5772/66798

Concrete cubic Compressive strength	Beam C30	Column C40	UHTCC
f_{cu}/MPa	31.95	45.78	40

Table 1. Material parameters of concrete and UHTCC.

Specimen	Volume-stirrup RatioV_f/%	Stirrup space	Axle pressure/kN	Axial compression Ratio n_1
Specimen 1	0.58	ϕ8@150	385	0.25
Specimen 2	1.17	ϕ8@100	540	0.35
Specimen 3	—	—	540	0.35

Table 2. Sample parameters.

Bar diameter	Yield strength f_y/MPa	Ultimate strength f_u/MPa
Φ6	319.55	490.56
Φ8	300.5	471.35
Φ16	386.55	572.92
Φ20	309.87	459.88

Table 3. Material parameters of steel.

2.2. Loading method

A schematic of loading modes was shown in **Figure 2**. All specimens were tested under a low-reversed cyclic load provided by a digital closed-loop controlled hydraulic loading system. The controlled value of axle pressure on the top of the column was adopted using a small hydraulic jack in load control, and low-reversed cyclic displacement controlled loading was applied on both free ends of the beam by means of a loading collar [4].

2.3. Test results

The final damage on the joint of UHTCC specimen was illustrated in **Figure 3**. The shear failure of the UHTCC beam-column joint occurs in the core area and this process includes four stages: initial crack, penetrating crack, ultimate state, and failure state. From the first stage to the end, the vertical crack first appeared in the inner end of the concrete beam. The cracks about 0.02 mm from the center of core area would appear with the continuous load. The fine inclined cracks would appear along the other diagonal direction when the reverse load was applied. With the increase of the magnitude of the load and the number of cycles, crack extended to the ends of the core area and significantly increased and finally formed a typical oblique X-type microcrack with 1 mm of the average crack spacing. According to the joints failure modes and fine cracks shown in **Figure 3**, the concrete surface in the core area did not spall, which was mainly due to the bridging stress provided by the PVA fiber in UHTCC enhanced the shear capacity of beam-column joints, so the specimen had multiple microcrack characteristic cracks under the ultimate load.

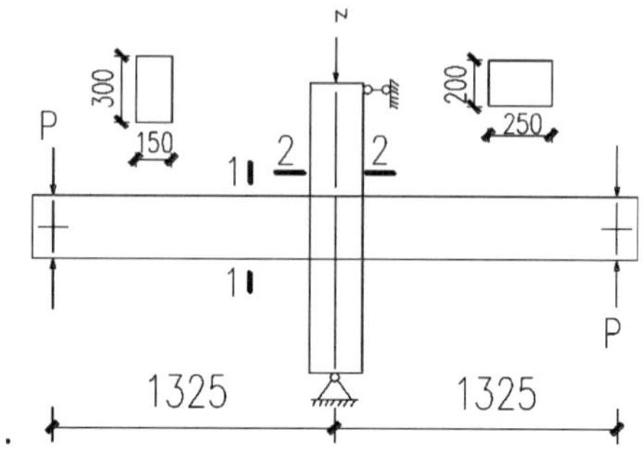

Figure 2. Loading method.

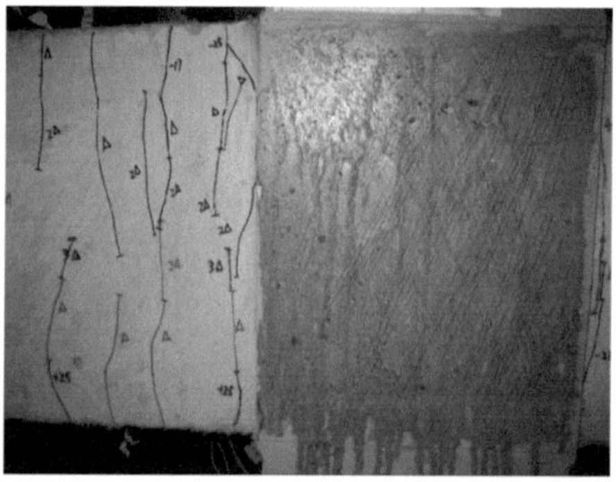

Figure 3. Damage on the joint.

3. Numerical study

3.1. The establishment of the model

ANSYS, a large-scale general finite element software across a range of disciplines, was used to simulate the response of beam-column joint. Depending on the size of the frame node

http://dx.doi.org/10.5772/66798

specimens and the symmetry of the specimen, one-fourth solid was modeled appropriately to reduce the time consumption of computing. The Link8 element, the Solid65 element, and the combin39 element were, respectively, used to simulate the steel, concrete (include UHTCC), and bond slip in the specimen. According to the position of the steel, mapping method was chosen to divide the grid on the geometric entities with the unit size of 50 mm; meanwhile, the sideline was defined as steel.

In addition, in order to better satisfy the engineering demand and analyze overall responses of the new beam-column joint element, the Open System for Earthquake Engineering Simulation (OpenSees) software was also selected to stimulate all the specimens. OpenSees was the computational platform using modern software techniques to provide a common analytical research framework for both structural and geotechnical engineering research study [5, 6]. It had advanced capabilities for creating and analyzing effects in structural and geotechnical engineering using built-in models and solution algorithms.

In OpenSees, the unconfined and confined concrete in beam and column were all stimulated by the "Concrete02" material model that reflected compression behavior through two segments in the rise and fall polyline curve. In general, to ensure a smooth limit-state function the steel material was modeled by the "Steel02" material with a bilinear model. Nonlinear beam-column, force-based nonlinear beam-column element, was used to model each of the members with 2D fiber sections. For the same reason, as shown in **Figure 1**, the cross-sections of column and beam were all discretized into different fibers in the in-plane direction.

3.2. The numerical modeling results

The comparison of hysteretic curves and skeleton curve between numerical simulation and test was shown in **Figure 4**. The following conclusion might be drawn for the damage features of UHTCC through the observation and analysis of experimental and numerical simulation.

The area of hysteresis loop was narrow and small when the displacement was small. However, with the increasing number and time of cyclic loading, inclined cracks in the center of the core repeatedly opened and closed. Finally, degradation of specimen stiffness was serious and hysteresis loop was shown with inverse S-shaped in **Figure 4**. The pinching effect from software simulation results were not obvious; this may be because there were less parameters to define unit Pinching4 in the OpenSees and lead the inaccurate reflection of pinching effect with the test curve steeper than hysteresis curve in the unloading phase.

From the hysteretic curves and skeleton curve, all the numerical simulation captured the deterioration characters of the resistance after yielding and gave reasonable prediction on the initial stiffness, unloading stiffness, and reloading stiffness of the joint; however, compared with the ANSYS simulation results, the simulation of OpenSees was closer to the test.

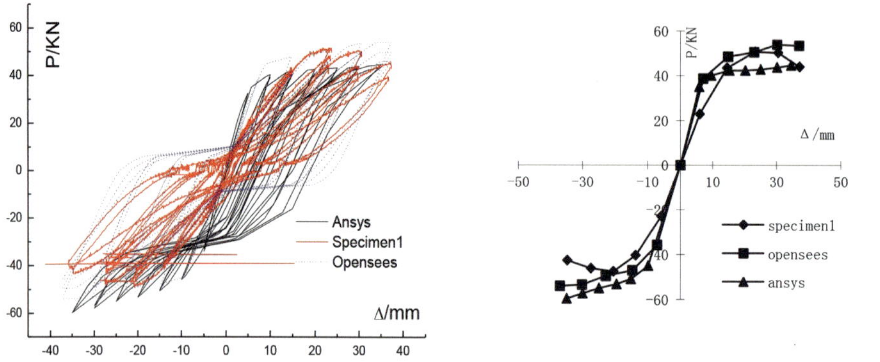

(a)

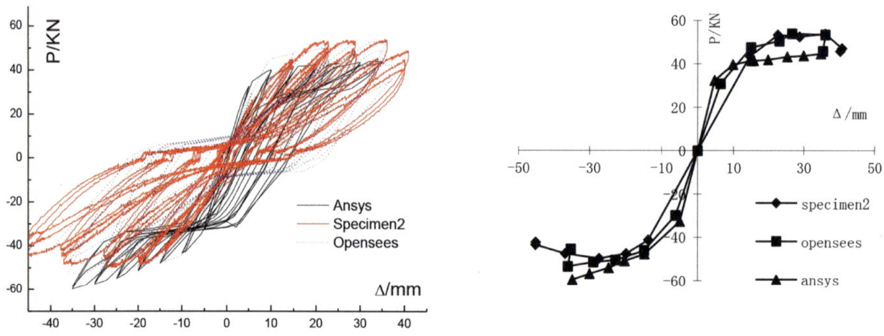

(b)

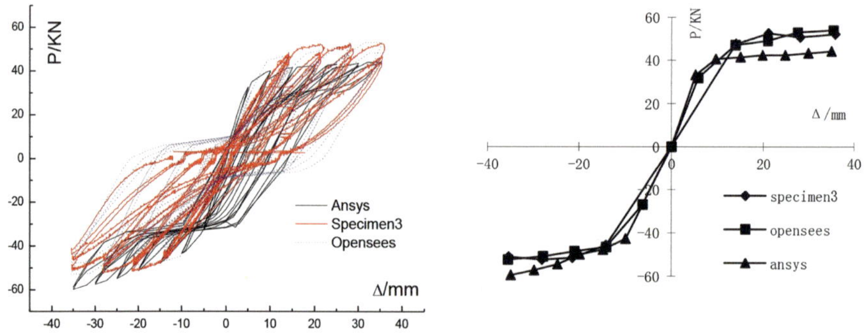

(c)

Figure 4. Comparison of hysteretic curves and skeleton curves: (a) Specimen1, (b) Specimen2, (c) Specimen3.

http://dx.doi.org/10.5772/66798

4. Conclusions

The following conclusions could be drawn through the research of the mechanical properties of the UHTCC beam-column joints.

The shear failure occurs in all the core area of UHTCC beam joints. From the four failure stages begin to end, the UHTCC material showed ultra-high toughness and superior ability of dispersing cracks with a number of fine cracks (average 1 mm) appearing in the center of the joint core.

The specimens with the change of stirrups and axial compression ratio had no significant effect on the shearing resistance capability by the UHTCC used in the center of the core joint that had excellent shearing resistant property to replace or partially replace the amounts of stirrups.

ANSYS and OpenSees were all capable of capturing the connection of UHTCC beam-column joints from initial cracking to destruction, including column failure and shear panel failure mechanisms. But obviously OpenSees simulation results were better than ANSYS with lower cost on calculation and modeling.

Acknowledgements

The author was grateful to Dr. Jun Su for providing the test data needed in this study. Acknowledgment also goes to Dr. Bohumir Strnadel and Dr. Sanhai Zeng for their precious suggestions.

Author details

Wei Li
Address all correspondence to: weilee903@outlook.com

Center of Advanced Innovation Technologies, VŠB-Technical University of Ostrava, Ostrava-Poruba, Czech Republic

References

[1] Jun Su, Shilang Xu. "Experimental study on seismic behaviors of UHTCC new beam-column joints with high axial compression." Journal of Huazhong University of Science and Technology (Natural Science Edition), 38 (2010): 53–56.

[2] Jun Su, Shilang Xu. "Research on performance of new-type frame joints of ultra-high toughness cementitious composites." Journal of Earthquake Engineering and Engineering Vibration. 30 (2010): 59–63.

[3] Xu-Lin Tang, et al. "Seismic behaviour of through-beam connection between square CFST columns and RC beams." Journal of Constructional Steel Research, 122 (2016): 151–166.

[4] Negar Elhami Khorasani, Maria EM Garlock, Spencer E. Quiel. "Modeling steel structures in OpenSees: Enhancements for fire and multi-hazard probabilistic analyses." Computers & Structures, 157 (2015): 218–231.

[5] Liping Kang, Roberto T. Leon, and Xilin Lu. "A general analytical model for steel beam-to-CFT column connections in OpenSees." Journal of Constructional Steel Research, 100 (2014): 82–96.

[6] Terje Haukaas. "Unified reliability and design optimization for earthquake engineering." Probabilistic Engineering Mechanics, 23(4) (2008): 471–481.

Section 4

Safety and Reliability in the Civil Engineering and Industry

Chapter 22

Effect of Size of Ignition Energy on the Explosion Behaviour of Selected Flammable Gas Mixtures

Miroslav Mynarz, Petr Lepík and Jakub Melecha

Additional information is available at the end of the chapter

http://dx.doi.org/10.5772/66799

Abstract

The determination of explosion indices of flammable gases is an important part of explosion prevention. Explosion indices could be influenced by initial temperature, initial pressure, humidity, ignition energy and others. This contribution deals with the effect of ignition energy size on explosion indices of flammable gases. For experimental measurements, three flammable gases were chosen—methane, propane and hydrogen. The chapter introduces the measurement results of the gas explosion parameters at various sizes of the ignition energy using explosion autoclave VA-20. Conclusions include the evaluation of influence of the ignition energy size on particular explosion indices.

Keywords: Explosion indices, 20-L apparatus, methane, Propane, Hydrogen, Ignition energy

1. Introduction

The explosion could happen wherever fuel, oxygen and sufficient ignition source appear together. The ignition energy is really significant. An explosive mixture is not ignited unless energy of the ignition source is sufficient. The ignition energy with a value of 10 J is used by default for the determination of explosion parameters of gases and vapours of flammable liquids. If the mixture is not ignited under given experimental conditions, it does not mean necessarily that the examined mixture is not explosive. When a higher ignition energy is used, the mixture could be ignited and high explosion indices could be reached (Mynarz et al., 2012).

Besides ignition source—the standard (EN 1127-1, 2011) defines 13 groups of ignition sources—duration time also matters. The explosion range is more wide with increasing size of the ignition energy— the lower explosive limit (LEL) is decreasing and the upper explosive limit (UEL) is growing. **Table 1** introduces the effect of the ignition energy on the methane explosive limits.

Ignition energy Ei (J)	LEL (vol.%)	UEL (vol.%)	Explosion range (vol.%)
1	4.9	13.8	8.9
10	4.6	14.2	9.6
100	4.3	15.1	10.8
10,000	3.6	17.5	13.9

Table 1. The effect of the ignition energy on the explosive limits of methane-air mixture (SAFEKINEX, 2002).

Increasing ignition energy affects the explosive limits but also increases maximum explosion pressure and the maximum rate of explosion pressure. The effect of the ignition energy is significant especially at the rate of explosion pressure rise.

The most common ignition sources suitable for measurement of explosive limits and explosion indices are inductive spark, chemical (pyrotechnic) igniter and fuse wire. Efficiency of each mechanism is different; various results could be reached with the above-mentioned ignition sources. This is manifested by results of measurements of the effect of initial pressure on hydrogen explosive limits using nickel fuse wire and electric spark, see **Figure 1**.

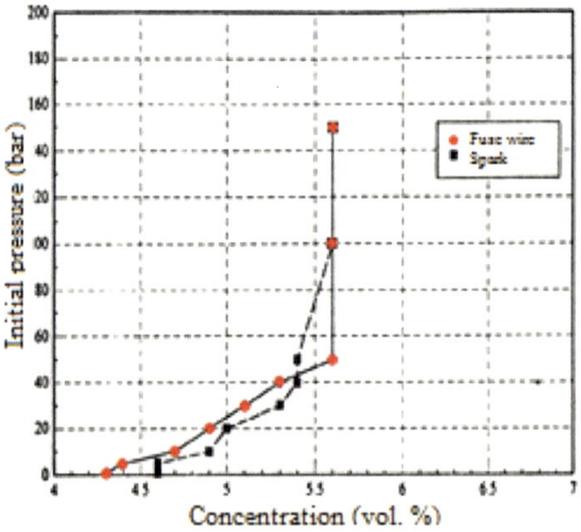

Figure 1. The effect of initial pressure on explosive limits with the use of nickel fuse wire and electric spark (Conrad and Kaulbars, 1995).

2. Tested samples

For experimental measurements of the effect of the ignition energy size on explosion indices, three samples of gases were chosen—methane, propane and hydrogen. Parameters of particular gases are shown in **Table 2**.

http://dx.doi.org/10.5772/66799

	Methane	Propane	Hydrogen
Physical state at 20°C/101.3 kPa	Gas	Gas	Gas
Molar mass (g/mol)	16.04	44.00	2.01
Melting point (°C)	−182	−188	−259
Boiling point (°C)	−161	−42	−253
Density (kg/m^3)	0.676	1.910	0.090
Critical temperature (°C)	−82.7	97.0	−239.9
Critical pressure (MPa)	4.60	4.25	0.13
Auto-ignition temperature (°C)	595	470	560

Table 2. Properties of tested gases (Material Safety Data Sheet-Methane, Propane, Hydrogen).

3. Experimental setup

The explosion autoclave VA-20 was used for experimental measurement of the effect of the ignition energy size on gas explosion indices. The setup is made for determination of the explosion indices of dust, gases and hybrid mixtures. The volume of the experimental double-coat chamber is 20 L (Kuhner Safety). **Figure 2** presents the scheme of the explosion autoclave VA–20.

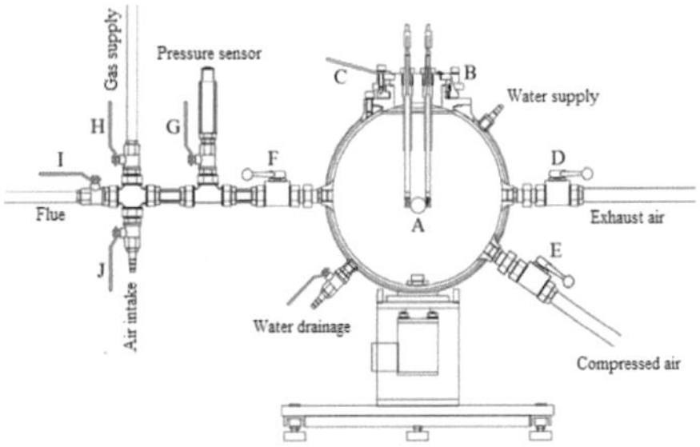

Figure 2. Scheme of 20-L apparatus.

4. Measurement results

Following chapters introduce the experimental results of measurement of explosion indices of methane, propane and hydrogen with air using the apparatus VA-20. The chemical igniter with ignition energies of 80, 160 and 240 J was used. The values of maximum explosion indices and lower explosive limit were determined in a range of minimum 0.5 vol.%.

4.1. Methane

Experimental results of the effect of the ignition energy on the explosion indices of methane are presented in **Table 3** and **Figures 3** and **4**. **Table 4** compares the maximum explosion pressure, maximum rate of explosion pressure rise and the lower explosive limit of methane for particular energies of ignition sources. Percentage changes related to the measurement with the lowest energy are also listed.

Concentration (vol.%)	4.5	5	6	7	8	9	10	11	12	13
					80 J					
p_m (bar)	0	2.2	4.3	5.3	6.1	6.8	7.2	6.8	6.3	5.8
$(dp/dt)_m$(bar s^{-1})	0	12	37	82	138	165	194	145	85	41
					160 J					
p_m (bar)	0	1.9	4.8	6.1	6.9	7.3	7.6	7.7	7.3	6.6
$(dp/dt)_m$(bar s^{-1})	0	12	58	155	209	228	280	218	146	76
					240 J					
p_m (bar)	0.1	0.1	3.0	4.7	6.1	6.7	7.2	7.6	7.3	6.9
$(dp/dt)_m$(bar s^{-1})	0	3	17	68	189	216	255	290	216	133

Table 3. Properties of tested gases (Material Safety Data Sheet—Methane, Propane, Hydrogen).

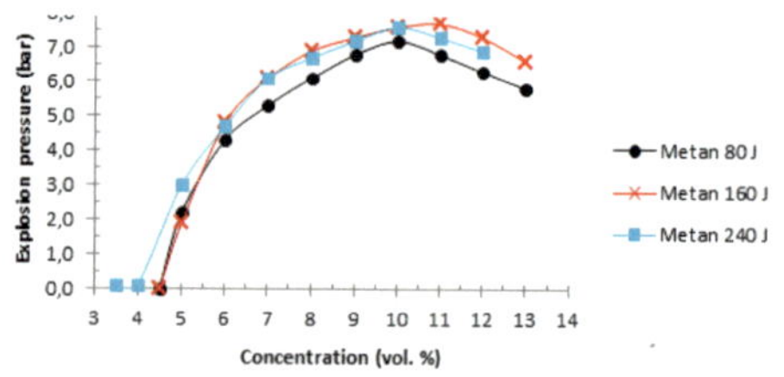

Figure 3. Graph of explosion pressure depending on methane concentration for various ignition energies.

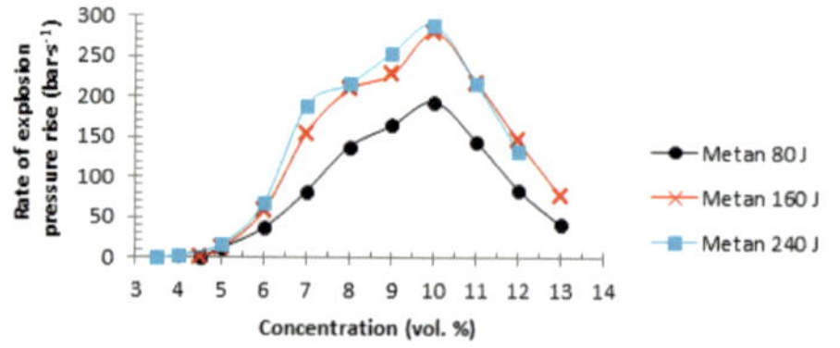

Figure 4. Graph of rate of explosion pressure rise depending on methane concentration for various ignition energies

http://dx.doi.org/10.5772/66799

	80 J	160 J	Methane percentage change (%)	240 J	Percentage change (%)
p_m (bar)	7.3	7.7	5.5	7.6	4.1
$(dp/dt)_m$ (bar s^{-1})	214	283	32.2	290	35.5
K_{Gmax} (bar m s^{-1})	58	77	32.2	79	35.5
LEL (vol.%)	4.5	4.5	0.0	3.5	−22.2

Table 4. Comparison of maximum explosion indices of methane.

4.2. Propane

Table 5 and **Figures 5** and **6** show experimental results of the effect of the ignition energy on the explosion indices of propane. **Table 6** compares the maximum explosion pressure, maximum rate of explosion pressure rise and the lower explosive limit of propane for particular energies of ignition sources. Percentage changes related to the measurement with the lowest energy are also listed.

Concentration (vol.%)	2	2.5	3	4	4.5	5	6	7
				80 J				
p_m (bar)	0	4.8	6.0	7.8	8.1	7.9	7.2	5.6
$(dp/dt)_m$ (bar s^{-1})	0	47	122	262	290	234	125	36
				160 J				
p_m (bar)	0	5.2	6.2	7.7	8.3	8.3	7.5	6.6
$(dp/dt)_m$ (bar s^{-1})	0	63	183	269	362	350	171	80
				240 J				
p_m(bar)	0.1	0.1	6.2	7.7	8.2	8.2	7.6	6.7
$(dp/dt)_m$ (bar s^{-1})	0	5	22	342	374	329	238	121

Table 5. Explosion indices of propane.

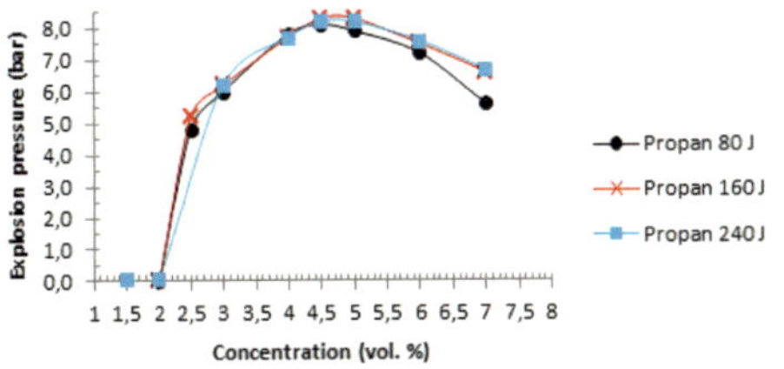

Figure 5. Graph of explosion pressure depending on propane concentration for various ignition energies.

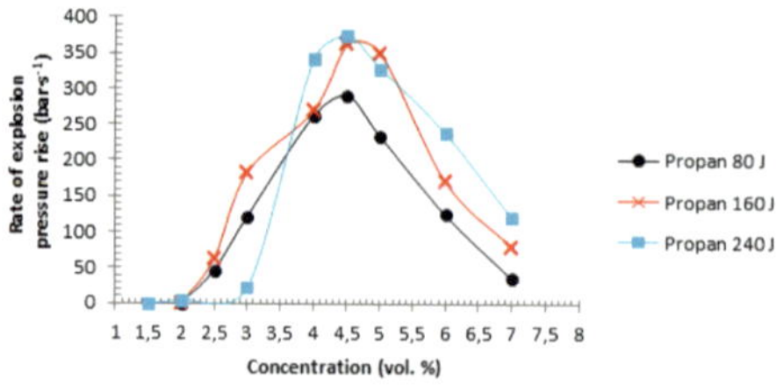

Figure 6. Graph of rate of explosion pressure rise depending on propane concentration for various ignition energies.

	80 J	160 J	Propane percentage change (%)	240 J	Percentage change (%)
p_m (bar)	8.2	8.3	1.2	8.2	0.0
$(dp/dt)_m$ (bar s^{-1})	305	366	20.0	377	23.6
K_{Gmax} (bar m s^{-1})	83	99	20.0	102	23.6
LEL (vol.%)	2.0	2.0	0.0	1.5	–25.0

Table 6. Comparison of maximum explosion indices of propane.

4.3. Hydrogen

Tables 7-A and **7-B** and **Figures 7** and **8** show experimental results of the effect of the ignition energy on the explosion indices of hydrogen. **Table 8** compares the maximum explosion pressure, maximum rate of explosion pressure rise and the lower explosive limit of hydrogen for particular energies of ignition sources. Percentage changes related to the measurement with the lowest energy are also listed.

Concentration (vol.%)	3.5	4	5	6	8	10	15
			80 J				
p_m (bar)	0	0.1	0.4	0.9	1.7	2.8	4.3
$(dp/dt)_m$ (bar s^{-1})	0	5	9	8	10	22	232
			160 J				
p_m (bar)	0	0.1	0.7	–	–	–	4.3
$(dp/dt)_m$ (bar s^{-1})	0	1	8	–	–	–	236
			240 J				
p_m (bar)	0.1	0.1	0.7	–	–	–	4.3
$(dp/dt)_m$ (bar s^{-1})	2	3	9	–	–	–	261

Table 7-A. Explosion indices of hydrogen.

http://dx.doi.org/10.5772/66799

Concentration (vol.%)	20	25	30	31	32	35
			80 J			
p_m (bar)	5.4	6.4	6.9	7.1	7.0	6.8
$(dp/dt)_m$ (bar s^{-1})	838	1566	1899	2166	2091	2105
			160 J			
p_m (bar)	–	6.4	7.0	7.0	7.0	6.8
$(dp/dt)_m$ (bar s^{-1})	–	1498	1880	2018	2203	2010
			240 J			
p_m (bar)	–	6.3	6.8	7.0	7.0	7.0
$(dp/dt)_m$ (bar s^{-1})	–	1546	2170	2183	2288	2295

Table 7-B. Explosion indices of hydrogen—part A.

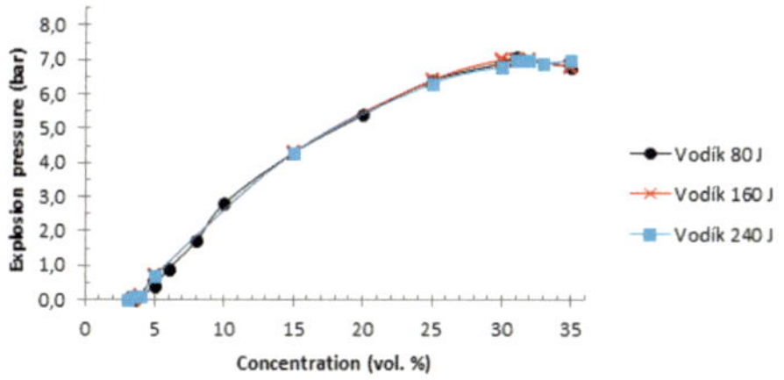

Figure 7. Graph of explosion pressure depending on hydrogen concentration for various ignition energies.

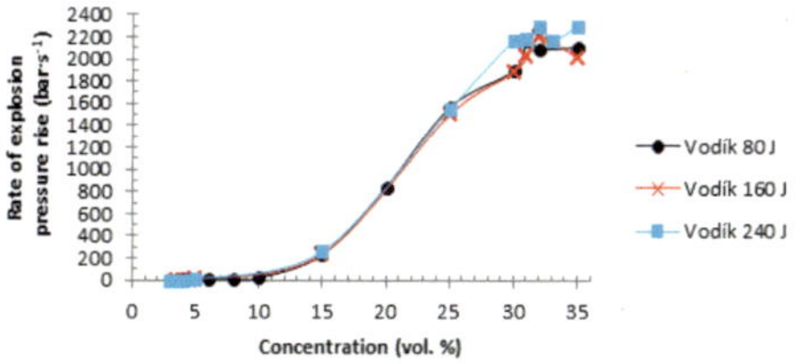

Figure 8. Graph of rate of explosion pressure rise depending on hydrogen concentration for various ignition energies.

	80 J	160 J	Hydrogen percentage change (%)	240 J	Percentage change (%)
p_m (bar)	7.1	7.0	−1.4	7.1	0.0
$(dp/dt)_m$ (bar s^{-1})	2324	2205	−5.1	2372	2.1
K_{Gmax} (bar s^{-1})	631	599	−5.1	644	2.1
LEL (vol.%)	3.5	4.0	14.3	3.5	0.0

Table 8. Comparison of maximum explosion indices of hydrogen.

5. Conclusion

While methane was measured, the rate of explosion pressure rise increased by 32.2% using double energy (160 J) and it increased by 35.5% using triple energy (240 J). Maximum explosion pressure increased by 5.5% using double energy (160 J) and it increased by 4.1% using triple energy (240 J). The lower explosive limit did not change at double energy (160 J). LEL decreased by 22.2% using triple energy (240 J).

While propane was measured, the rate of explosion pressure rise increased by 20.0% using double energy (160 J) and it increased by 23.6% using triple energy (240 J). Maximum explosion pressure increased by 1.2% using double energy (160 J) and it had not changed using triple energy (240 J). The lower explosive limit did not change at double energy (160 J). LEL decreased by 25% using triple energy (240 J).

While hydrogen was measured, the rate of explosion pressure rise decreased by 5.1% using double energy (160 J) and it increased by 2.1% using triple energy (240 J). Maximum explosion pressure decreased by 1.2% using double energy (160 J) and it had not changed using triple energy (240 J). The lower explosive limit increased by 14.3% at double energy (160 J). LEL did not change using triple energy (240 J).

According to experimental data, the inference was made that the size of ignition energy affects especially the rate of explosion pressure rise and the lower explosive limit. Its effect on explosion pressure is only minimal.

Acknowledgement

This paper was financially supported by the project of grant ministry of interior the Czech Republic under the Id. No. VI20152019047, entitled "Development of the rescue destructive bombs for the disposal of statically damaged buildings".

Author details

Miroslav Mynarz*, Petr Lepík and Jakub Melecha

*Address all Corresponding to: miroslav.mynarz@vsb.cz

Faculty of Safety Engineering, VSB-Technical University of Ostrava Lumirova, Ostrava-Vyskovice, Czech Republic

References

Conrad, D., Kaulbars, R., 1995. Pressure dependence of the explosive limits of hydrogen, Chem.-Ing.-Tech. 67.

http://dx.doi.org/10.5772/66799

EN 1127-1 ED.2. 2011. Explosive atmospheres – Explosion prevention and protection – Part 1: Basic concepts and methodology. Brussel: CEN – European Committee for Standardization.

Kuhner Safety: 20-L Apparatus. In: [online]. [cit. 2015-06-02]. http://safety.kuhner.com/en/product/apparatuses/safety-testing-devices/id-20-l-apparatus.html

Material Safety Data Sheet—Hydrogen. In: [online]. [cit. 2016-04-22].http://prodkatalog.linde-gas.cz/international/web/lg/cz/prodcatlgcz.nsf/RepositoryByAlias/BL8335/$file/BL8335.pdf

Material Safety Data Sheet—Methane. In: [online]. [cit. 2016-04-22]. http://www.catp.cz/BL/BL8321.pdf

Material Safety Data Sheet—Propane. In: [online]. [cit. 2016-04-22]. http://prodkatalog.linde-gas.cz/international/web/lg/cz/prodcatlgcz.nsf/RepositoryByAlias/BL0104/$file/BL0104.pdf

Mynarz, M., Lepík, P., Serafín, J., 2012. Experimental determination of deflagration explosion characteristics of methane-air mixture and their verification by advanced numerical simulation, Twelfth International Conference on Structures under Shock and Impact, Kos, Greece, WIT Transactions on The Built Environment, Vol. 126, pp. 169–178. ISBN: 978-1-84564-612-7, ISSN: 1746-4498 (print).

Project SAFEKINEX: Report on experimental factors influencing explosion indices determination, 2002. Deliverable No. 2. Federal institute for materials research and testing (BAM). In: [online]. [cit. 2016-04-10].

Chapter 23

The Heat Radiation of Wooden Facing on Facades

Dana Chudová, Adam Thomitzek and Martin Trčka

Additional information is available at the end of the chapter

http://dx.doi.org/10.5772/66800

Abstract

The chapter deals with the heat radiation of wood during fire. The aim is to verify the theoretical assumption about the heat radiation of spruce wood. For the purpose of verification, the temperatures and radiation were measured under laboratory conditions. The results were compared with the theoretical calculation.

Keywords: wood, heat radiation, temperature, heat flux density

1. Introduction

The natural material, wood, has been newly and increasingly used in the construction industry in the Czech Republic. In the world, particularly in the USA and Canada, there are buildings from wood (i.e. wooden building) common. Under the conditions prevailing in the Czech Republic, there has been an increase in the number of wooden buildings in recent years, especially with regard to their cost and rapid implementation in the construction. Ever more popular is the use of wood as the cladding material even though the negative qualities of wood as a low biological resistance and high flammability are pointed out. The situation is not so tragic from the perspective of fire-safety engineering. Although it is a flammable material, the wood has fire resistance when it is properly used. In the construction industry, it can be used safely with the proper implementation of construction and compliance with other regulations. To prevent the spread of fire from the burning object, the safety distance between the objects should be defined. Around the burning object, there is fire danger zone where there is the risk of transmission of fire, thanks to heat radiation. In the Czech Republic, as well as in many places in the world, there exists some codex of standards that address this problem. The Czech technical standards for the area of safety distances and their settings are based on the principle of restricting the fall of flammable structures and limiting heat flux density from the adjacent building. The aim of the article, based on a series of experiments, is to evaluate that standard values of a heat flux density are realistic for the calculation of safety distances from outside wooden wall.

1.1. Theoretical analysis

According to Jönsson et al. [1], the fire spreading between buildings is caused mostly due to radiation. However, the influence of radiation can cause an ignition of the material at a much longer distance than by convection–thus a direct contact of the flame, as was described by McGuire [2] already in 1965. The characteristic value for radiation when there is an ignition of wood is 33.5 kW.m^{-2} for the auto-ignition (i.e. the inflammation with the absence of sources of ignition) and 12.5 kW.m^{-2} for the ignition in the presence of sources of ignition, that can be, for example, sparks as presented by Barnett [3]. These values are commonly used worldwide.

Carlson [4] states that factors influencing the transmission of the radiation between the burning object and vulnerable object are the effect of the flame from holes, the emissivity of the flame, the configuration factor and the intervention of fire fighters. From the basic physical relationship of heat transport, $E = \varepsilon\sigma T4$, it is obvious that the temperature has the greatest influence on the size of the radiated energy. As given in Refs. [5, 6], the rates of burning depend on the wood species, its density, its moisture content and the type of facades and their ways of using. Clark, in 1998 [7], conducted a series of experiments that showed the temperature of ignition of the material is caused by radiant heat. Clark also dealt with the effect of wood moisture on its radiation. He found out that higher wood moisture has an effect on the value of heat flux and on the ignition timing. Clark points out on the study deals with the influence of wood moisture carried out by Janssens in 1991 that has been found that the heat flux is approximately between 10 and 14 kW.m^{-2}.

Most of the studies show three situations how a fire spreading occurs along the surface of the facade. The first one is a case where the object catches fire from outside (so along the facade); it occurs either by radiation or by the action of the flames from the adjacent building. The second situation of the initiation is due to the sources of ignition that are from the nearby buildings such as fire dustbins, cars etc. The third one is the fire inside the building, while there is spreading of fire out of the window. Therefore, the most building regulations in the European countries limit the using wood and its products on low buildings–maximally on three-storey buildings. There is a small percentage of limitation on the total surface of the facade for using wooden facing. According to the European harmonized classification, for facing materials, the material should be of class B-S3, D2, while the wood-based panels are generally classified as D-s2, d0 [8]. There is the national standard for fire protection called ÖNORM in Austria, for example ONR 22000 Fire Safety in High–Rise Buildings or ÖNORM B 3806 Requirements for fire behaviour of building products (building materials) valid from 01.07.2005. There are also the OIB Guidelines of Austrian Institute of Construction Engineering that deal with general requirements and stability of fire, behaviour of construction materials when on fire, requirements for fire resistance of building materials etc. According to the standard ÖNORM B 3800-5, the aim of the protection of object against the fire spreading is:

Fire behaviour of building materials and components–part 5: Fire behaviour of facades: The fire cannot spread along the surface, large parts cannot fall down and people cannot be endangered.

1.2. Solving the issue in Czech conditions

In the Czech Republic, the walls that are having a heat flux density on the surface higher than 15 $kW.m^{-2}$ [9] are considered as fully or partially open danger surface and the safety distance depends on them. The fire danger zone is formed around burning objects where there is a risk of transmission of fire due to heat radiation or falling of construction parts of burning objects. The size of fire danger zone is defined by the safety distance from open danger surface of the fire section of the object. The fire danger zone cannot interfere over the border of building land, but it can interfere in the public spaces such as streets, squares, parks, etc. In this space could be other objects only when the peripheral walls of these objects dispose with an open danger surface with the species DP1 or when the walls have surface covering of other products wide at least 20 mm and with the fire classification A1 or A2. The peripheral wall with insulation must have these types of modifications with a flame spread index (FSI) of 0 $mm.min^{-1}$.

Fully open danger surface of peripheral wall or its part is defined as the surface with heat flux density greater than 60 $kW.m^{-2}$, at the time required for fire resistance for outer side of the peripheral wall. This fully open danger surface could also be considered if there is no fire resistance and outer side of walls is made from materials of the fire classification E or F. The amount of heat released is higher than 150 $MJ.m^{-2}$. From the calculation of heat flux density, another classification of peripheral wall could be established. The same applies for peripheral wall species DP1 or DP2 that have materials from outer side of fire classification B to D with the amount of the released heat higher than 350 $MJ.m^{-2}$.

A partly open danger surface is the surface of peripheral wall or its part that reports the heat flux density from 15 to 60 $kW.m^{-2}$ at time required for fire resistance for outer side of the peripheral wall.

The main criteria for determining the fully or partly open danger surface are the amounts of released heat from the products (materials) and the heat flux density on outer side of the peripheral wall. The amount of heat Q [MJ] released from 1 m^2 of flammable products of outer side of the peripheral walls is dependent on the calorific value of products H_i [$MJ.kg^{-1}$] and basis weight M_i [$kg.m^{-2}$]. This amount of heat is calculated using Eq. (1) [10]:

$$Q = \sum_{i=1}^{j} M_i \cdot H_i [\mathrm{MJ} \cdot \mathrm{m}^{-2}] \tag{1}$$

The determination of heat flux density I [$kW.m^{-2}$] from the burning surface of the peripheral wall may take into account the speed of burning surface m_v [$kg.m^{-2}$-min^{-1}] and this heat transfer [9]. This heat flux density is calculated using Eq. (2):

$$I = 0.35 \cdot \frac{m_v \cdot H_i}{60} [\mathrm{kW} \cdot \mathrm{m}^{-2}] \tag{2}$$

The total heat flux density is multiplied by the coefficient, 0.35. This coefficient shows the quotient of radiant component in heat transfer. A burning speed (m_v) is experimentally determined or it is used in the tabular values.

The released heat from 1 m^2 of the outer side of the peripheral wall of flammable products is determined with Eq. (1) [10]. The calculation is applicable for peripheral walls from spruce wood.

$$Q = 0.02 \cdot 470 \cdot 17 = 159.8 \text{ MJ} \cdot \text{m}^{-2} \tag{3}$$

$$Q = 0.02 \cdot 470 \cdot 17 = 159.8 \text{ MJ} \cdot \text{m}^{-2} \tag{4}$$

The released heat that is higher than 150 MJ-m^{-2} but less than 350 MJ-m^{-2}, it is typical for the walls that are determined as partly open danger surface. However, it is still necessary to take into account a heat flux density. This heat flux density is calculated according to Eq. (2).

$$I = 0.35 \cdot \frac{0.45 \cdot 17}{60} = 44.63 \text{ kW} \cdot \text{m}^{-2} \tag{5}$$

1.3. Experimental verification

To verify the calculated values, a set of tests under laboratory conditions was conducted. The first seven tests were unsuccessful mainly because of insufficient sources of ignition and an improper disposition of the sample. Therefore these data are not mentioned. This chapter mentions only the data from two measurements that do not have the identified deficiencies from previous measurements.

2. Materials and methods

The measurements were realized under the same conditions at the temperature 20°C, humidity and pressure. Spruce wood with 10.3% of humidity was used as the test material. The test sample with dimensions of 1400 mm × 450 mm was assembled from the laths with the width of 45 mm and a thickness of 20 mm (see **Figure 1**).

The back side of the sample was joined by continuous surface of boards and with 10 mm of space between particular boards. The test sample was installed on the wall in a vertical position and at 30 mm above the floor. On the floor was placed the source of ignition with a volume of about 300 ml and there was a mixture of flammable liquids (n-heptane (V = 50 ml) and a mixture of hydrocarbons–petroleum (V = 250 ml)).

The temperature and radiation during the testing were measured. The thermocouples of type K were used for the process of temperature measurement. They were distributed as follows: the first thermocouple was placed at a height of 650 mm and another at 250 mm above the other on the axis of the sample (see **Figure 2**).

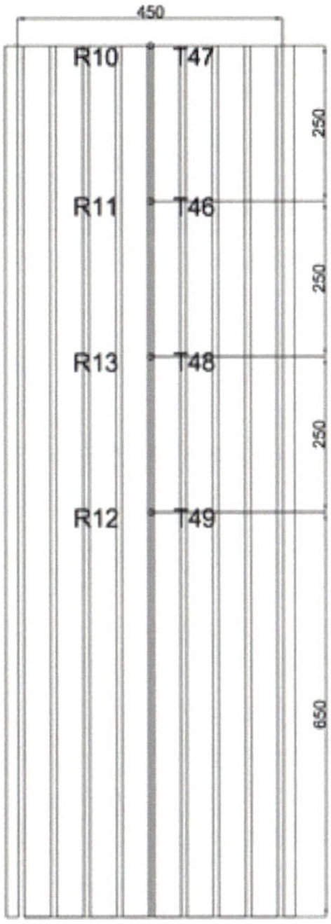

Figure 1. Experimental setup.

Figure 2. Radiometer setup.

The radiation was measured using radiometers Hukseflux SBG01 with a range of 0–100 kW.m^{-2}, which were placed at the same positions of height as thermocouples and 300 mm away from the sample. To score the process was used the measuring system ALMEMO 5690-2. The basic calibration of sensors of the heat flux was carried out at the full measuring range of the system. The initial calibration accuracy is ±3%. There are other errors which are caused by non-linearity, convection and radiation balance. The radiometer SBG01 shows the error caused by non-linearity of signal ±4.5% in the measurement range up to 44 kW.m^{-2}.

3. Experimental results

In **Figure 3**, there is appreciable process of heat radiation on a particular radiometer. The first growth is associated with burning of the sample as well as with a source of ignition. The values at the even combustion of sample without an ignition source for the verification of values of heat radiation with the theoretical calculation were used. The highest values were achieved by radiometer R12 which was placed at the lowest position at a distance of 650 mm from the edge of the sample. The maximum value of heat radiation was 43.81 kW.m^{-2} at this radiometer and it was reached at the time 3:16 min after the ignition of the source. A maximum value of heat radiation was also measured at the radiometer R13 positioned 250 mm above, almost at the same time. The maximum value of heat radiation at the other radiometers that were placed at a height of 1150 and 1400 mm was reached about 10–20 seconds later. This was caused by the absence of combustion at the upper part of the sample. There are differences always about 10 kW.m^{-2} at the radiometers. The temperature increase at that time corresponding to the theoretical assumption can be seen in **Figure 4**.

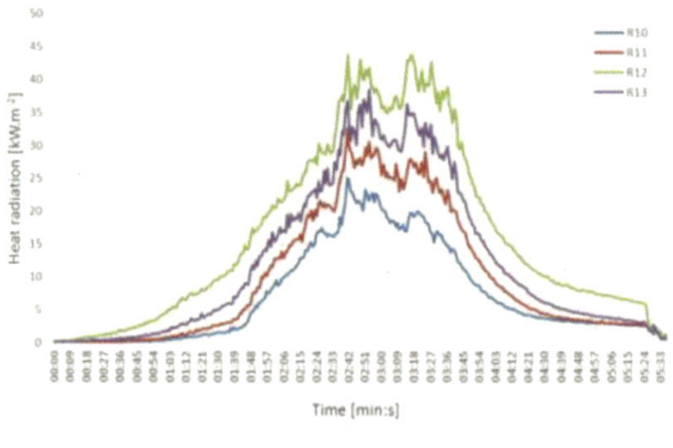

Figure 3. The values of heat radiation on various radiometers during test no. 01.

An initial increase of temperature corresponds to the combustion of the sample together with the source of ignition. Therefore there are used data of 3 min for realistic description of the temperature during fire. The highest temperature about 740°C was measured at the thermocouple T48 and T49 at the time 3:20 min. The lowest temperatures were measured at the thermocouple T47 which was placed at a height of 1400 mm, and that it also shows the effect of the absence of combustion in the upper part of the sample.

http://dx.doi.org/10.5772/66800

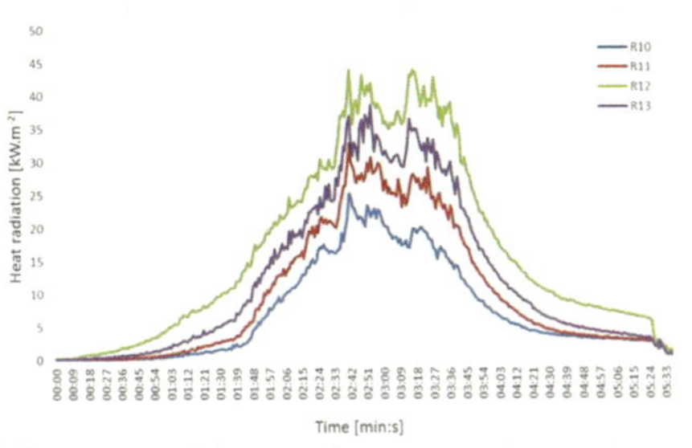

Figure 4. The measured temperature on thermocouples during test no. 01.

During the second test, as it is obvious in **Figure 5** and **Figure 6**, a lower heat radiation was achieved, while the temperature at the thermocouples was higher by about 50°C. Processes of heat radiation and temperatures that are a little different in time are also seen. In the second test, there was a gradual increase of radiation already from the first min and a rapid increase of values already before the second min. The maximum of heat radiation 36 kW.m^{-2} was measured at the thermocouple R12 that was placed in the axis on the lowest place, at the time 2:49 min when the temperature reached almost 800°C. There was a significant increase of radiation at the time 3:28 min at the thermocouple R13 and R11 that was caused by a rapid movement of the flames due to changes in the ventilation of a test chamber.

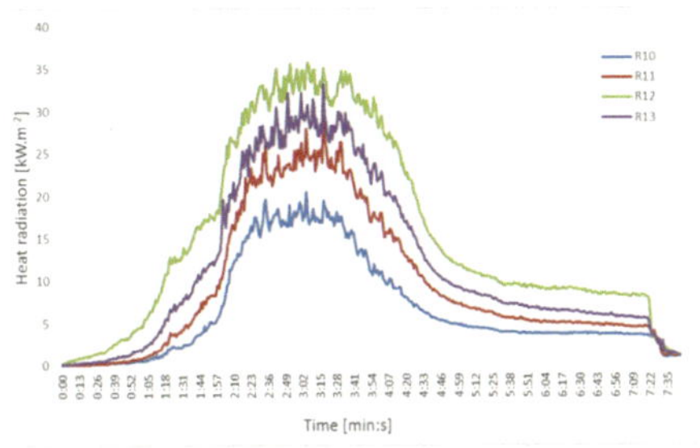

Figure 5. The values of heat radiation on various radiometers during test no. 02.

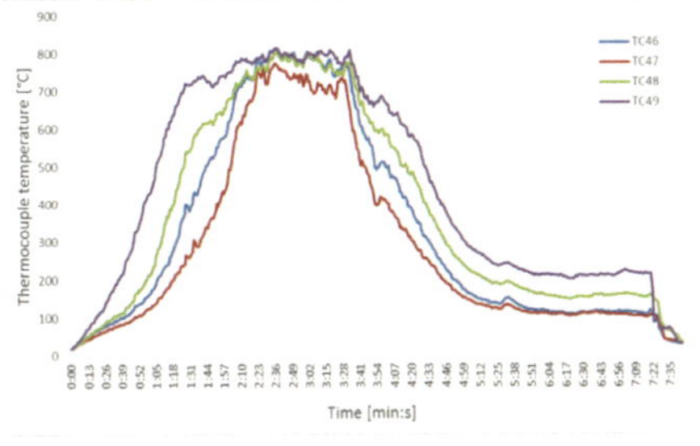

Figure 6. The measured temperature on thermocouples during test no. 02.

4. Conclusion

The aim of the article was to verify the theoretical assumption about heat radiation of spruce wood. The authors also wanted to test and possibly upgrade the information that has been obtained from the measurements. The calculated value of heat flux density is 44.63 kW.m^{-2} assuming that during a diffuse combustion with good access of air there is 35% of the heat shared from the flame of radiation. Maximum value of heat flux density, which was measured, is 43.81 kW.m^{-2}. This maximum value is influenced by an action of heat that is released from source of ignition. It means that measured heat radiation achieves a calculated value only at the most adverse measured place. These values are affected by the measurement error that is affected by a measuring technique which was used to take measurements of values at lower values than is the half of the measuring range that affects the measurement errors. It has not been possible to use the heat flux sensors with a lower measuring range, without the verification of the actual values of heat flux density, as it may cause damage to the equipment. There is also difficult to achieve the uniform spreading of flames on the surface in condition of sample. In fact, it is assumed that flames come from windows during the under-ventilated fire. An initiation of the facade is caused by very intensive action of flames. Test conditions were limited by a possibility to use powerful source of ignition it the test chamber. The sample was modified in the way to ensure better burning-off (by using the vertical spaces). Actually the wooden facing gradually dries on the facades and the burning surface on old facades is significantly higher than on the new facades. It means that an artificial increase of surface partially stimulate the actual condition of the wooden facing. Based on the tests, it is possible to conclude that the calculated values of the heat flux density are closed to real values. It is necessary to focus on a measurement at different combination of shapes and samples for further experiments. Also it is necessary to consider realization of the large size test.

Author details

Dana Chudová*, Adam Thomitzek and Martin Trčka

*Address all correspondence to: dana.chudova@vsb.cz

VŠB-Technical University of Ostrava, Faculty of Safety Engineering, Ostrava, Lumírova, Ostrava-Výškovice, Czech Republic

References

[1] Jönsson R., et al. Fire Building Regulations Theory and Practice , LTH-Brandteknik, Stockholm, 1994

[2] McGuire J.H., Fire and the Spatial Separation of Buildings, Fire Technology Vol. 1, No. 4, 1965, p. 278.

[3] Barnett C.R., Fire Separation between External Walls of Buildings, Fire Safety Science, Proceedings of the 2nd International Symposium (pp. 841–850), 1988.

[4] Carlson, E. External Fire Spread to Adjoining Buildings. Report 5051. Lund University. The faculty of engineering. Lund 1999.

[5] Lignum Dokumentation Brandschutz: 7.1, Auβenwäande, Konstruktion und Bekleidung, www.lignum.ch. Zurich 2007.

[6] Hakkairainen, Tuula a Tuuli Oksanen. Fire safety assessment of wooden facades. Fire and Materials [online]. 2002, 26(1), 7–27. DOI: 10.1002/fam.780. ISSN 0308-0501.

[7] Pilar Giraldo, M. et al. Computer-Simulation Research on Building Facade Geometry for Fire Spread Control In Buildings with Wood Claddings. World Conference on Timber Engineering. Auckland 2012.

[8] Giraldo M.P., Avellaneda J., Lacasta A.M., Rodríguez V., Computer-Simulation Research on Building-Façade Geometry for Fire Spread Control in Buildings with Wood Claddings. In: Conference: The 2012 World Conference on Timber Engineering, Auckland, New Zealand, Vol. 2, 2012

[9] ČSN 730810 Fire protection of buildings – General requirements 5/2009. Czech Office for Standards, Metrology and Testing (ÚNMZ). Prague.

[10] ČSN 730802 Fire protection of buildings – Non-industrial buildings 6/2009. Czech Office for Standards, Metrology and Testing (ÚNMZ). Prague.

Chapter 24

The Calculation Method of Safety Degree and Its Application in Coal Mine Enterprises

Nie Baisheng, Huang Xin, Wang Longkang,
Yu Hongyang and Li Xiang Chun

Additional information is available at the end of the chapter

http://dx.doi.org/10.5772/66800

Abstract

In order to evaluate the situation of safety production of coal mine enterprises effectively, quantitative analysis is necessary and very important. Safety degree of coal mine enterprises based on the concept of safety degree is defined and the method of calculating quantitatively the safety degree is put forward. The validity of this method is verified by empirical research in view of micro- and macroanalyses. In view of micro analysis the safety degree is derived with the calculation method based on information of one coal mine. The safety degree of this coal mine went through rapid increase period, stable period, and slow increase period. Macroresearch results show that the situation of safety production of coal mine enterprises in China has significantly been improving and the level of safety degree also has been increasing year by year since 1979, the year when the policy of reform and opening began. The reasons are the advancement of technology, strengthening of safety management and education, increasing of safety investment, and perfection of policies, laws, and regulations. These achievements can provide quantitative method for assessing the status of coal mines.

Keywords: coal enterprise, safety, safety degree, empirical researches

1. Introduction

China is one of the largest producers of coal in the world. The coal production in China accounted for 46.9% of the total coal production around the world. But the coal consumption in China accounted for 50.6% of the total coal consumption around the world (BP Group, 2015).

At the same time, the coal industry is considered as the most dangerous industry in China. And the number of new occupational patients tops all industry (Liao et al., 2009). All kinds of danger and risk exist during coal production. And not only personal casualty but also stoppages in production of coal mine may be induced by accidents, which cause huge loss to the coal enterprise (Xu, 2014; Mahdevari et al., 2014). So it is imperative to study the safety issues from the quantitative point of view. Many scholars did lot of researches. The quantitative methods for safety analysis include micro-level Markov models (Knegtering and Broracher, 1999), computer-aided fault tree synthesis method (Wang et al., 2002), dynamic fault tree method (Čepin and Marko, 2002), and the decision tree method of incident management, and so on (Baumont et al., 2000). For example, safety technology investment model for assessing quantitatively the enterprise's risks and potential threats was put forward (Bojanc et al., 2012). The safety level of traffic system was evaluated by SIL probability model, and its hazards were found out (Beugin et al., 2007). With a comprehensive method of quantitative analysis on energy security, the safety degree was assessed from five dimensions (Benjamin and Mukherjee, 2011). Furthermore, quantitative research was also applied to analyze coal mine accidents, and thus improvement measures were taken to ensure safe production (Paul and Maiti, 2007). According to the time of accident occurrence and intervals of mechanical failure, a model to analyze safety issues in coal mines was established and the study showed that accidents are related to reliability of mechanical equipment and management effectiveness (Vivek et al., 2011). Also, hazards and probability of accidents in the coal mine production system are found out by statistical methods, and multiple probability of accident by severity of damage can be conducted as a risk factor of the system (Denby and Kizil, 1992; Hatton and Whateley, 1995; H.S.B., 2005). A coal mine macro, meso, and micro dynamic warning system which is based on portable examination instrument, risk information card, and wireless communication network was also put forward (Wang et al., 2016).

In China the death toll is high when compared with other countries, but in recent years the safety status has improved and the death toll has been on the decline. In this chapter the safety degree of coal mine enterprises will be defined and quantitative calculation methods will be put forward based on the death toll and the number of injured. The safety status of coal mine enterprise in China will be assessed and the key factors affecting the safe production in coal mine enterprises will be analyzed.

2. The concept of safety degree of coal mine enterprise

For safety degree, there is no uniform concept, and most are defined from the perspective of the safety state of things. Related definitions include: describing the probability of things in a safe state; the safety level of the system; and the degree objective from danger. The safety degree was defined as a situation of production safety in enterprises by analyzing the relationship between safety level and safety degree (Huang et al., 1999; Golbraikh et al., 2003).

http://dx.doi.org/10.5772/66800

By comparative analysis of the definition of safety degree, the safety degree of coal mine enterprise is defined as: the probability where there are no casualties and economic losses suffered by coal mine enterprises in a certain period of time. The concept is a quantitative form on safety production, which reflects the safety situation of the enterprises.

3. Calculation method of coal mine enterprise's safety degree

The safety degree of coal mine enterprises is the result of internal factors' interaction and is the quantification of coal mine safety situation. The range of coal mine enterprise's safety degree $S \in [0, 1]$, that is, the absolute unsafely degree is 0 and the absolute safety degree is 1. The reverse concept of S is risk degree, which is the probability of accidents in coal mine enterprises, $S = 1 - R$ (R is the risk degree).

3.1. Calculation of safety degree in view of staff system in coal mine enterprises

It is not easy to count the safety degree of staff system in practice because there is a measuring standard. The safety degree of staff system can be obtained by estimating method based on the number of casualties in coal mine enterprises. Whether unsafe acts can cause injury or not is random. Also, a lot of unsafe acts may be not counted that do not cause consequence. According to the Heinrich accident triangle rule (Heinrich, 1980), serious injury and death:injuries:no injuries = 1:29:300, so the later (no injuries) is 10 times of the total of the two formers (serious injury and death, injuries). Thus, the total number of violations can be attained.

The safety degree of staff system can be expressed by the following formula:

$$S_H = 1 - \frac{n}{N} \tag{1}$$

where S_H is the safety degree of staff system, dimensionless; N is the number of enterprises staff in one statistical year; and n is the number of casualties in one statistical year, where

$$n = (injuries + deaths) \cdot (1 + 10) \tag{2}$$

3.2. Calculation of the safety degree of coal mines enterprises

The theories of accident consequence chain show that the direct reasons of accident were unsafe act of staff and unsafe condition of logistics system. Therefore, the system's situation can be reflected by an integrated study of unsafe act and unsafe condition. According to the research of a renowned Japanese scholar, 88% of factors in an accident is contributed by human's unsafe act, 10% is attributed by unsafe condition of things, and other reasons account for 2%. Accordingly, the weight of human safety degree is 88%, the weight of logistics system 10%, and the weight of other factors is 2%.

Then, the total safety degree of coal mine enterprises is:

$$S = 0.88 \cdot S_H + 0.1 \cdot S_M + 0.02 \cdot S_O \tag{3}$$

4. Some common mistakes

Empirical research includes the micro and macro level. For micro level, one coal mine is taken into account and for macro level the information of the total coal mines of China is used. The safety degree of the cases will be calculated and safety status will be analyzed.

5. Empirical study in view of micro level

5.1. The original data of one coal mine

In view of micro level, one coal mine is used for study. The coal mine is located in the Shandong Province and is an old mine with more than 30 years of operation. For years, lots of coal was produced, but various accidents also caused some irreparable losses. During the periods 1974–2005, more than 11,600 injuries of workers are reported, of which 329 persons were seriously injured and 219 people lost their lives. The accidents and casualties for every calendar year are shown in **Table 1** (Hu, 2006) and **Figures 1** and **2**.

5.2. Calculation of the safety degree of staff system

We can calculate the safety degree of staff system according to formula (1) and the data in **Table 1**. The accurate number of no injury is not known, but it is likely to cause an injury. The safety degree of the staff system can be got through transform method,

$$\frac{n}{N} = \frac{\omega + 10\omega}{1000} \tag{4}$$

where ω is the casualty rate per thousand persons.

Due to lack of statistics of logistics system such as operating rates of machinery and equipment, and production lines and roadway repair, the safety degree of logistics system and the total safety degree cannot be calculated. But a large number of studies show that unsafe act of coal mines is one of the main causes of accidents and at least 80% of coal mine accidents were caused by unsafe act. Therefore, the safety degree of staff system can reflect the total safety degree of the enterprises by at least 80%.

5.3. Analysis of the safety situation of coal mine

We can get the trend chart of safety degree according to the data in **Table 2**. **Figure 3** shows that the safety degree of this coal mine went through rapid increase period, stable period, and slow increase period, which indicates the improvement of situation since 1994.

http://dx.doi.org/10.5772/66800

Year	Raw coal production (ton)	The annual average number of employed	The fatality rate per millions tons	The fatality rate per thousand persons	Serious injury rate per thousand persons	The injuries rate per thousand persons	The casualties rate per thousand persons
1974	225,000	6316	6.00	1.58	2.69	82.01	83.60
1975	1,643,802	5172	8.51	2.71	3.67	65.35	68.06
1976	1,317,570	5503	6.83	1.64	1.27	58.33	59.97
1977	1,334,789	5441	7.49	1.84	2.21	68.74	70.58
1978	1,322,312	5708	6.05	1.40	0.88	41.70	43.10
1979	1,083,721	4640	3.69	0.86	0.65	49.78	50.65
1980	1,220,258	4160	7.46	0.72	1.44	77.64	78.37
1981	1,342,292	3819	3.72	131	2.62	60.75	62.06
1982	654,939	3707	4.58	0.81	1.89	96.57	97.38
1983	443,283	3803	4.51	0.53	1.58	54.17	54.69
1984	667,629	5163	7.48	0.97	1.36	47.45	48.42
1985	1,195,572	5561	1.17	0.36	2.34	69.05	69.41
1986	1,322,312	5745	3.78	0.87	0.52	53.26	54.13
1987	1,419,797	5774	2.11	0.52	0.87	48.67	49.19
1988	1,471,427	5790	3.39	0.86	1.21	48.01	48.88
1989	1,163,186	6469	3.03	0.62	0.77	43.90	44.52
1990	1,465,714	6426	5.45	1.24	0.47	55.56	56.80
1991	1,216,473	6435	6.57	1.24	0.62	49.41	50.66
1992	1,608,392	6443	3.10	0.78	2.17	64.80	65.65
1993	1,847,960	6793	8.65	2.36	1.47	68.16	70.51
1994	2,020,607	6897	3.46	1.01	1.88	62.92	63.94
1995	1,782,517	7142	3.93	0.98	1.68	47.04	48.03
1996	1,643,530	7144	2.43	0.56	1.68	47.17	47.73
1997	1,816,656	839	2.20	0.48	1.55	36.79	37.26
1998	1,841,486	8140	3.26	0.74	1.11	35.01	35.5
1999	1,806,549	8105	2.77	0.62	3.21	33.68	34.30
2000	1,802,670	8405	4.44	0.95	0.83	24.27	25.24
2001	1,556,855	8227	2.63	0.49	0.24	16.29	16.77
2002	1,536,553	7890	2.04	0.38	0.51	15.34	15.72
2003	1,315,052	7865	4.02	0.63	0.38	12.21	1.84
2004	1,231,467	7913	5.14	0.88	0.13	11.37	12.26
2005	1,174,920	6977	1.99	0.29	0.14	9.03	9.32

Table 1. Statistical table of accident and casualty rates of calendar year.

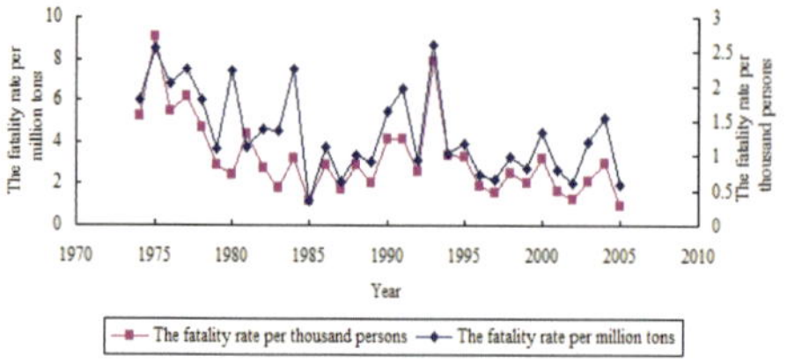

Figure 1. The fatality rate per thousand persons and per million tons.

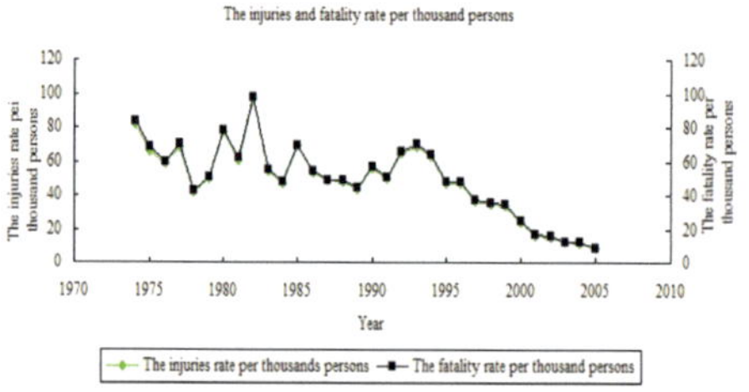

Figure 2. The casualty rate and injuries per thousand persons.

Year	The safety degree of flow systems	Year	The safety degree of flow systems
1974	0.91641	1990	0.94320
1975	0.93194	1991	0.94935
1976	0.94003	1992	0.93442
1977	0.92942	1993	0.92948
1978	0.95690	1994	0.93607
1979	0.94936	1995	0.95198
1980	0.92164	1996	0.95227
1981	0.93794	1997	0.96273
1982	0.90262	1998	0.96425
1983	0.94530	1999	0.96570
1984	0.95158	2000	0.97478
1985	0.93059	2001	0.98322
1986	0.94587	2002	0.98428
1987	0.95081	2003	0.98715
1988	0.95113	2004	0.98775
1989	0.95548	2005	0.99068

Table 2. The safety degree of staff systems in certain coal mine.

http://dx.doi.org/10.5772/66800

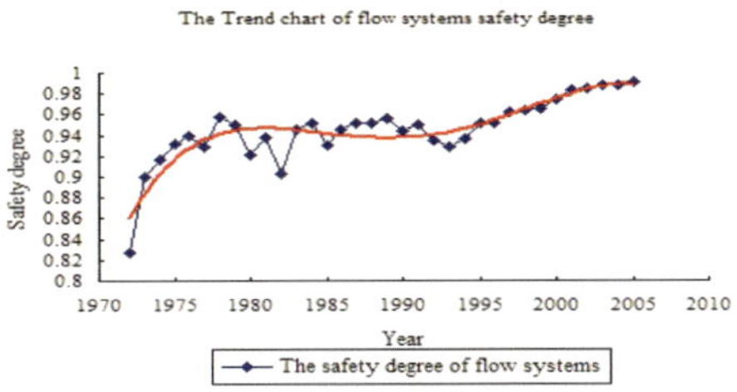

Figure 3. The trend chart of safety degree in certain coal mine.

6. Empirical study on the macro level

6.1. The accident statistics of China's coal mine industry

The safety degree of China's coal mine industry can be expressed through the number of casualties, injury, and potential injury of the accidents. According to statistics, the death rate per 100,000 persons is used to reflect safety situation, as shown in **Table 3** (Chen, 2012).

Year	The numbers of deaths	The death rates of per 100,000 persons
1964	1350	6.49
1965	1104	4.81
1966	1556	6.28
1967	1431	5.65
1968	1687	6.46
1969	2017	6.99
1970	3027	9.03
1971	3766	9.91
1972	3597	8.83
1973	4079	9.53
1974	3722	8.29
1975	4736	9.65
1976	4948	9.26
1977	5637	10.15
1978	6001	9.07
1979	5566	8.11
1980	5165	7.04
1981	5162	6.77
1982	4873	6.13

Year	The numbers of deaths	The death rates of per 100,000 persons
1983	5567	6.73
1984	5872	6.43
1985	6912	6.98
1986	6888	6.45
1987	7049	6.32
1988	6902	5.97
1989	7625	6.69
1990	7360	5.65
1991	6412	4.85
1992	5992	4.85
1993	6244	4.41
1994	7239	4.98
1995	6907	4.64
1996	6556	4.26
1997	7083	2.51
1998	6302	4.03
1999	6469	4.14
2000	5796	3.87
2001	5670	3.66
2002	6995	4.36
2003	6434	4.01
2004	6027	3.76
2005	5986	3.73
2006	4746	2.96
2007	3786	2.36
2008	3215	2.01
2009	2631	1.64
2010	2433	1.52
2011	1973	1.23

Table 3. The number of deaths in China's coal mine enterprises from 1964 to 2011.

6.2. Calculation of safety degree of China's coal mine industry

Due to the lack of the number of wounded and unsafe act in the statistics, we can estimate the number of injuries and unsafe act by applying the Heinrich accident triangle rule: serious injuries and death:slight injuries:no injury = 1:29:300. In formula (1):

http://dx.doi.org/10.5772/66800

$$\frac{n}{N} = \frac{30\omega + 300\omega}{100,000} \tag{5}$$

where ω is the death rate per 100,000 persons.

Table 4 shows the safety degree of China's coal mine industry (see **Figure 4**).

Year	Safety degree	Year	Safety degree	Year	Safety degree
1964	0.978575	1980	0.976778	1996	0.985956
1965	0.984116	1981	0.977676	1997	0.985130
1966	0.979272	1982	0.979775	1998	0.986708
1967	0.981371	1983	0.977777	1999	0.986347
1968	0.978686	1984	0.978784	2000	0.987228
1969	0.976929	1985	0.976952	2001	0.987927
1970	0.970188	1986	0.978717	2002	0.985600
1971	0.967290	1987	0.979151	2003	0.986755
1972	0.970857	1988	0.980312	2004	0.987593
1973	0.968535	1989	0.977939	2005	0.987677
1974	0.972632	1990	0.981343	2006	0.990230
1975	0.969443	1992	0.983999	2007	0.992206
1976	0.969443	1992	0.983996	2008	0.993382
1977	0.966512	1993	0.985445	2009	0.994584
1978	0.970067	1994	0.983561	2010	0.994991
1979	0.973252	1995	0.984687	2011	0.995938

Table 4. The safety degree of staff systems in certain coal mine.

6.3. Analysis of safety production of China's coal mine industry

The trend chart of safety degree of China's coal mine industry can be derived from the data in **Table 4**. The trend chart shows that safety degree of China's coal mine industry has a sharp reduction from 1964 to 1971, has been stable from 1972 to 1978, but the overall trend has increased after 1979, which indicates that the safety production situation of China's coal mine enterprises improved. Its reason may be that the policy of reform and opening began and the economy developed sharply. The improvement of the safety production situation reflects the important role of the advanced technologies and safety management, while coal mine enterprises improved the work environment by strengthening the safety investment and improved employees' quality by strengthening safety training, which ultimately improved the safety degree of the staff system and promoted the improvement of enterprises' overall safety degree. In addition, the related policies, laws, and regulations for coal mines in China have played a significant role in promoting the safety production.

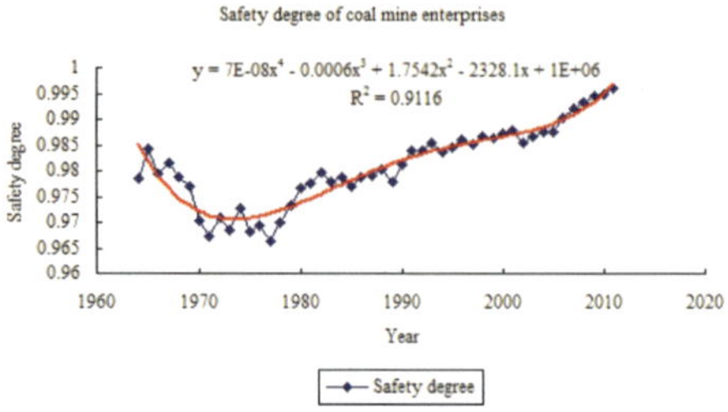

Figure 4. Trend chart of safety degree of China's coal mine enterprises in calendar year.

7. Conclusion and discussion

The safety degree of coal enterprise is defined, the calculation methods of safety degree are put forward, and empirical researches on the micro and macro view are done according to this method. Studies show that the calculation method of safety degree is valid and the safety degree reflects the situation of safety production of coal mine enterprises to a large extent and it is significant to quantify the safety problems of coal mines. By analysis of the results of empirical research it can be concluded that the reasons for the increase of the safety degree are due to the advancement of technology, strengthening of safety management and education, increasing of safety investment, and perfection of policies, laws, and regulations.

Acknowledgements

The authors gratefully acknowledge funding by Fundamental Research Funds for the Central Universities (no. 2009kz03) and the Innovation Foundation of CUMTB for PhD Graduates (no. 800015Z602).

Author details

Nie Baisheng[1,2*], Huang Xin[1,2], Wang Longkang[3], Yu Hongyang[1,2] and Li Xiang Chun[1,2]

*Address all correspondence to: bshnie@cumtb.edu.cn

1 School of Resources and Safety Engineering, China University of Mining and Technology (Beijing), Beijing, China

2 State Key Lab of Coal Resources and Safe Mining, China University of Mining and Technology (Beijing), Beijing, China

3 School of Management, China University of Mining and Technology (Beijing), Beijing, China

http://dx.doi.org/10.5772/66800

References

Golbraikh, et al. 2003. Rational selection of training and test sets for the development of validated QSAR. Journal of Computer-Aided Molecular Design 17, 241–253.

Denby, M.S. Kizil, 1992. Application of expert systems in geotechnical risk assessment for surface coal mine design. International Journal of Rock Mechanics and Mining Science & Geomechanics Abstracts 2(2), 110.

Knegtering, A.C. Broracher, 1999. Application of micro Markov models for quantitative safety assessment to determine safety integrity levels as defined by the IEC61508 standard for functional safety, Reliability Engineering and System Safety 66, 171–175.

BP Group, 2015. BP Statistical Review of World Energy, Accessed July 30, 2015, http://www.bp.com/zh_cn/china/reports-and-publications/_bp_2015.html.

E.S. Huang, R. Samudrala, J.W. Ponder, 1999. Abinitio fold prediction of small helical proteins using distance geometry and knowledge-based scoring functions. Journal of Molecular Biology 290, 267–281.

G. Baumont, F. Ménage, J.R. Schneiter, A. Spurgin, A. Vogel, 2000. Quantifying human and organizational factors in accident management using decision tree: the HORAAM method. Reliability Engineering and System Safety 70, 113–124.

G. Hu, 2006. Research on the cause of coal mine accidents and safety management countermeasures of WALi. China University of Mining & Technology, Beijing (in Chinese).

H. Heinrich, 1980. Industrial Accident Prevention. 5th ed. New York: McGraw-Hill.

H. Liao, Q. Sun, Z. Zhang, J. Chen, J. Li, J. Guo, 2009. Analysis for cross-sectional survey data of occupational hazards in coal mine. Journal of Safety Science and Technology 5(5), 152–156.

H.S.B., 2005. Duzgun: Analysis of roof fall hazard and risk assessment for Zonguldak coal basin underground mines. International Journal of Coal Geology 64(2), 104–115.

J. Beugin, D. Renaux, L. Cauffriez, 2007. A SIL quantification approach based on an operating situation model for safety evaluation in complex guided transportation systems. Reliability Engineering and System Safety 92, 1686–1700.

J. Chen, 2012. Statistical Analysis on China Coal Mine Accident and Forecasting Based on Optimal Combination Prediction Model. TaiYuan University of Technology (in Chinese).

K.S. Benjamin, I. Mukherjee, 2011. Conceptualizing and measuring energy security: A synthesized approach. Energy 36, 5343–5355.

L. Wang, B. Nie, J. Zhang, et al., 2016. Study on coal mine macro, meso and micro safety management system. Perspectives in Science 7, 266–271.

M. Čepin, B. Marko, 2002. A dynamic fault tree. Reliability Engineering and System Safety 75, 83–91.

P.S. Paul, J. Maiti, 2007. The role of behavioral factors on safety management in underground mines. Safety Science 45, 449–471.

R. Bojanc, B. Jerman-Blazic, M. Tekavcic, 2012. Managing the investment in information security technology by use of a quantitative modeling. Information Processing and Management 48, 1031–1052.

S. Mahdevari, K. Shahriar, A. Esfahanipour, 2014. Human health and safety risks management in underground coal mines using fuzzy TOPSIS. Science of the Total Environment, 488–489, 85–99.

S. Xu, 2014. Study on occupational safety and health mechanism for school from social responsibility point of view. China Safety Science Journal 24(6), 129–134.

V.K. Vivek, J. Maiti, P.K. Ray, 2011. A methodology for evaluation and monitoring of recurring hazards in underground coal mining. Safety Science 49, 1172–1179.

W. Hatton, M.K.G. Whateley, 1995. Risk assessment applied to coal tonnage estimation in the United Kingdom. International Journal of Rock Mechanics and Mining Science & Geomechanics 32(6), 276.

Y. Wang, T. Teague, H. West, S. Mannan, 2002. A new algorithm for computer-aided fault tree synthesis. Journal of Loss Prevention in the Process Industries 15, 265–277.

Chapter 25

The Elastic Deformation of Soil Around Models of Rigid Slab and Raft

Kamil Burkovic and Martina Janulikova

Additional information is available at the end of the chapter

http://dx.doi.org/10.5772/66802

Abstract

Load tests and the results of the numeric modeling represent this chapter. Two kinds of foundation models were tested. Tests started during the year 2014 in the Faculty of Civil Engineering, VŠB–Technical University of Ostrava. Test equipment STAND, located in the campus of the Technical University, is primarily designed to test the interaction of the foundations and the subsoil. Presented are the data processed experimental measurement of the adjacent terrain around the foot and raft. The measured data will be verified by theoretical calculations using the MIDAS GTS based on the method of finite elements. Tests of several foundation constructions were carried out. The aim of this test was to verify the behavior of raft foundation by comparing to equivalent surface foundation and a pilot. Three types of foundation constructions were examined. Reinforced concrete foundation slab, raft foundation (made of reinforced concrete foundation slab supported by drilled reinforced concrete pilot), and a separate reinforced concrete drilled pilot. All these types of foundation constructions were constructed as models, in a reduced scale, approx. 1:10. The size had to be adjusted due to limited capacity of the testing device and financial reasons. The measurements were carried out by the STAND device in the area of VŠB-TU Ostrava. The values of load (vertical point force) and vertical deformations (subsidence) were measured with the individual tested models. Beside the main task, concerning the measuring of the behavior of the foundation constructions, there was also carried out measurement of the behavior of the adjacent terrain. The aim of this chapter is to compare the behavior of adjacent terrain near the model of rigid slab and the model of the raft. The measurement results are compared with the results of numerical modeling.

Keywords: slab-soil system, finite element method, settlement, measurement, foundation

1. Measurement

STAND is a device (see **Figure 1**) which serves to research the interactions between the foundation constructions and the subsoil (Buchta et al., 2016; Cajka, 2014; Cajka et al., 2016a,b; Mynarcik et al., 2016). The loading equipment (hydraulic presses) can be freely anchored to the cross members. The position of the presses can be arbitrarily changed in the range of the ground area. The subsoil is a homogeneous clay with high plasticity and hard to firm consistency in the depth down to 5.0 m. All the measured models were subjected to a load test by vertical point force which was transmitted in the construction by steel boards of a hydraulic press. The load values (vertical point force) and vertical deformation (subsidence) of the individual tested models and foundation constructions were measured. Measurement of the raft was carried out in August 2014, and measurement of separate slab in July 2014. The hydraulic press was anchored by steel attachments and washers to the construction of the STAND. The beams for installation of the meters track were located on both sides of the press. Beams for installation of displacements sensors were supplemented by cross beams for adjacent terrain deformation measurements. For the measurement of the raft four displacement sensors were used, soil was measured by seven sensors (see **Figure 2**), to a distance of 600 mm.

Figure 1. The STAND–measuring equipment.

Four displacement sensors were used for the measurement of the slab; soil was measured by eight sensors. Seven sensors were placed to a distance of 600 mm. The last sensor was at a distance of 1.0 m from the foundation (see **Figure 3**). Cylinder pressure sensor and the displacement sensors were connected to the data bus. The measurements were carried out in 5-min cycles after 20 kN. The measurements were completed when there was a break of the plate and no more high pressure level in the hydraulic press could be achieved. After measuring the pressure in the press was released and creep of subsoil was measured. Deformation of the land around the raft (see **Figure 4**) and the slab (see **Figure 5**) are drawn in the following graphs. The graphs record the data of the individual sensors during the load tests. The graphs in **Figure 6** and **Figure 7** show pushing of the surrounding soil at the edges of the foundation, roughly up to 200–250 mm away (by raft) and roughly up to 100–150 mm away (by slab). Behind this line the soil is lifted by the push of the subsoil.

http://dx.doi.org/10.5772/66802

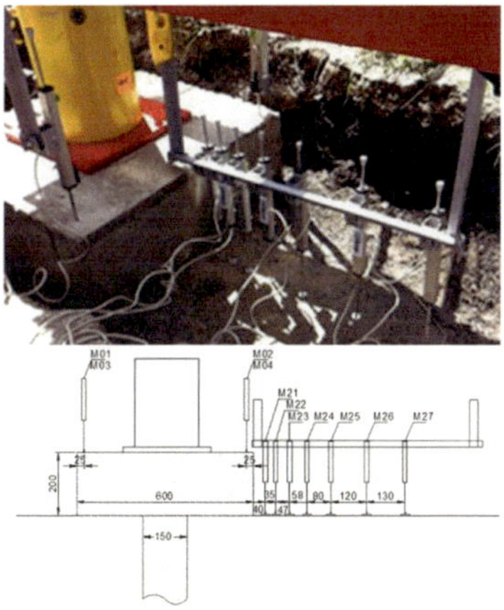

Figure 2. Raft model measurement with displacement sensors.

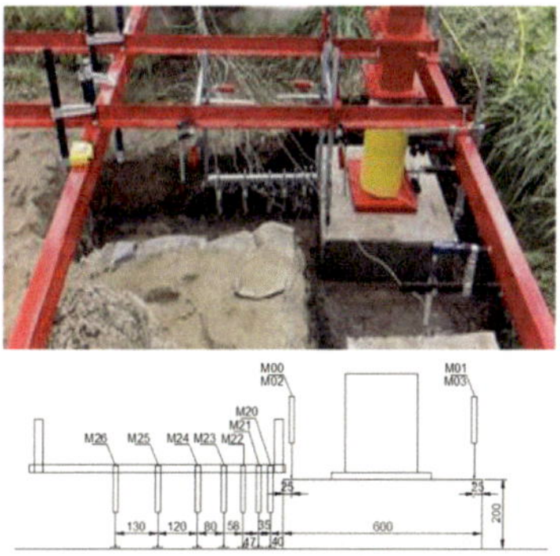

Figure 3. Slab model measurement with displacement sensors.

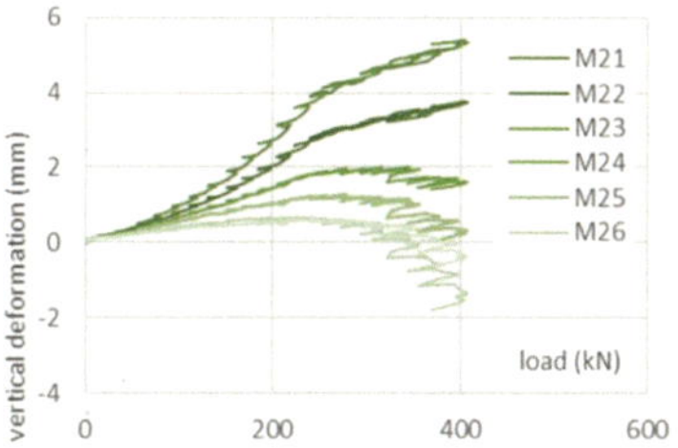

Figure 4. Load-deformation dependence of the raft.

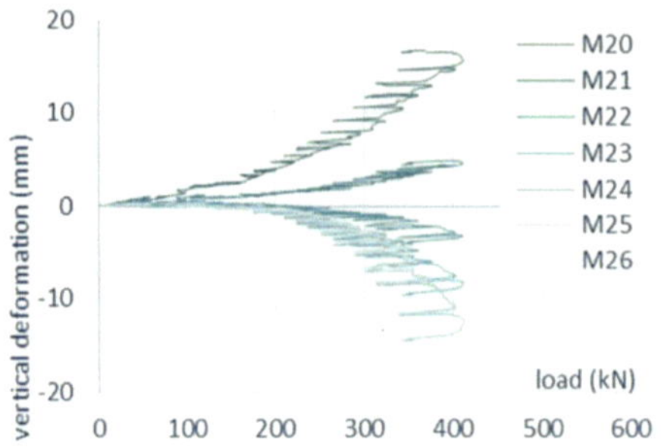

Figure 5. Load-deformation dependence of the slab.

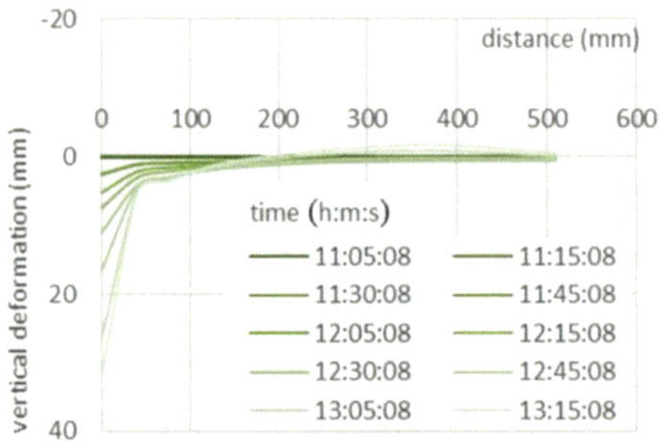

Figure 6. Deformation-time-distance–raft.

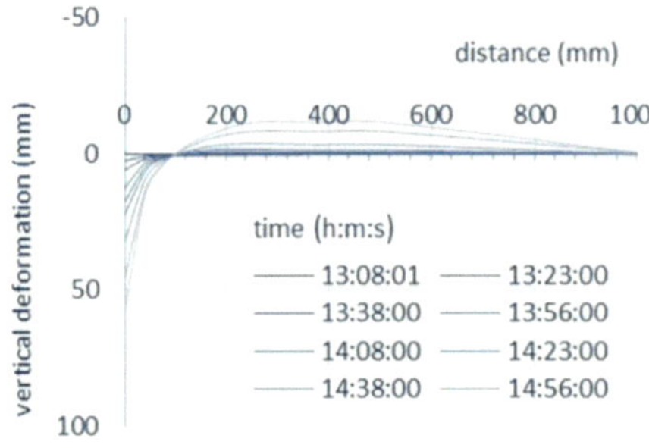

Figure 7. Deformation-time-distance–slab.

2. Mathematic modeling of the deformation of the terrain

The analysis of the behavior of the foundation slab and raft was carried out by the final element method, using software MIDAS GTS NX. A 3D model of the slab and the subsoil was made (Hrubesova et al., 2015). The model was used for the bedrock "Mohr-Coulomb," which

http://dx.doi.org/10.5772/66802

is usually used for analysis of the elastic behavior of the bedrock. The modulus of elasticity of concrete was considered at 27.0 GPa, and the compressive strength at 20 MPa. The point load entering the calculations corresponds with the force use data in the load test. It ranged equally from 0 to 400 kN. As the force was transmitted by the steel slab board measuring 0.2 × 0.2 m, equal surface load rating from 0 to 10 MPa was used (Burkovic and Duris, 2015). The calculation was carried out excluding the influence of the groundwater.

3. Slab foundation

Elastic behavior of the foundation slab material is assumed. The concrete slab and the subsoil were made by a hybrid network of final elements with automatically generated contact points between the slab and the subsoil. The calculation run in two phases (construction stage). In the first phase, the calculation of the weight of the slab itself and the subsoil was carried out and the deformations were then reset at zero.

The following phase of nonlinear analysis was carried out in 10 iteration steps. Clearly, schemes of vertical deformations of the terrain of separate iterative steps were received from the calculations in **Figure 8**. Graphs of deformations for comparison of the measured values were made from the adjacent points in **Figure 9**.

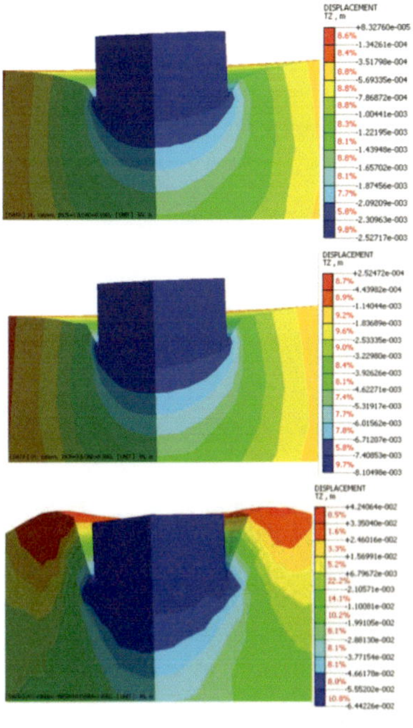

Figure 8. Deformation of the terrain–steps 1, 3, and 10.

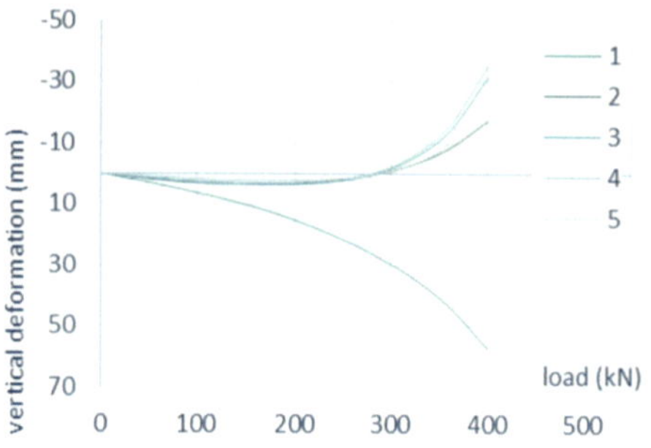

Figure 9. Load-deformation dependence–slab.

4. Raft foundation

Feet and bedrock were made as the 3D element (box). The pilot was modeled as a 1D element (truss). The 3D element of the bedrock has been divided into a finite number of elements in the system (hybrid mesher). The material was used for shoe model of elastic behavior. The concrete was used for the pilot model of the behavior of an elastic material, which does not use with material nonlinearity (pile element) for defining the sheath of friction piles. The calculation took place in two stages (construction stage). The first phase was carried out with the calculation of the own weight of the shoe and the subsoil and deformation were subsequently cleared. The subsequent phase of nonlinear analysis was carried out in 10 iterative steps. Understandable schemes of vertical deformations of the terrain of separate iterative steps were received from the calculations in **Figure 10** and **Figure 11**. Graphs of deformations for comparison of the measured values were made from the adjacent points in **Figure 12**.

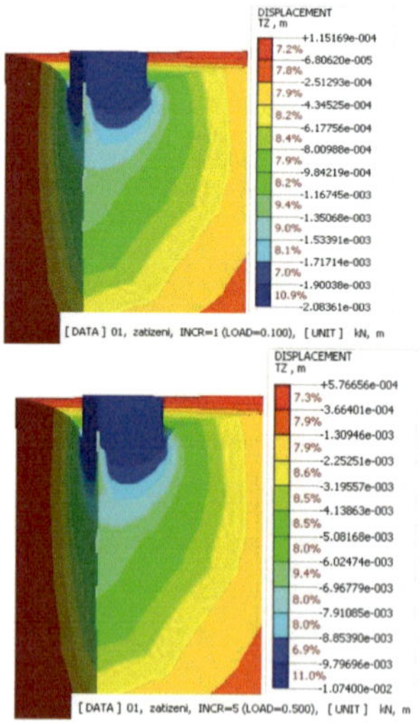

Figure 10. Deformation of the terrain–steps 1 and 5.

http://dx.doi.org/10.5772/66802

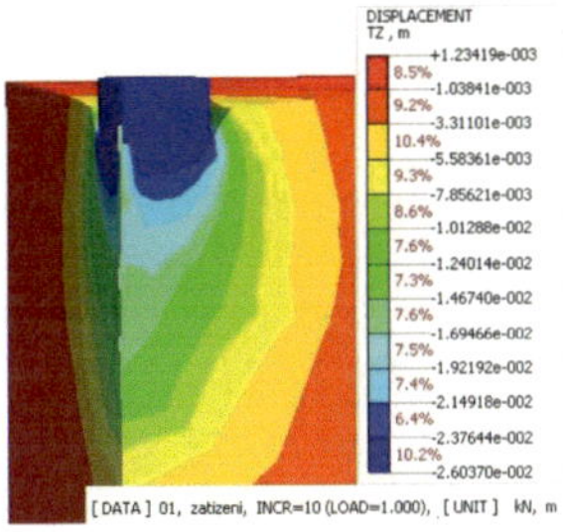

Figure 11. Deformation of the terrain–step 10.

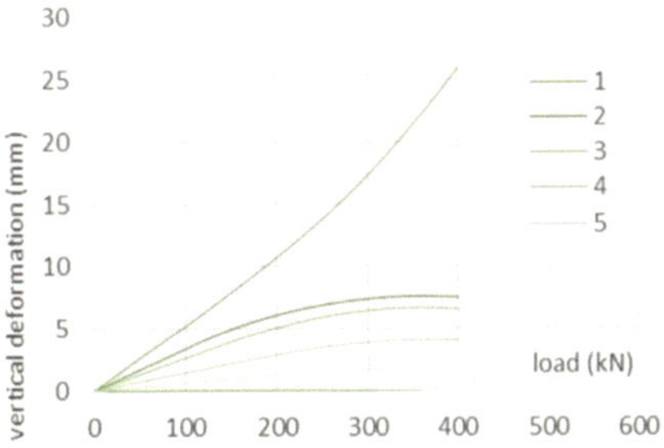

Figure 12. Load-deformation dependence–raft.

5. Comparison of measurements and conclusion

Understandable comparing graph was set from the data gained by the load test and the values gained from the calculations. You can compare the progress of the measured maximum values of adjacent terrain deformation. **Figure 13** shows recorded waveforms of the maximum deformation of adjacent terrain, slab, and raft. Furthermore, you can compare the values measured and calculated values of settlement adjacent terrain at the slab and raft in **Figure 14** and **Figure 15**.

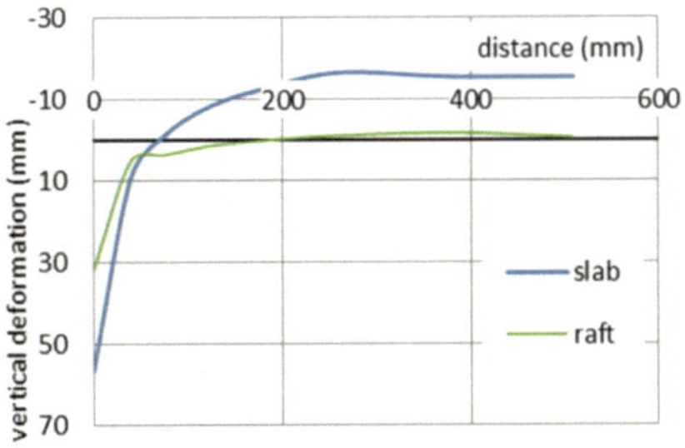

Figure 13. Maximum deformation of adjacent terrain.

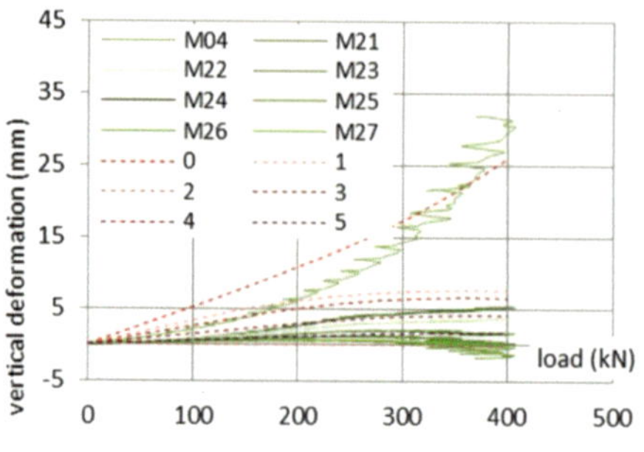

Figure 14. Comparison of results–slab.

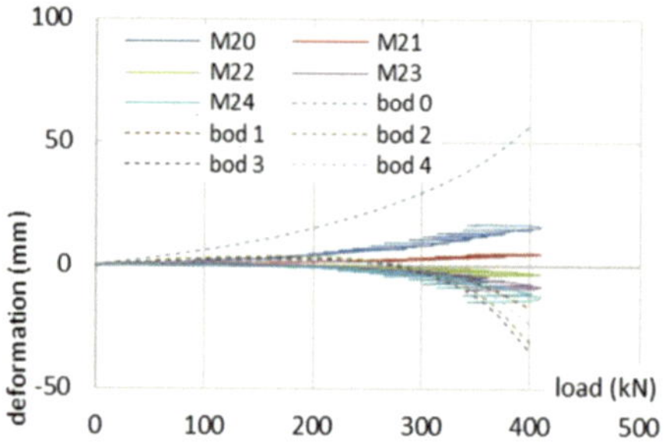

Figure 15. Comparison of results–raft (right).

These disproportions are caused by incorrect entering values as well as selected number of final elements and, above all, the behavior of the concrete foundation slab at the end of the experiment. The slab behaves in the calculation as a solid body.

Acknowledgments

The works were supported from sources for conceptual development of research, development, and innovations for 2016 at the VSB-Technical University of Ostrava that were granted by the Ministry of Education, Youths and Sports of the Czech Republic.

Author details

Kamil Burkovic* and Martina Janulikova

*Address all correspondence to: kamil.burkovic@vsb.cz

Faculty of Civil Engineering, VSB-Technical University of Ostrava, Ostrava, Poruba, Czech Republic

http://dx.doi.org/10.5772/66802

References

Vojtech Buchta, Martina Janulikova, Roman Fojtik. Experimental tests of reinforced concrete foundation slab. Procedia Engineering, 114, 530–537, 2016.

Radim Cajka. Comparison of the calculated and experimentally measured values of settlement and stress state of concrete slab on subsoil. Applied Mechanics and Materials, 501–504, 867–876, 2014.

Radim Cajka, Petr Mynarcik, Jana Labudkova. Experimetal measurement of soil-prestressed foundation interaction. International Journal of GEOMATE, 10, 2101–2108, 2016.

Petr Mynarcik, Jana Labudkova, Jiri Koktan. Experimental and numerical analysis of interaction between subsoil and post-ensioned slab-on-ground. Jurnal Teknologi, 78, 23–27, 2016.

Eva Hrubesova,Hynek Lahuta, Lukas Duris, Mohammed Jaafar. Mathematical modeling of foundation-subsoil interaction. 15th International Multidisciplinary Scientific GeoConference SGEM 2015, SGEM2015 Conference Proceedings, 2, 437–444, 2015.

Kamil Burkovic, Lukas Duris. Experimental Modeling and Verification of Data Models for Foundation Elements. Proceedings of the Fifteenth International Conference on Civil, Structural and Environmental Engineering Computing. Civil-Comp Press, Stirlingshire, UK, 108, 2015.

Radim Cajka, Jana Labudkova, Petr Mynarcik. Numerical solution of soil–foundation interaction and comparison of results with experimental measurements. International Journal of GEOMATE, 11, 2116–2122, 2016.

Radim Cajka, Jana Labudkova. Numerical modeling of the subsoil-structure interaction. Key Engineering Materials, 691, 333–343, 2016.

Jana Labudkova Radim Cajka. Comparison of the results from analysis of nonlinear homogeneous and nonlinear inhomogeneous half-space. Procedia Engineering, 114, 522–529, 2015.

Chapter 26

Influence of Contact Stress Model on the Stability of Bridge Abutment

Radim Cajka, David Pustka and David Sekanina

Additional information is available at the end of the chapter

http://dx.doi.org/10.5772/66803

Abstract

This chapter deals with the behaviour of an abutment pier on subsoil subjected to flood changes. The floods increase the cross-section of the river bed and change the properties of the foundation soil under the foundation. First, the soil saturates with water. Then, fine-grained particles will wash away and finally parts of the basement rock will be washed off. Finite element method has been used for the calculation of the interaction between the foundation and the subsoil. The foundation has been modelled in a 2D environment using spatial components. For the subsoil, an element with effects of an elastic foundation has been used. The stiffness of the bedrock has been characterized by the *C* parameter. The chapter describes situations related to the collapse of the structure.

Keywords: bridge, abutment pier, basement rock, floods, soil-structure, interaction

1. Introduction

Unflagging growth of anthropogenic activities has been causing changes in the Earth's climate. These changes have led to the changes of weather in comparison to the past. Changes in weather frequently have brought increased values of loads (e.g. due to wind, snow and water) which can significantly influence reliability (see, e.g. Tikalsky et al., 2005; Pustka et al., Raizer, 2009; Briaud et al., 2014; Králik and Králik, 2014; Markova et al., 2014; Pustka, 2014; Janas et al., 2015; Pustka, 2015; Koteš et al., 2016) of (civil) engineering structures. To assure required level of reliability of these structures, it is necessary to deal with this issue. Climate's changes have brought, among others, heavier precipitations which have led to excessive water flows or even to floods. This unexpected flows of water can significantly damage bridge structures crossing these watercourses (see, e.g. Cajka and Manasek, 2005; Link et al., 2008; Pasiok and

Stilger-Szydlo, 2010; Burns et al., 2011; Wang et al., 2011; Yu et al., 2011; Khosronejad et al., 2012; Collins et al., 2013; Lin et al., 2014; Afzali, 2015; Ehteram and Meymand, 2015; Klinga and Alipour, 2015; Fael et al., 2016; Mohamed et al., 2016). In association with this growing risk, a study examining effects of scour to a bridge abutment was elaborated.

In the following model, an example of a bridge pier (Strasky et al., 2001; Navratil, 2004; CNI, 2005; Parke and Nigel, 2008; Navratil and Zich, 2013; Sucharda and Brozovsky, 2013) is considered. To analyse interaction between the basement rock and foundation (see, e.g. CNI, 1988; CNI, 2004; Cajka et al., 2011; Cajka, 2013a,b,c; Cajka et al., 2014; Unlu et al., 2013; Hrubesova et al., 2015; Lahuta et al., 2015; Hrubesova et al., 2016; Cajka et al., 2016a,b; Labudkova and Cajka, 2016) a parametric study has been created. In the study, the finite element method on elastic subsoil has been utilised. The floods increase the cross-section of the river bed and change the properties of the foundation soil under the foundation (see, e.g. Ettema et al., 2000). In the first stage, the soil saturates with water. In the second stage fine-grained particles will wash away. In the third stage, parts of the basement rock will be washed off.

2. Model example of an abutment pier

2.1. Assumptions of calculation

For the calculation of interaction between the foundation and basement finite element method has been used (FEM consulting, 2002). The foundation has been modelled in a 2D environment using spatial components. For the basement rock, an element with effects of an elastic foundation has been used. The *C* parameter represents properties of the basement rock.

2.2. Subsoil model

The most efficient way for solutions of interaction tasks is a 2D model of the basement rock. Such model represents correctly, through a surface model, deformation properties of the whole mass of the foundation soil. The physical properties are expressed by means of subsoil parameters. The set of the interaction parameters is marked briefly as *C*. The parameters are allocated directly to structure components that are in the contact with the basement rock. The parameters describe the properties that influence the stiffness matrix. To simplify the situation, the *C* parameter can be imagined as the supporting by means of a dense liquid $\gamma = C_{1z}$ (MN m^{-3}) or by means of a set of vertical springs with an infinite density. From the physical point of view, there is not any difference. In case of extreme simplification, the *C* parameter can be imagined as Winkler's elastic foundation model.

2.3. Modelling and description of the structure

As a material for the foundation concrete C16/20 has been considered. Dimensions of the abutment pier are evident from **Figure 1**. The pier has been loaded by the horizontal load-carrying structure of the bridge (forces R_{gk} and R_{qk}). The load developed by the soil and random load of the road that influences the back face of the pier structure, have been introduced by H_k force (see **Figure 1**).

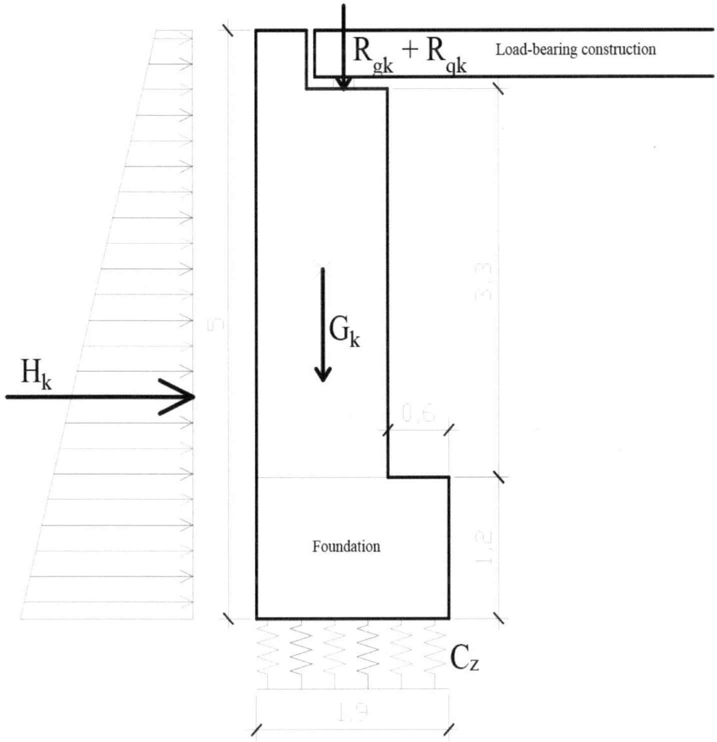

Figure 1. Scheme of the abutment pier with considered loads.

As far as the structure of the abutment pier is concerned, the foundation structure has been used only for the calculation. The loading of the whole upper construction has been re-calculated and simplified. Only the vertical loading and bending moment in the centre of gravity of the stem have been taken into consideration. The basement rock has been modelled using the C_z parameter. For purposes of the calculation, the following reference value has been used: C_z = 25 MN m^{-3}. This rough value is given by characteristic of gravel with fine-grain particles and by the loading and deformation for a specific type of the basement rock. The interaction has been solved for several cases: the value of C_z has changed because of the lower stiffness of the basement rock that was caused by the washing off of the fine-grain particles. In another case, the washing off of the basement rock has been taken into consideration. Finally, the combination of the both cases above has been investigated. **Figure 2** shows the foundation with considered distributions of the basement rock stiffness C_z.

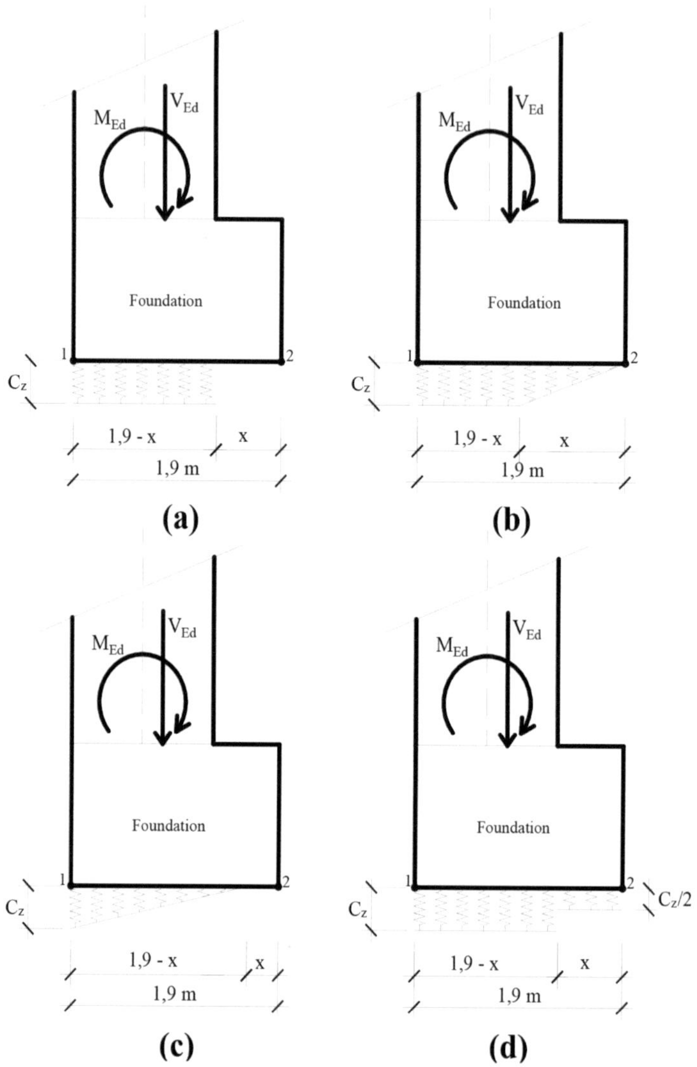

Figure 2. The foundation with considered distributions of the basement rock stiffness C_z.

2.3.1. Partial loss of contact between the foundation and basement rock

The flow of water washes away the basement rock. This reduces the contact surface resulting in increase of the stress in the foundation joint. Because of the non-homogeneous distribution of the tension in the foundation joint, the settlements in points 1 and 2 (see **Figure 2**) are

http://dx.doi.org/10.5772/66803

different. Consequently, the foundation joint rotates. **Table 1** shows the settlements of the pier in the points 1 and 2 and the total rotation. Assumed deformation of the foundation is shown in **Figure 3**. Rotation is calculated according to Eq. (1):

$$\varphi = arctg\frac{\Delta w}{b} \tag{1}$$

x [m]	Origin (1.9 – x) [m]	w_1 [mm]	w_2 [mm]	$\Delta w = w_2 - w_1$ [mm]	Rotation of foundation [deg]	Max. stress on foundation surface [MPa]
0.0	1.9	6.92	11.98	5.06	0.152	0.299
0.1	1.8	5.50	14.95	9.46	0.285	0.361
0.2	1.7	3.68	19.07	15.39	0.464	0.435
0.3	1.6	1.35	24.81	23.46	0.707	0.526
0.4	1.5	–1.66	32.88	34.54	1.042	0.638
0.5	1.4	–5.57	44.37	49.94	1.506	0.778
0.6	1.3	–10.71	60.99	71.70	2.163	0.955
0.7	1.2	–17.56	85.45	103.01	3.109	1.182
0.8	1.1	–26.83	122.29	149.12	4.506	1.479
0.9	1.0	–39.65	179.31	218.96	6.632	1.877
1.0	0.9	–57.82	270.62	328.44	10.004	2.426

Table 1. Deformation of the foundation for the case 'a'.

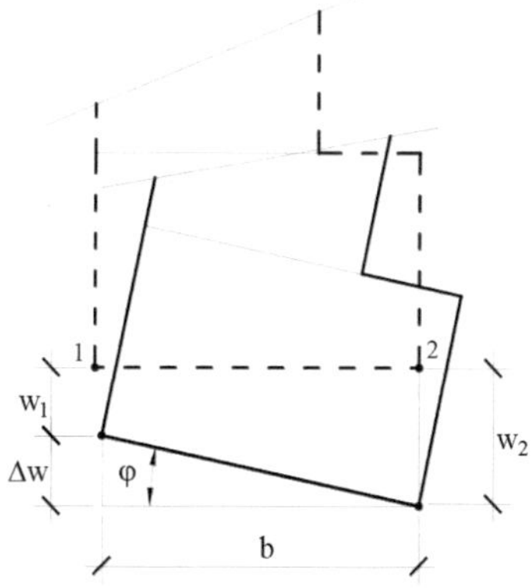

Figure 3. Assumed deformation of the foundation.

Figure 2 shows the x-coordinates related to considered distribution of the basement rock stiffness C_z. When the contact surface is reduced to a certain level, tensile forces are generated. Elements, where tensile stress appeared, have been excluded from the calculation. **Figure 4**

shows the chart for the calculation where the tension in the contact surface is taken into account (case (a*)). For case (a), an iteration method has been used.

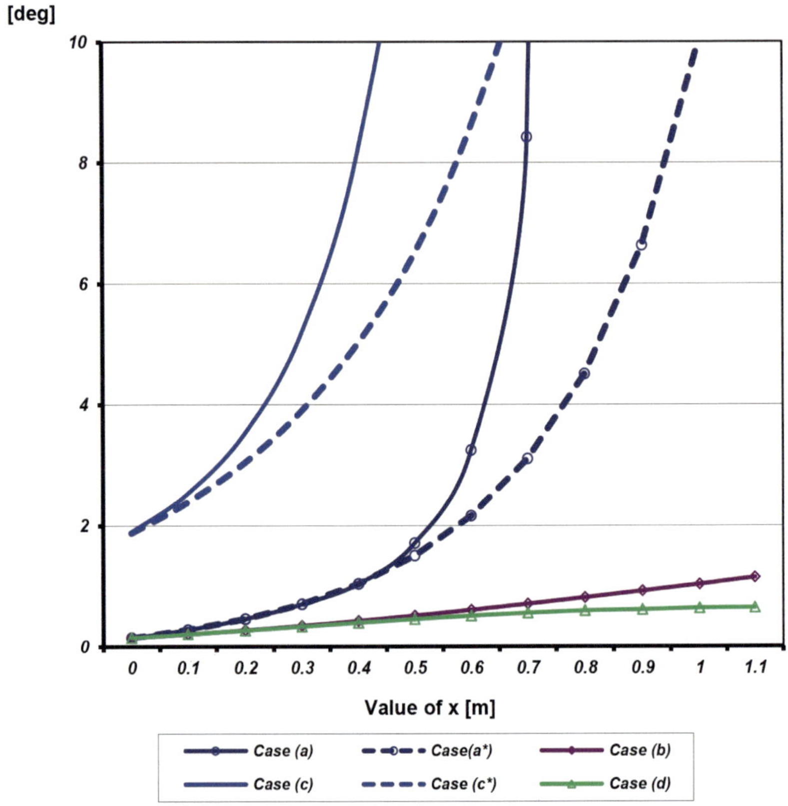

Figure 4. Dependency of the rotation of the foundation surface for x values.

2.3.2. *Gradual decrease in the stiffness of the basement rock*

In case (b) (see **Figure 2**), the interaction parameter C_z decreases gradually. The development of the C_z values is constant up to the place that is, in all likelihood, affected by water penetration. From that point onwards, the stiffness is linear up to the point 1 where the stiffness of the basement rock is assumed to be zero. Resulting values are listed in **Table 2**. The development of the values is shown in **Figure 4**.

2.3.3. *Gradual washing-away of soil and washing-off of fine-grain particles*

Combination of both the previous situations represents the case 'c'. Here, the C_z is considered to be constant below the point 1 (see **Figure 2**). The soil is washed off gradually, thus decreasing C_z. Resulting values are listed in **Table 3**. From the chart in **Figure 4**, it is evident

http://dx.doi.org/10.5772/66803

that the tensile stress in the contact surface appears as early as in the first phase. The procedure has been similar to that used in case (a). An iteration method has been used for case (c). The case (c*) describes the situation where the basement rock is subjected to the tension.

x [m]	Origin (1.9 – x) [m]	w_1 [mm]	w_2 [mm]	$\Delta w = w_2 - w_1$ [mm]	Rotation of foundation [deg]	Max. stress on foundation surface [MPa]
0.0	1.9	6.92	11.98	5.06	0.152	0.299
0.1	1.8	6.29	13.30	7.02	0.212	0.321
0.2	1.7	5.60	14.81	9.21	0.278	0.344
0.3	1.6	4.88	16.52	11.64	0.351	0.364
0.4	1.5	4.12	18.42	14.31	0.431	0.381
0.5	1.4	3.33	20.54	17.21	0.519	0.396
0.6	1.3	2.53	22.85	20.32	0.613	0.405
0.7	1.2	1.74	25.36	23.63	0.712	0.410
0.8	1.1	0.96	28.05	27.09	0.817	0.409
0.9	1.0	0.22	30.90	30.69	0.925	0.401
1.0	0.9	–0.35	33.78	34.13	1.029	0.392

Table 2. Deformation of the foundation for the case 'b'.

x [m]	Origin (1.9 – x) [m]	w_1 [mm]	w_2 [mm]	$\Delta w = w_2 - w_1$ [mm]	Rotation of foundation [deg]	Max. stress on foundation surface [MPa]
0.0	1.9	–3.57	59.18	62.75	1.893	0.38
0.1	1.8	–9.10	74.74	83.84	2.530	0.427
0.2	1.7	–18.68	98.10	116.78	3.526	0.49
0.3	1.6	–36.13	135.23	171.35	5.181	0.574
0.4	1.5	–70.31	199.51	269.82	8.192	0.691
0.5	1.4	–141.30	319.79	461.09	14.184	0.861
0.6	1.3	–323.69	602.49	926.18	30.374	1.182
0.7	1.2	–669.93	1253.43	1923.36	91.693	2.009

Table 3. Deformation of the foundation for the case 'c'.

2.3.4. Step decrease in the parameters of the basement rock

Because the soil is saturated with water and fine-grain particles have been washed off, the stiffness will decrease (see **Figure 2**). In contrast to the calculation with the linear distribution (case 'b'), a step division of C_z has been chosen. When modelling by means of two parameters, the entering of values is simpler and faster. When modelling the linear development, the entering of values is more complex and C_z is different for each element. **Table 4** and **Figure 4** give the values for case 'd.

x [m]	Origin (1.9 – x) [m]	w_1 [mm]	w_2 [mm]	$\Delta w = w_2 - w_1$ [mm]	Rotation of foundation [deg]	Max. stress on foundation surface [MPa]
0.0	1.9	6.92	11.98	5.06	0.152	0.299
0.1	1.8	6.29	13.30	7.02	0.212	0.321
0.2	1.7	5.65	14.75	9.09	0.274	0.342
0.3	1.6	5.06	16.27	11.21	0.338	0.369
0.4	1.5	4.52	17.84	13.31	0.401	0.372
0.5	1.4	4.09	19.39	15.30	0.461	0.38
0.6	1.3	3.77	20.87	17.10	0.516	0.382
0.7	1.2	3.59	22.24	18.65	0.562	0.379
0.8	1.1	3.56	23.45	19.88	0.600	0.372
0.9	1.0	3.69	24.17	20.48	0.618	0.36
1.0	0.9	3.96	25.29	21.33	0.643	0.346

Table 4. Deformation of the foundation for the case 'd'.

3. Conclusion

Figure 4 summarises the results of the conditions described above. Also, the chart shows the rotation of the foundation surface. **Table 1–4** can be used to determine the values for a specific case and to determine the maximum stress that appears in the contact surface. The structure collapses if the basement rock plasticizes and the load-carrying capacity is lost. According to the limiting rotation requirements by CNI (1988), the ratio $\Delta w/b = 0.003$ applies to the concrete foundation structure. The rotation angle is $\phi = 0.17°$. It follows from the calculation that the structure does not meet this requirement when the x-parameter (case 'b') decreases below the foundation surface 0.1 m. This is the beginning of the condition when the fine-grained particles start washing away. Most adverse results occur in the case 'c' when the lower stiffness of the basement rock is combined with the loss of contact with the basement rock. Because of the lost contact between the foundation and basement rock, the stress re-distributes and tensile stress appear in the contact surface. It is clear from the chart that there is a difference in the calculations (case 'a' and case 'c') where the tensile stress is, or is not, considered for the contact surface. The situation where the tensile stress exists is marked with an asterisk. The results are absolutely different. Therefore, the tensile stress in the foundation surface should not be taken into account.

Conflict of interest

The authors declare that there is not conflict of interest.

http://dx.doi.org/10.5772/66803

Acknowledgement

The work was supported from sources for conceptual development of research, development and innovations 2016 at the VŠB-Technical University of Ostrava that were granted by the Ministry of Education, Youths and Sports of the Czech Republic. In this undertaking, theoretical results gained in the project GAČR 16-08937S were partially exploited.

Author details

Radim Cajka, David Pustka* and David Sekanina

*Address all correspondence to: david.pustka@vsb.cz

Faculty of Civil Engineering, VSB-Technical University of Ostrava, Ostrava, Poruba, Czech Republic

References

S. H. Afzali. New model for determining local scour depth around piers. Arabian Journal for Science and Engineering, 1–9, 2015, DOI: 10.1007/s13369-015-1983-4.

J. L. Briaud, P. Gardoni, and C. Yao. Statistical, risk, and reliability analyses of bridge scour. Journal of Geotechnical and Geoenvironmental Engineering, 140 (2), art. no. 0000989, 2015, DOI: 10.1061/(ASCE)GT.1943-5606.0000989.

S. E. Burns, S. K. Bhatia, C. M. C. Avila, B. E. Hunt, B.E., (eds.). Scour and Erosion–Proceedings of the Fifth International Conference on Scour and Erosion (ICSE-5). American Society of Civil Engineers (ASCE), 2011.

R. Cajka, P. Manasek. Building structures in danger of flooding. In: IABSE Symposium Report. International Association for Bridge and Structural Engineering, New Delhi, India. WOS: 000245746100072, 2005.

R. Cajka, V. Krivy, D. Sekanina. Design and development of a testing device for experimental measurements of foundation slabs on the subsoil. Transactions of the VŠB – Technical University of Ostrava, Civil Engineering Series, 11(1), 1–5, 2011, DOI: 10.2478/v10160-011-0002-2.

R. Cajka. Accuracy of stress analysis using numerical integration of elastic half-space. Applied Mechanics and Materials, 300–301, 1127–1135, 2013a, DOI: 10.4028/www.scientific.net/AMM.300-301.1127

R. Cajka. Analysis of stress in half-space using Jacobian of transformation and gauss numerical integration. Advanced Materials Research, 818, 178–186, 2013b, DOI:10.4028 /www.scientific.net/ AMR.818.178.

R. Cajka. Analytical derivation of friction parameters for FEM calculation of the state of stress in foundation structures on undermined territories. Acta Montanistica Slovaca, 18(4), 254–261, 2013c.

R. Cajka, K. Burkovic, V. Buchta and R. Fojtik. Experimental soil–concrete plate interaction test and numerical models. Key Engineering Materials, 577–578, 33–36, 2014.

R. Cajka, J. Labudkova and P. Mynarcik. Numerical solution of soil–foundation interaction and comparison of results with experimental measurements, International Journal of Geomate, 11(1), 2116–2122, 2016.

R. Cajka, P. Mynarcik, and J. Labudkova. Experimental measurement of soil-prestressed foundation interaction, International Journal of GEOMATE, 10(4), 2101–2108, 2016.

CNI. ČSN 73 1001 Foundation of structure. Subsoil under shallow foundations, Prague, 1988 (in Czech).

CNI. ČSN EN 1997-1 Geotechnical design–Part 1: General rules, Prague, 2004 (in Czech).

CNI. ČSN EN 1992-2 Design of concrete structures–Concrete bridges–Design and detailing rules, Prague, 2005 (in Czech).

J. Collins, M. Steele, D. Wilkes, D. Ashurst and B. Harvey. Investigation into highway bridge damage and failures during the November 2009 Cumbria flood event. In: Forensic Engineering–Informing the Future with Lessons from the Past–Proceedings of the Fifth International Conference on Forensic Engineering organised by the Institution of Civil Engineers and held in London, UK. ICE Publishing, pp. 49–60, 2013.

M. Ehteram, A. M. Meymand. Numerical modeling of scour depth at side piers of the bridge, Journal of Computational and Applied Mathematics, 280, 68–79, 2015, http://dx.doi.org/10.1016/j.cam.2014.11.039.

R. Ettema, R. Arndt, P. Roberts and T. Wahl (eds.). Hydraulic Modeling–Concepts and Practice. American Society of Civil Engineers (ASCE), ISBN 0-7844-0415-1, 2000.

C. Fael, R. Lança and A. Cardoso. Effect of pier shape and pier alignment on the equilibrium scour depth at single piers, International Journal of Sediment Research, 1–7, 2016, http://dx.doi.org/10.1016/j.ijsrc.2016.04.001.

FEM consulting, 2002. Nexis 32 rel.3.50 - Soilin. Manual for software, SCIA group.

E. Hrubesova, H. Lahuta, L. Duris and M. Jaafar. Mathematical modeling of foundation-subsoil interaction. In: International Multidisciplinary Scientific GeoConference Surveying Geology and Mining Ecology Management, SGEM, 2 (1), pp. 437–444, 2015.

E. Hrubesova, J. Marsalek and J. Holis. The influence of inaccuracies of soil characteristics on the internal forces in the retaining wall. In: Advances and Trends in Engineering Sciences and Technologies–Proceedings of the International Conference on Engineering Sciences and Technologies, ESaT 2015, pp. 69–74, 2016.

P. Janas, M. Krejsa, V. Krejsa and R. Briš. Structural reliability assessment using direct optimized probabilistic calculation with respect to the statistical dependence of input variables. In: Safety and Reliability of Complex Engineered Systems–Proceedings of the 25th European Safety and Reliability Conference, ESREL 2015, pp. 4125–4132, 2015.

A. Khosronejad, S. Kang, and F. Sotiropoulos. Experimental and computational investigation of local scour around bridge piers, Advances in Water Resources, 37, 73–85, 2012, http://dx.doi.org/10.1016/j.advwatres.2011.09.013.

J. V. Klinga, A. Alipour. Assessment of structural integrity of bridges under extreme scour conditions. Engineering Structures, Volume 82, 55-71, http://dx.doi.org/10.1016/j.engstruct.2014.07.021, 2015.

P. Koteš, J. Vičan, and M. Ivašková, M. Influence of reinforcement corrosion on reliability and remaining lifetime of RC bridges. Materials Science Forum, 844, 89–96, 2016, DOI: 10.4028/www.scientific.net/MSF.844.89.

J. Králik and J. Králik Jr. Failure probability of NPP communication bridge under the extreme loads. Applied Mechanics and Materials, 617, 81–85, 2014, DOI: 10.4028/www.scientific.net/AMM.617.81.

J. Labudkova and R. Cajka. Experimental measurements of subsoil-structure interaction and 3D numerical models, Perspectives in Science, 7, 240–246, 2016, http://dx.doi.org/10.1016/j.pisc.2015.11.039.

H. Lahuta, E. Hrubesova, L. Duris, L. and T. Petrasova. Behaviour subsoil of slab foundation under loading, In: International Multidisciplinary Scientific GeoConference Surveying Geology and Mining Ecology Management, SGEM, 2 (1), pp. 119–126, 2015.

Ch. Lin, J. Han, C. Bennettand R. L. Parsons. Case History Analysis of Bridge Failures due to Scour. In: Proceedings of the International Symposium of Climatic Effects on Pavement and Geotechnical Infrastructure 2013. American Society of Civil Engineers (ASCE), pp. 204–2016, 2014.

O. Link, F. Pfleger and U. Zanke, U. Characteristics of developing scour-holes at a sand-embedded cylinder, International Journal of Sediment Research, 23(3), 258–266, 2008, http://dx.doi.org/10.1016/S1001-6279(08)60023-2.

J. Markova, M. Holicky, M. Sykora, and J. Kral. Probabilistic assessment of traffic loads on bridges. In: Safety, Reliability and Risk Analysis: Beyond the Horizon–Proceedings of the European Safety and Reliability Conference, ESREL 2013, pp. 2613–2618, 2014.

Y. A. Mohamed, G. M. Abdel-Aal, T. H. Nasr-Allah and A. A. Shawky. Experimental and theoretical investigations of scour at bridge abutment, Journal of King Saud University–Engineering Sciences, 28(1), 32–40, 2016, http://dx.doi.org/10.1016/j.jksues.2013.09.005.

J. Navratil. Structural analysis of bridges, legitimate conservatism and obsolete theories. Concrete Engineering International, 8(1), 17–19, 2004.

J. Navratil and M. Zich. Long-term deflections of cantilever segmental bridges. Baltic Journal of Road and Bridge Engineering, 8(3), 190–195, 2013, DOI: 10.3846/bjrbe.2013.24.

G. Parke, H. Nigel. ICE Manual of Bridge Engineering (2nd Edition). ICE Publishing, Thomas Telford Ltd, 1 Heron Quay, London E14 4JD, UK 2008.

R. Pasiok, and E. Stilger-Szydlo. Sediment particles and turbulent flow simulation around bridge piers, Archives of Civil and Mechanical Engineering, 10(2), 67–79, 2010, http://dx.doi.org/10.1016/S1644-9665(12)60051-X.

D. Pustka, R. Cajka, P. Marek and L. Kalocova. Multi-components load effect analysis on a slender reinforced concrete column using probabilistic SBRA method. In: EASEC-11– Eleventh East Asia-Pacific Conference on Structural Engineering and Construction. Taiwan, 19, pp. 334–335, 2008.

D. Pustka. Probabilistic reliability analysis of high-performance reinforced concrete beam using Matlab software. International Journal of Mechanics, 8, 101–111, 2014.

D. Pustka. Probabilistic approach to stability analysis of a reinforced concrete retaining wall. Advanced Materials Research, .1079–1080, 248–251, Trans Tech Publications, Switzerland, 2015, doi:10.4028/www.scientific.net/AMR.1079-1080.248.

V. Raizer. Reliability of Structures. Analysis and Applications. Backbone Publishing Company, Fair Lawn, USA, ISBN 978-09742019-7-9, 2009.

J. Strasky, J. Navratil and S. Susky. Applications of time-dependent analysis in the design of hybrid bridge structures. PCI Journal, 46 (4), 56–74, 2001.

O. Sucharda and J. Brozovsky. Bearing capacity analysis of reinforced concrete beams. International Journal of Mechanics, 7(3), 192–200, 2013.

P.J. Tikalsky, D. Pustkaand Pavel Marek. Statistical variations in chloride diffusion in concrete bridges. ACI Structural Journal, 102(3), 481–486, 2005.

T. Unlu, H. Akcin, and O. Yilmaz. An integrated approach for the prediction of subsidence for coal mining basins. Engineering Geology, 166, 186–203, 2013, DOI: 10.1016/j.enggeo.2013.07.014.

C.Y. Wang, J.H. Cheng, H.P. Shihand J.W. Chang. Ring columns as pier scour countermeasures, International Journal of Sediment Research, 26(3), 353–363, 2011, http://dx.doi.org/10.1016/S1001-6279(11)60099-1.

X. Yu, J. Tao and X. Yu. Comparison Study on Computer Simulations for Bridge Scour Estimation. In: Proceeding of Georisk 2011–Geotechnical Risk Assessment & Management. American Society of Civil Engineers (ASCE), pp. 1125–1132, 2011.

Chapter 27

Comparison of Properties of Concretes with Different Types and Dosages of Fibers

Vlastimil Bilek, Jan Hurta, Petra Done and
Libor Zidek

Additional information is available at the end of the chapter

http://dx.doi.org/10.5772/66804

Abstract

Concretes with PP fibers 12 mm, construction polymer fibers 25 mm, 3D steel fibers 25 mm, and steel microfibers 12 mm were prepared in dosages 0.5 and 1%. The mechanical properties (compressive strength, bending strength, fracture properties, and modulus elasticity) and the frost resistance of these concretes were tested and they are discussed. The behavior of these concretes is also discussed using graphs load vs. deflection. As bad results of frost resistance are sometimes recorded for concrete with fibers, this property is also evaluated. As was expected, mechanical properties are enhanced with the addition of suitable fibers. Frost resistance is usually comparative with concrete without fibers, but in the case of concrete with 1% of steel fibers, it is reduced.

Keywords: fiber-reinforced concrete, high-performance concrete, fibers, fracture, frost resistance

1. Introduction

Nowadays, fiber-reinforced concrete (FRC) is increasingly used for different applications—pavements, industrial floors, or high-performance concrete applications. The applications in high-performance concrete elements and structures are especially important because structures from HPC are subtle and there is not enough space for flexural reinforcements—stirrups—inside them. The question is which type of fibers and which dosage to apply. Although there has been some good experience with synthetic structural fibers [1], steel fibers are most widely used in these applications. The effect of the steel fiber is affected especially by the shape of the fibers; a more detailed study was performed, for example, by Pajak and Ponikiewski [2]. Corrugated or hooked fibers show the best effect. The next question is what the correct testing method is for the estimate of the fiber effect. Usually, prisms with the cross-section 150 × 150 mm are recommended [4]. But the subtle constructions have smaller dimensions. Yhang et al. [3] used 100 × 100 mm prisms for the testing of FRC with microwires. In this paper, prisms with the cross-section 80 × 80 mm are used because they correspond to produced element dimensions [1, 5]. In Ref. [4], a four-point bending test is recommended on an unnotched prism whose volume is nearly 16 L. In this paper, notched prisms 80 × 80 × 480 mm (volume 3.1 L) are used with the central notch and they are tested in a three-point bending. As expected, the results are different from that of Ref. [4] and not objective from the point of view of concrete properties, but they are more similar to a practical situation from the point of view of the arrangement of fibers.

The structural design of these subtle constructions is often very difficult and the best way to evaluate the properties are tests in a 1:1 scale, see also Refs. [5, 6].

There are different kinds of fibers for different purposes. For example, steel wires or structural synthetic fibers are designed for similar purposes, but it is clear that their performance will be different. The possibility to substitute different types of fibers with other ones and optimize their dosage was the main purpose of this paper.

2. Experimental procedure

2.1. Materials

The concrete was designed as a high-performance concrete for class C70/85 XF1 (see EN 206). Ordinary Portland cement CEM I 42.5 R produced in cement plant Mokra was used. Metakaolin was added for enhancing the workability of the mixture and also for enhancing strengths. Commercial polycarboxylether superplasticizer, drinkable water, and commonly produced aggregates—sand 0/4 and crushed granite (Litice nad Orlici quarry)—were also used. Water to binder ratio $w/(c+m) = 0.27$.

Different types of fibers were used—polypropylene 12 mm long fibers in dosage $0.5\%_{vol}$, so-called structural synthetic polymer 25 mm long fibers—0.5 and $1.0\%_{vol}$, 3D steel fibers 30 mm long 0.5, and $1.0\%_{vol}$ and steel microfibers 13 mm long 0.25 and $0.5\%_{vol}$. Properties of these fibers are presented in **Table 1**.

		Length	Diameter	Tensile strength	Modulus of elasticity
	Designation in this paper	[mm]	[mm]	[MPa]	[GPa]
PP micro fibers	PP	12	0.09	460	3.5
Structural fibers	S	25	1	650	5
Steel 3D fibers	D	30	0.54	1200	
Steel Microfibers	M	13	0.21	2750	200

Table 1. Type and properties of used fibers.

Concretes were mixed in a laboratory mixer; the volume of each batch was 30 L. Workability was measured using reverse Abrams cone. After mixing the concrete was placed into steel molds. After demolding at the age of 22–24 hours, specimens were stored in water $t = (20 \pm 2)°C$.

2.2. Specimens

Cubes 100 mm were used for compressive strengths tests at the age of 28 days. Prisms 80 × 80 × 480 mm were made for the testing of fracture properties and for the testing of frost resistance. A notch to depth cca 28 mm (1/3 of high) was cut into the beams 220 mm from the one end of the beam at the age of 28 days. Fracture tests in accordance with the Karihaloo and Nallathambi effective crack model [7] were performed on the notched beam (span 400 mm). The fracture toughness K_{IC} is the main result of these tests. Toughness G_c was also computed. Fracture work W_F was computed from the load-deflection curve in accordance with RILEM recommendation [8, 9]. The modulus of elasticity in a three-point bending on notched beam E as well as the modulus of rupture f_r (flexural strength on the notched beam) are the partial results of these tests. After the fracture test, a part of the broken beam, whose length is approximately 260 mm, was used for the test of flexural strength f_b (span 220 mm). Other beams 80 × 80 × 480 mm were exposed to 125 freezing and thawing cycles (FT-cycles) in accordance with Czech norm CSN 73 1322—frost resistance of concrete. One cycle represents 4 hours in the freezer at the temperature −20°C and 2 hours in water +20°C. After the cycles, values of flexural strength, modulus of rupture, and modulus of elasticity were measured and compared with the values at the age of 28 days. Activity indexes I_b, I_r, and I_E were calculated as a ratio value of *values comparative beams/values of frosted beams*.

3. Results and discussion

3.1. Mechanical properties

The recorded results are shown in **Table 2**. The first property is workability. It is evident that the concrete which was obtained with the cone flow cca 350 mm was completely different from the originally very flowable, self-compacting concrete with the cone flow 850 mm. This was the case of PP fibers, but the dosage of these fibers was five times higher than normally recommended. This experiment was performed for the comparison and the information about the impact on workability in concretes with enhanced fire resistance, thanks to PP application. In all these cases, the workability can be regulated with a higher dosage of superplasticizers but this method was not used in this case.

The compressive strength f_c is the highest for the content of fiber 0.5% or 0.25 for M-fibers. These concretes are well compacted and fibers can work well. The value of f_c is not affected by the fiber materials; all structural fibers show similar results in dosage 0.5%.

The flexural strengths f_b and f_r show different courses. The differences are demonstrated in **Figures 1–3**. In **Figure 1**, the load-deflection curve for the reference concrete and the concrete with 0.5% of S-fibers is shown. The maximum value of load is nearly the same, but the postpeak regime is different—there is some residual strength for FRC. The load-deflection

		ref.	PP 0.5	S 0.5	S 1.0	D 0.5	D 1.0	M 0.25	M 0.5
cone *f.* [mm]		**850**	**350**	**530**	**340**	**790**	**610**	**470**	**360**
f_c	[MPa]	93.7	89.1	102.5	95.0	105.2	95.1	101.4	98.4
		±2.8	±3.3	±2.2	±3.1	±1.7	±15.5	±1.5	±7.5
f_b	[MPa]	7.4	7.6	9.45	8.4	10.4	12.5	8.7	11.3
		±0.6	±0.2	±0.4	±0.7	±0.9	±0.4	±0.5	±0.7
f_r	[MPa]	6.8	7.0	7.8	7.0	7.9	8.6	7.0	9.4
		±0.3	±0.3	±0.1	±0.4	±0.2	±0.1	±0.2	±0.6
E	[GPa]	39.4	36.9	39.8	37.3	38.6	41.0	36.2	37.8
		±1.1	±0.5	±0.9	±1.4	±2.1	±0.8	±0.9	±1.1
K_{IC}	[MPam$^{1/2}$]	1.62	1.65	1.58	1.54	1.68	2.26	1.78	1.92
		±0.1	±0.2	±0.1	±0.1	±0.11	±0.8	±0.32	±0.25
G_c	[Jm^{-2}]	67.6	75	63	63	74	125	92	100
		±6	±18	±7	±5	±7	±10	±37	±30
W_F	[Jm^{-2}]	141	231	620	1215	5025		193	332
		±9	±26	±100	±100	±680		±10	±95
I_b	[%]	100	96	84	92	82	61	81	76
I_r	[%]	89	91	91	94	95	120	80	87
I_E	[%]	104	102	89	86	104	84	86	88

Table 2. Workability, mechanical properties, and indexes of frost resistance of concretes with fibers.

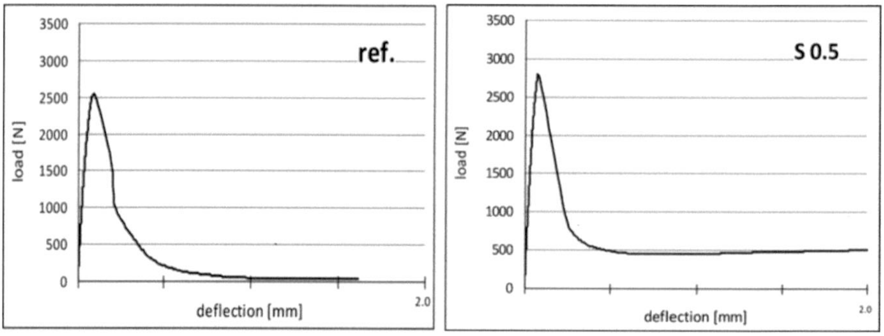

Figure 1. Typical load-deflection curves of concrete "ref" and S 0.5.

curve for the content of 1% S-fibers is very similar, only the residual strength is higher—it does not show here. The highest values of load are reached for steel fibers, especially for the steel fibers with a special shape—hooked ends—3D fibers (see **Figure 2**). These fibers hold in the matrix very well and they enhance f_b and f_r. D-fibers are long enough, which means that the

http://dx.doi.org/10.5772/66804

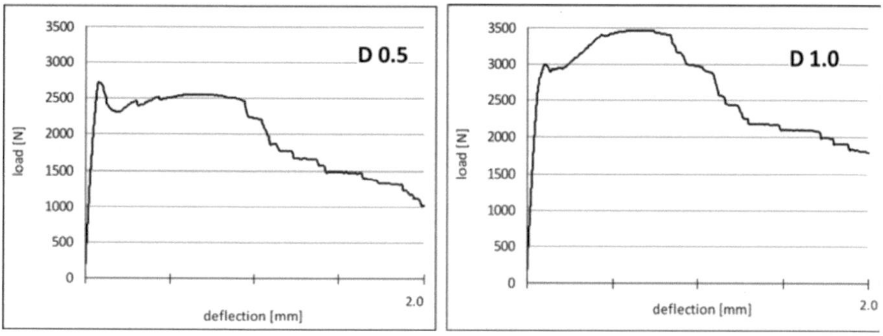

Figure 2. Typical load-deflection curves of concrete D 0.5 and D 1.0.

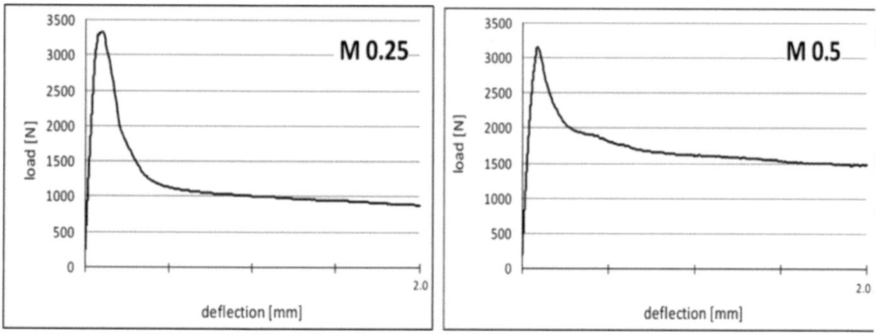

Figure 3. Typical load-deflection curves of concrete M 0.25 and M 0.5.

concrete shows a significant residual strength after the first peak. The results for 0.5 and 1% differ especially in the course curve after the first peak. These results are very similar to those of Ref. [2]. Micro steel fibers (**Figure 3**) work well for the big amount; the improvement of f_b and f_r is the most conspicuous and the dosage of fibers is 1/2 of the previous—only 0.25 and 0.5%, but they are short and they are pulled out early.

The values of the modulus of elasticity are nearly the same—with respect to the standard deviation. Also fracture toughness K_{IC} and toughness Gc is similar for all of the mixtures. The load from the first peak was considered the value of the maximum recorded load. The effect of fibers on the improvement of concrete without microcracks is considered in this case.

3.2. Frost resistance

The indexes of frost resistance are shown in **Figure 4**. The first columns are the indexes in terms of flexural strength f_b, the second ones in terms of flexural strengths on notched prisms f_r, and the third ones in terms of the modulus of elasticity E measured in a three-point bending on

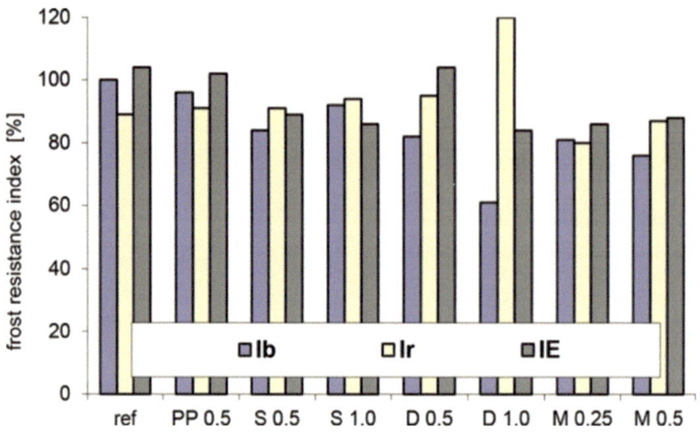

Figure 4. Frost resistance indexes I_b, I_r and I_E expressed in terms of f_b, f_r, and E.

prisms with a central notch. The f_b is affected especially by the properties of the surface layer while the f_r is affected by the properties of the central regions of the cross-section [9].

The best values are recorded for the concrete without fibers. All FRC fulfill the requirements of CSN 73 1322 – $I > 75\%$, except the concrete with 1% of D-fibers in terms of f_b. Concrete M 0.5 shows I_b on the boundary of acceptable value. The experience proves that FRC with wires has worse frost resistance. In terms of f_r and E the frost resistance is good. This probably means that the surface layer of the concrete is affected by the water which penetrates around the wires into the concrete and freezes later. This result—significant reduction of frost resistance—can be a consequence of the small size of the specimen, but it is especially important for the subtle or thin construction from FRC.

4. Conclusions

From the above-mentioned results, some general conclusions can be drawn:

1. Fire-reinforced concretes show similar or higher compressive strength in comparison with reference concrete.
2. All fibers enhance flexural strength, and the most conspicuous increase is recorded for steel fibers.
3. All fibers affect the postpeak course of the load-deflection diagram. Only steel fibers make it possible to reach a higher load after the first peak. But in this case, the microcracks are developed and bridged with fibers. These fibers are not protected by concrete and they can be destroyed by the attack of some aggressive media. Durability of the concrete can be a problem.

http://dx.doi.org/10.5772/66804

4. The frost resistance of concretes with steel fibers can be reduced—probably thanks to the penetration of water along the fibers. This is especially important for the subtle cross-sections. This can also be one of the reasons why to perform tests using small size specimens, too.

5. On the basis of the published results, the concrete with 0.5 of D-fibers was selected as the best possibility for the production of subtle concrete elements. Tests in a 1:1 scale confirmed rightness of this choice [6].

This investigation is being continued with the target to compare the results recorded in small size specimens with those of preliminary norm requirements [4].

Acknowledgements

This outcome has been achieved with the financial support of the project GACR No. 16-08937S, "State of stress and strain of fiber reinforced composites in interaction with the soil environment."

Author details

Vlastimil Bilek, Jan Hurta*, Petra Done and Libor Zidek

*Address all correspondence to: jan.hurta@vsb.cz

Department of Building Materials and Diagnostics, Faculty of Civil Engineering, VSB-Technical University of Ostrava, Ludvika Podeste, Ostrava-Poruba, Czech Republic

References

[1] Fiala, C., Hájek, P., Bílek, V.: Optimised balcony precast elements from fibre concrete. *Proceeding of International Conference Central Europe towards Sustainable Building (CESB 10)*, Hajek, P. et al. (eds), Prague, 2010, pp. 379–382, ISBN 978-80-247-3624-2.

[2] Pajak, M., Ponikiewski, T.: Effect of the shape of steel fibers on the mechanical properties of reinforced self-compacting concrete. *Cement, Wapno, Beton*. 2013, Vol. 18. No. 6, pp. 335–342, ISSN-1425-8129.

[3] Zhang, X., Ruiz, G., Tafira, M., Cendon, D.: Loading rate effect on the fracture behaviour of three different steel fiber-reinforced concretes. *9th International Conference FraMCoS-9*, V. Sauoma, J. Bolander and E. Landis (Eds), DOI: 10.21012/FC9.057.

[4] Hanzlová, J., Veselý, V., Vodička, J.: *New Czech Standards for Fibre-Reinforced Concrete*, BETON TKS, Czech. 2016, pp. 3–6.

[5] Fiala, C., Hejl, J., Bílek, V., Růžička, J., Vlach, T., Novotná, M., and Hájek, P.: Experimental verification of subtle frame components prototypes from high performance concrete for energy efficient buildings. *Solid State Phenomena*. Vol. 249, pp. 301–306, ISSN: 1662-9779, DOI: 10.4028/www.scientific.net/SSP.249.301.

[6] Vlach, T., Laiblová, L., Fiala, C., Novotná, M., Ženíšek, M., Hájek, P.: Eccentricity influence on bearing capacity of subtle column using numerical analysis and experimental verification. *Experimental Stress Analysis*, Český Krumlov, (CD-ROM), 2015. pp. 473–476, ISBN 978-80-01-05735-3.

[7] Karihaloo, B.L., Nallathambi, P.: An improved effective crack model for determination of fracture toughness of concrete, *Cement and Concrete Research*. Vol. 19, pp. 603–610.

[8] Elices, M., Guinea, G.V., Planas, J.: On the measurement of concrete fracture energy using three-point bend test. *Materials Structures*, Vol. 30, No. 1997, pp. 375–376.

[9] RILEM TC-50 FMC (Recommendation): Determination of fracture energy of mortar and concrete by means of three-point bend test on notched beams. *Materials Structures*. 1985, Vol. 18, No. 107, pp. 285–290.

Chapter 28

Aseismic Study on Mountain Tunnels in High-Intensity Seismic Area

Gao Bo, Wang Shuai-shuai, Wang Ying-xue and Shen Yu-sheng

Additional information is available at the end of the chapter

http://dx.doi.org/10.5772/66805

Abstract

The chapter discusses the antiseismic and shock absorption study on the mountain tunnels in high-seismic intensity areas using numerical analysis and shaking table test for recent years and proposes the seismic challenges of tunnel design in Sichuan-Tibet Railway. The aseismic design of the tunnel entrance and the inner part in the fracture zone are presented according to the previous research results.

Keywords: mountain tunnels, high-intensity area, aseismic design, antiseismic, shock absorption

1. Introduction

There have been many investigations on the underground structures damages during strong earthquakes in the world (Konagai, 2005; Sharma and Judd, 1992), and after the Wenchuan earthquake, our research team investigated the earthquake disaster of the mountain tunnels located in the Du (Dujiangyan)-Wen (Wenchuan) highway (Wang et al., 2009). It has been investigated that the mountain tunnel suffered serious damages at the tunnel portal due to widespread landslides and rockfalls, and the major damage of the inner part of the tunnels were concentrated in the poor geological sections due to forced displacement, when tunnel crossed the fracture zones and active fault. The serious mountain tunnel damage during Wenchuan earthquake inspired Chinese researchers and engineers to pay attention to the antiseismic and shock absorption research of the mountain tunnel.

In this chapter, the studies of our research group on the antiseismic and shock absorption of mountain tunnels in high-intensity seismic area in recent years are presented. The first part of this chapter introduces the studies of antiseismic and damping design of the mountain tunnels in high-intensity seismic zone located in the Ya (Yaan)-Xi (Xichang) highway. The second part introduces the challenges in the aseismic design of mountain tunnels located in Chuan (Chengdu)-Zang (Lhasa) high-speed railway.

2. Brief description of Yaxi Expressway

As shown in **Figure 1**, Yaxi Expressway begins in Yaan and ends in Lugu Town, designed as a four lane highway with speed of 80 km/h, and the line has a total length of 240 km, which passes through the mountains with a tunnel ratio as 55%. Climbing from the margin of the Sichuan basin to Hengduan mountainous highlands, the Yaxi Expressway crosses through the China Southwest Geological disaster prone deep canyon area, along the line the terrain conditions are extremely precipitous, with extremely complex geological structure, changeable climate, and fragile ecological environment. The engineering construction conditions are very hard, and it is very difficult to ensure operation safety, so the highway project is considered as one of the highest contents of science and technology expressway in mountainous area with the worst natural environment.

Figure 1. Yaxi Expressway.

Yaxi Expressway crosses 12 earthquake fault zone, and the PGA reaches from 0.15 to 0.4 g, which is the largest ground motion parameter in the highway design and construction in China at present. The tunnels of Yaxi Expressway are all located in high-intensity seismic zone, especially the Le Bukoragi tunnel, which passes through the Anning River active fault zone with an amplitude ground motion of 0.4 g. Investigations have shown that the mountain tunnel entrance and the inner part across the fault tend to sustain serious damage subjected to strong earthquake motions (Wang et al., 2009), so it is essential and important to investigate the deformation mechanism of the mountain tunnel.

In this chapter, the aseismic studies of our group on the dynamic response of the mountain tunnels are presented for recent years, and the antiseismic and shock absorption measures in the tunnel design are introduced.

3. Aseismic study on tunnel portal of Yaxi Expressway

In this section, the research results on the dynamic stress and deformation mechanism of tunnel entrance achieved by numerical simulation and shaking table test in recent years are introduced.

3.1. Numerical analysis on the dynamic response of the tunnel entrance

The numerical models of Cheyang tunnel and Xudianzi tunnel entrance in the dynamic time history analysis by FLAC3D are shown in **Figure 2** and **Figure 3**.

The bending moments of the Cheyang tunnel liners in the entrance at 8.7 second subjected to vertical shear waves are shown in **Figure 4**, when the maximum bending moment is in the left arch foot position. The distribution of the bending moment of the tunnel cross-section indicates that the dynamic stress concentration is in the arch foot position where the tunnel transverse shape mutates.

As shown in **Figure 5**, most elements of the slope in the tunnel entrance are in tension plastic state, especially these elements around the cavities are in tension plastic state during the strong ground motion.

Figure 2. Analysis model of Cheyang tunnel.

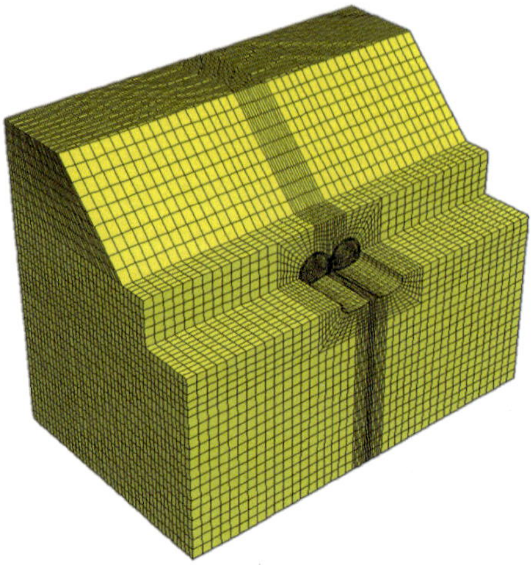

Figure 3. Analysis model of Xudianzi tunnel.

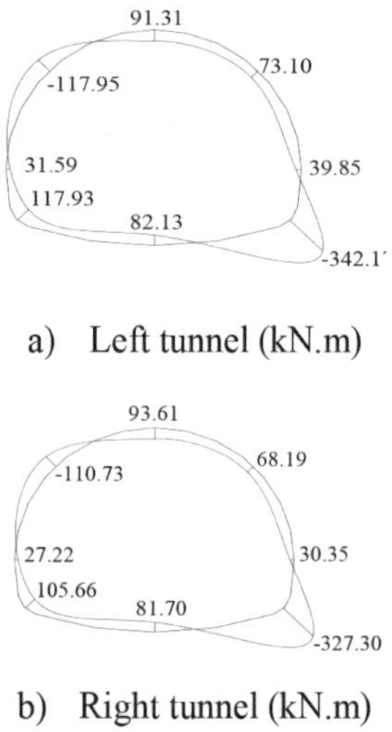

Figure 4. The bending moments of Cheyang tunnel in the entrance at 8.7 second.

The analysis results demonstrate that the bending moment of the liner in the entrance is much larger than that of the inner part, as the peak bending moments along the Xudianzi tunnel presented in **Figure 6**, it is shown that the maximum peak bending moment is in the position of tunnel entrance, and the associated value decreases as the distance to the tunnel entrance increases, and when the distance reaches four times the tunnel diameter, the tunnel entrance has little effect on the dynamic stress of the inner part.

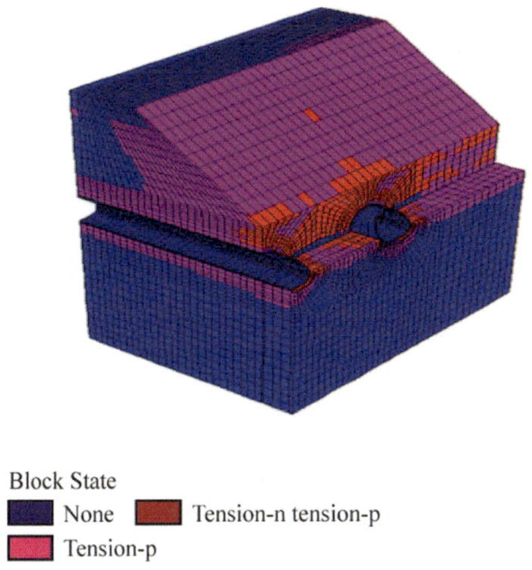

Figure 5. The tension plastic zone of Cheyang tunnel entrance.

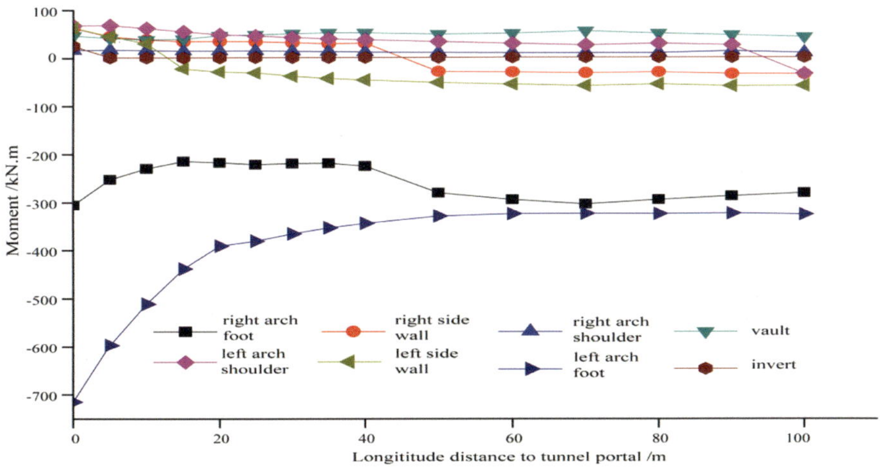

Figure 6. The peak bending moments of Xudianzi tunnel.

According the analysis in the previous sections, it is important to keep the stability of the tunnel entrance slope in the aseismic design.

3.2. Antiseismic and shock absorption study

Studies have shown that grouting the surrounding rock (Shen et al., 2014; Wang et al., 2014) and covering the tunnel with a soft layer (Kim and Konagai, 2001; Wang et al., 2015) are effective measures for mitigating seismic damage to tunnels, as shown in **Figure 7** and **Figure 8**.

The dynamic analysis of Xudianzi tunnel entrance for two kinds of structures with grouting and covering damping layer were achieved using FLAC3D, the analysis models are presented in **Figure 7** and **Figure 8**.

The bending moments distribution of the section in the entrance are presented in **Figure 9** and **Figure 10**, the bending moment values of the arch foot positions can be found larger than that of other positions. The maximum bending moment decreases with grouting as shown in **Figure 9**, and the associated value is effectively reduced by covering the tunnel with a soft isolation layer as seen in **Figure 10**.

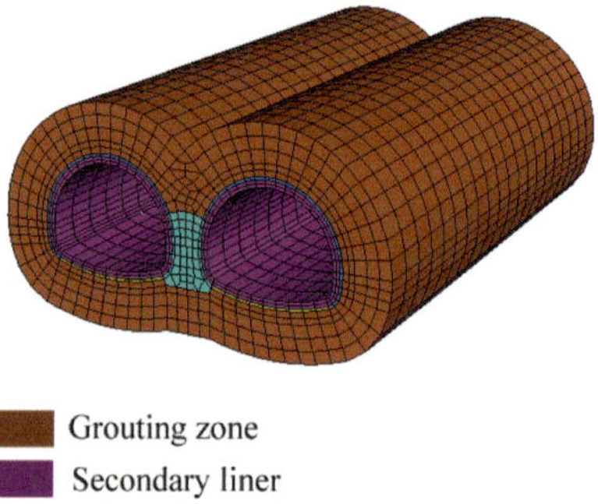

Figure 7. Grouting the surrounding rock.

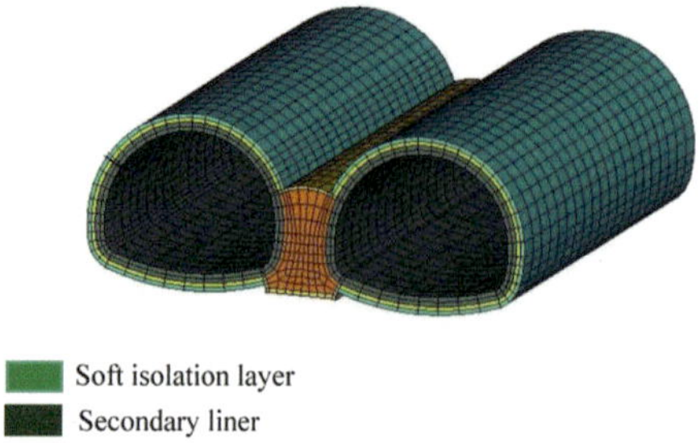

Figure 8. Covering the tunnel with a soft isolation layer.

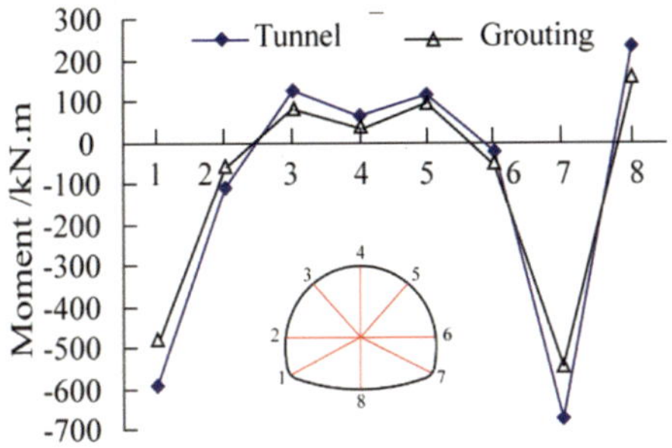

Figure 9. The bending moments with grouting.

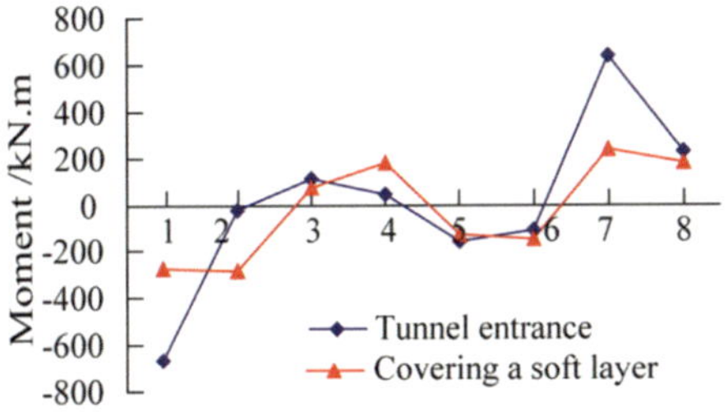

Figure 10. The bending moments with covering a soft isolation layer.

3.3. Shaking table model test on tunnel entrance

Two large scale shaking table model tests of the portals of two parallel tunnels were carried out, with the model geometry similarity ratio of 25, to investigate the dynamic response of tunnel liners and the interaction between surrounding rock and tunnel structure subjected to vertical shear waves in 2007. The model test items and methods of test are introduced in the research paper (Sun et al., 2011), and two cases are introduced here: case 1 for general tunnel entrance, and case 2 for covering the tunnels with soft isolation layer.

The distribution of ground cracks of tunnel entrance models subjected to vertical shear waves are presented in **Figure 11** and **Figure 12**.

The ground cracks are concentrated in the slope near the portal, as shown in **Figure 11**, and these shear cracks in the tunnel arch shoulders position are mostly caused by the interaction between the tunnel structure and surrounding rock under shear waves; however, the distribution of the ground cracks on the top surface of the entrance are in different directions of "X" shape.

Figure 11. The distribution of ground cracks.

Figure 12. The distribution of ground cracks with soft isolation layer.

The ground cracks are in tunnel arch shoulder position when the tunnels are covered with soft isolation layer, illustrated in **Figure 12**. The number of the model ground cracks for case 2 is significantly reduced by covering the tunnel with soft isolation layer.

The damage patterns of the liner models of the right tunnel are drawn in **Figure 13** and **Figure 14** after vibration, **Figure 13** shows that the side wall of general tunnel liner was badly damaged and fallen down after strong ground motion, and the damage of the tunnel liner with covering a soft damping layer is obviously reduced as illustrated in **Figure 14**.

3.4. Summary

The numerical analysis and model test results show that the dynamic response of the tunnel entrance is larger than the inner part subjected to shear waves, and the liner damage can be effectively eliminated by covering a soft isolation layer.

Figure 13. The fracture of the liner (right tunnel) in case 1.

Figure 14. The fracture of the liner with soft isolation layer (right tunnel) in case 2.

4. Aseismic study on tunnel in fault and fracture zone

The tunnel earthquake investigations (Konagai, 2005; Sharma and Judd, 1992; Wang et al., 2009) have shown that tunnels crossing fault and fracture zone suffered serious damage during earthquake.

In this section, the shaking table model test and numerical analysis results are introduced to investigate the dynamic response of the Le Bukoragi tunnel in fault and fracture zone.

4.1. Shaking table model test on tunnel across fault fracture zone

To investigate the dynamic stress and deformation mechanism of the tunnel in fault and fracture zone, two large scale model tests were carried out: case 1 for the general tunnel, and case 2 for the tunnel across fault fracture zone.

The longitudinal sections of the two models are shown in **Figure 15** and **Figure 16**. The similarity ratios and the material physical parameters of the surrounding rock and liner are presented in the research paper (Sun et al., 2011; Wang et al., 2015). The longitudinal width of the fault zone is 10 cm, by contrast with the prototype width reaching up to 3 m, and the physical parameters of the rock model in the fault fracture zone are listed in **Table 1**.

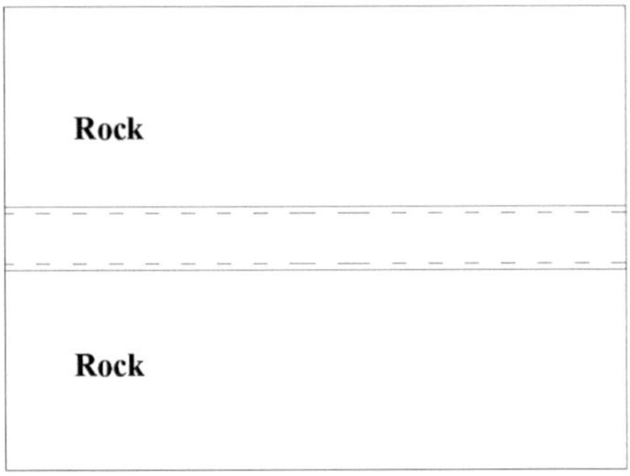

Figure 15. Longitudinal section of the general tunnel in case 1.

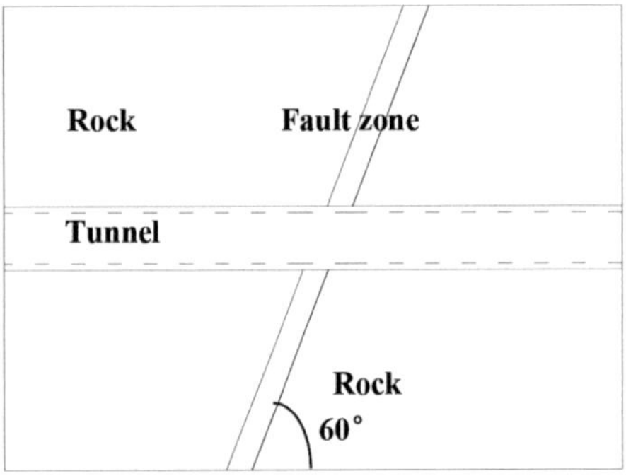

Figure 16. Longitudinal section of the tunnel across fault zone in case 2.

Parameter	Rock prototype	Rock prototype	Rock model	Units
Cohesive strength	85	45	2.4	kPa
Fraction angle	17	1	19.2	°
Young modulus	0.9×10^3	45	30.1	MPa
Density	1.9	1.5	1.3	g/cm^3

Table 1. Rock parameters of the fault zone.

As shown in **Figure 17**, the longitudinal cracks appear in the vault, arch foot, side wall, and invert positions of the liner after vertical shear wave excitation, caused by bending moment and compressive strain. As illustrated in **Figure 18**, one circumferential shear crack appears in the fault fracture zone caused by shear strain, which shows that the dynamic response of the liner will be greater when the tunnel passes through the fault fracture zone, due to the stiffness difference between the hard rock and fracture zone.

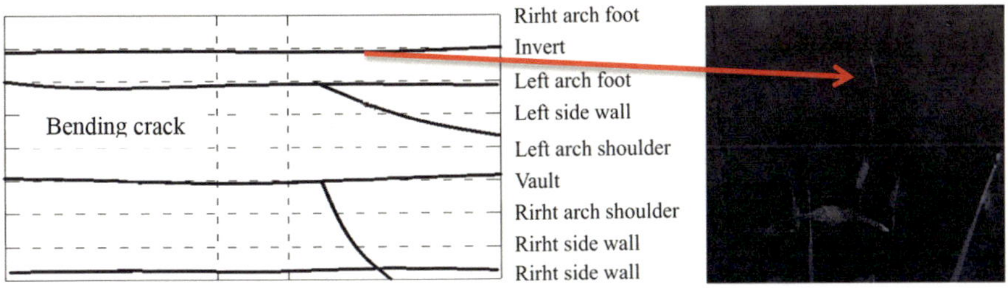

Figure 17. The cracks of the general tunnel in case 1.

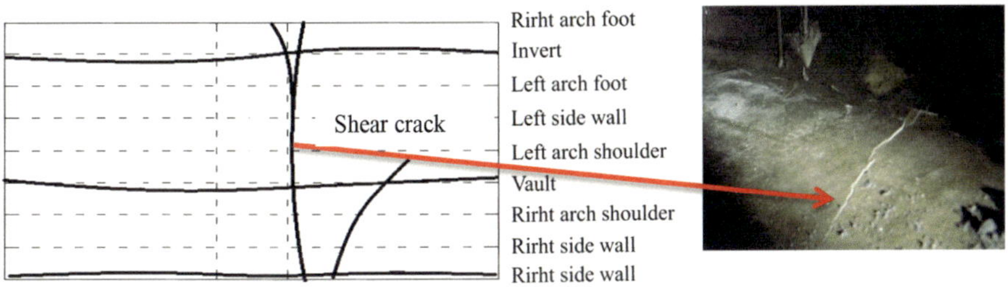

Figure 18. The cracks of the tunnel across fault zone in case 2.

4.2. Numerical analysis on jointed tunnel across fault fracture zone

It is possible to see that the stiffness difference between the soft rock in the fracture zone and the hard rock has an obvious effect on the tunnel response according to shaking table model test results. Here, the numerical analysis of the tunnel across the fault and fracture zone is introduced, and the analysis model using FLAC3D can be seen in **Figure 19**.

Researches (Shahidi and Vafaeian, 2005) have shown that the structure design using deformation joints to tolerate the longitudinal differential displacements can effectively reduce the damage of the tunnel, so two analysis cases were performed: case 1 for general tunnel without joints, and case 2 for the design structure with deformation joints along the longitudinal tunnel.

As shown in **Figure 20**, it is indicated that the displacement amplitudes of the vault in the fault zone are greater than other parts along the tunnel, which shows that the dynamic response of the tunnel in the fault fracture zone becomes larger for the differential stiffness between the soft and hard rock, as illustrated in **Figure 21**, the amplification effect of the bending moments of the linear vault are obvious in the fault fracture zone.

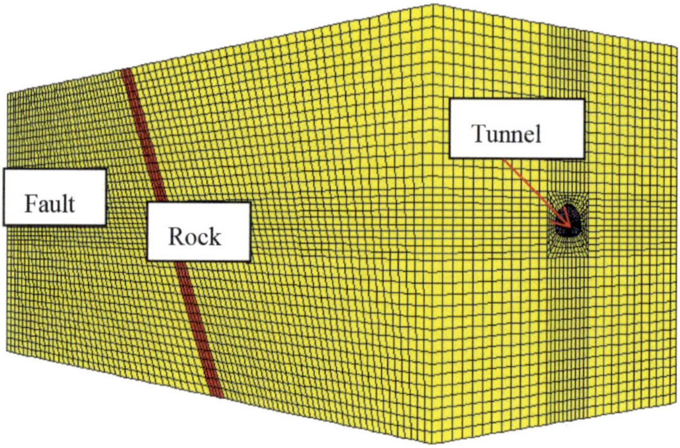

Figure 19. The numerical model of tunnel across fault zone.

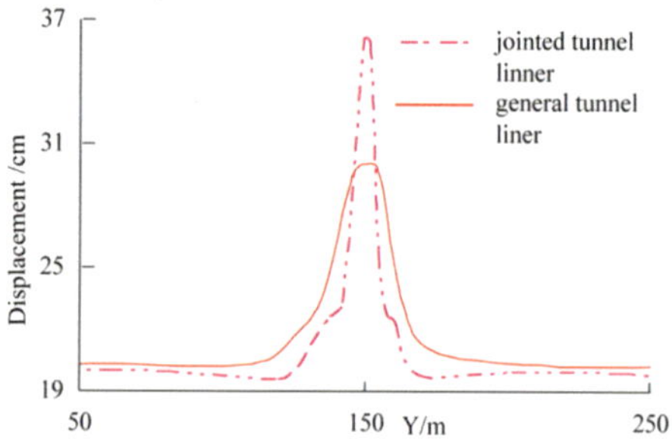

Figure 20. The displacement amplitude of vault of tunnel across fault zone.

As illustrated in **Figure 20**, it proves that the displacement amplitudes of the jointed tunnel in case 2 are greater than that of the general tunnel in case 1 in the fault fracture zone, while the associated values of the joined tunnel in other positions are smaller than that of the tunnel without joints. It can be seen from **Figure 21** that the moments amplitudes of the joined tunnel are smaller than the general tunnel, due to the better deformation capacity of the joined tunnel.

4.3. Summary

The research results by the shaking table tests and numerical analysis shows that the dynamic response of the tunnel becomes larger due to the differential stiffness between the fault

http://dx.doi.org/10.5772/66805

fracture zone and hard rock, and the dynamic response of the tunnel decreases with joints along the tunnel.

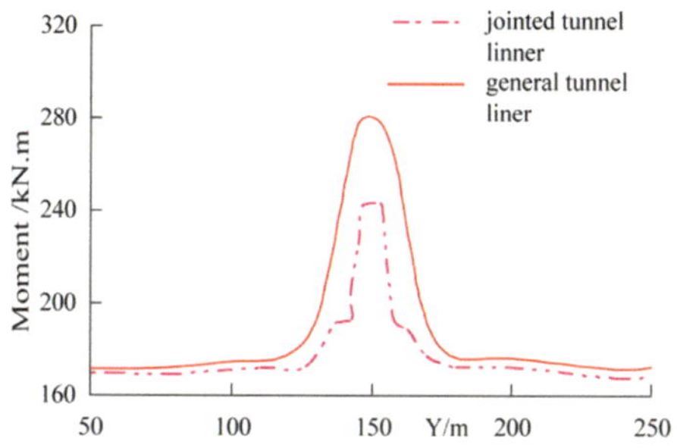

Figure 21. The bending moment amplitude of vault of tunnel across fault zone.

5. The challenges in the aseismic design of Sichuan-Tibet Railway tunnels

As shown in **Figure 22**, the Sichuan-Tibet Railway passes through several crust splicing tapes and active faults (Shang et al., 2005), especially for Chengdu to Kangding section with strong earthquake rupture in history. Studies have shown that the strong earthquake vibration, and the formation of secondary disasters such as landslides faulting, landslides, avalanches, and other damage are great challenges for the line selection and structure design (Konagai, 2005; Shang et al., 2005; Sharma and Judd, 1992).

As described, the tunnel portal and the inner part passing through the fault and fracture zone always suffered serious damage, and it is very important to reduce the risk and damage from strong earthquake and its secondary disasters.

The sections of the tunnel in which the antidamping design must be concentrated are presented, and according to the antiseismic and shock absorption research results, the following principles are shown in **Figure 23**:

(i) The primary dangers at tunnel entrances are landslides and rockfalls, and efforts should be concentrated to improve the stability of the slope, as illustrated in **Figure 23**, such as bolting and shotcreting, reducing the slope gradient by slope cutting, and improving the thickness of the tunnel backfill.

(ii) The settlement joints are proposed to set for improving the deformation capacity of the tunnel in the entrance, and it is also recommended when the tunnel passes through the fault and fracture zone.

(iii) The surrounding rock should be grouting, and it is suggested to cover the tunnel with soft isolation layer to reduce the dynamic stress of the liner in the tunnel portal.

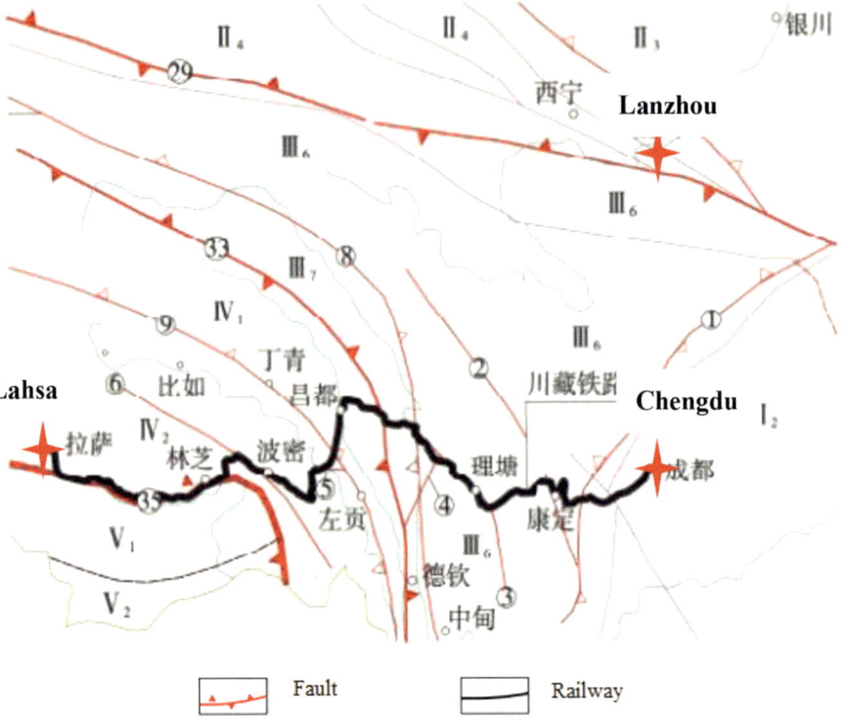

Figure 22. The distribution of faults along the Sichuan-Tibet Railway (Zhang et al., 2016).

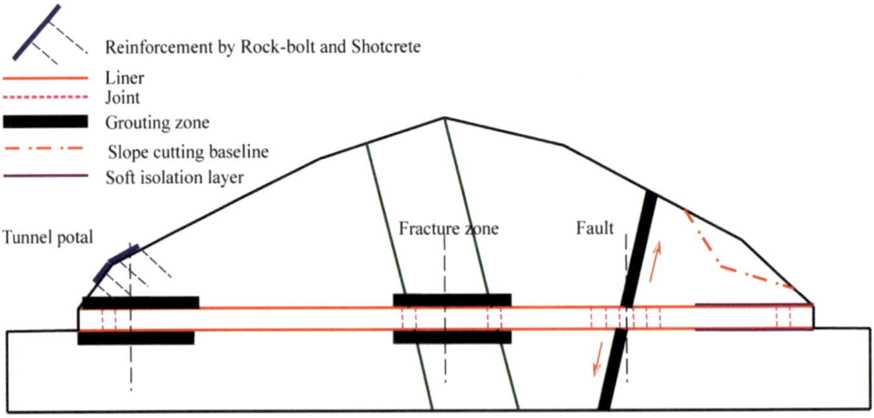

Figure 23. The aseismic design schematic diagram along the tunnels.

http://dx.doi.org/10.5772/66805

6. Summary

The chapter introduces the research results on the anti-seismic and shock absorption studies of our group of mountain tunnels in high seismic intensity areas in recent years, and discusses the challenges in the design of tunnels in Sichuan-Tibet Railway. It shows that the dynamic response of the tunnel in the portal and the fracture and fault zone becomes larger, and the dynamic response of the tunnel in the entrance can be effectively mitigated by grouting the surrounding rock and covering soft isolation layer, and the dynamic stress of the tunnel passing through the fracture and fault zone decreases with joined structure design.

Conflict of interest

The authors declare that there is no conflict of interest.

Acknowledgements

This work was supported by the National Natural Science Foundation of China (Grant no. 51108056 and 51178398).

Author details

Gao Bo*, Wang Shuai-shuai, Wang Ying-xue and Shen Yu-sheng

*Address all correspondence to: prgaobo@vip.sina.com

Civil Engineering School, Southwest Jiaotong University, Sichuan, China

References

D. Kim and K. Konagai. Key parameters governing the performance of soft tunnel coating for seismic isolation, pp. 1333–1343, 2001.

K. Konagai. Data archives of seismic fault-induced damage. Soil Dyn Earthq Eng 25, 559–570, 2005.

A. R. Shahidi and M. Vafaeian. Analysis of longitudinal profile of the tunnels in the active faulted zone and designing the flexible lining (for Koohrang-III tunnel), pp. 213–221, 2005.

Y. Shang, H. Parka and Z. Yang. Engineering geological zonation using interaction matrix of geological factors: an example from one section of Sichuan-Tibet Highway. Geosci J 9, 375–387, 2005.

S. Sharma, W.R. Judd. Underground opening damage from earthquakes. Eng Geol 30, 263–276, 1992.

T. Shen, Bo Gao, X. Yang and S. Tao. Seismic damage mechanism and dynamic deformation characteristic analysis of mountain tunnel after Wenchuan earthquake. Eng Geol 180, 85–98, 2014.

T. Sun, Z. Yue, B. Gao, Q. Liand and Y. Zhang. Model test study on the dynamic response of the portal section of two parallel tunnels in a seismically active area. Tunn Undergr Sp Tech 26, 391–397, 2011.

Z. Wang, B. Gao, Y. Jiang and S. Yuan. Investigation and assessment on mountain tunnels and geotechnical damage after the Wenchuan earthquake. Sci China Ser E: Technol Sci 52, 546–558, 2009.

Z. Z. Wang, Y. J. Jiang, C. A. Zhu and T. C. Sun. Shaking table tests of tunnel linings in progressive states of damage. Tunn Undergr Sp Tech 50, 109–117, 1015.

Z. Song, G. Zhang, L. Jiangand and G. Wu. Analysis of the characteristics of major geological disasters and geological alignment of Sichuan – Tibet Railway. Railway Stand Des 60(1), 14–19, 2016 (in Chinese).

S. Wang, B. Gao, Y. Shen, et al. Study on the mechanism of resistance and damping technology of deep soft rock tunnels subjected to incident plane SH waves. China Civil Eng J 47, 280–286, 2014 (in Chinese).

S. Wang, B. Gao, C. Sui, et al. Mechanism of shock absorption layer and shaking table tests on shaking absorption technology of tunnel across fault. Chinese J Geotech Eng 37(6), 1086–1092, 2015 (in Chinese).

Chapter 29

The Stability Analysis of Foundation Pit Under Seepage State Based on Plaxis Software

Hu Qizhi, Liu Zhou, Song Guihong and
Zhuang Xinshan

Additional information is available at the end of the chapter

http://dx.doi.org/10.5772/66806

Abstract

The stability of excavation engineering is closely related to groundwater, so it is important to study the impact of seepage flow on the stability of foundation pit. The work is based on the percolation theory and principles of strength reduction. The computation were done with use of the Plaxis software. There were studied simulations which included the seepage state and simulations which didn't include this effect. In order to studythe influence of seepage on the stability of foundation pit, there was computed the stability coefficient by using the strength. The results show that, when the seepage stability was not considered, the coefficient is 30% larger than when considering the seepage. Therefore, when designing and calculating the excavation, the seepage should be considered when checking stability if there is groundwater.

Keywords: foundation pit, seepage, Plaxis, stability, strength reduction

1. Introduction

For excavation, especially the excavation in high water table areas, there are head differences inside and outside the pit, besides, under that case, groundwater seepage will occur at the pit head difference. Pore pressure inside and outside the hole and the change of the effective stress caused by groundwater seepage are serious threats to the stability of the excavation engineering (Sheng, 2008). Research shows that 60% of pit accidents are directly or indirectly related to groundwater (Qiang et al., 2007). Thus, the analysis of excavation stability must be attached with great importance to the groundwater and groundwater seepage.

Seepage problems of excavation are mainly related to the role of groundwater flow in the rock excavation of soil pores or cracks and other media. Domestic researchers have done a lot of research on the impact of water and explained different aspects of seepage stability on the excavation and other related computing problems, and they have achieved certain project benefits (Zhuanzheng et al., 2012; Cheng et al., 2009; Chang et al., 2002; Huangchun and Xiaonan, 2001). Seen from the engineering point of view, it has greater applicability when handling finite element seepage problems. However, different finite element software and soil constitutive model would result in different calculated results. Based on this point, in order to explore the impact of groundwater seepage on pit excavation engineering further, the application of geotechnical engineering software Plaxis specific engineering examples and comparative analysis should both consider with seepage and without it, and apply strength reduction method to calculate the stability factor for these two different states.

2. Strength reduction

Strength reduction was first proposed by J. M. Duncan; he pointed out that the safety factor can be defined as the extent of soil shear strength reduction when the slope has just reached the critical failure state (Quan et al., 2008). By gradually reducing the shear strength index, the c value is divided by a corresponding reduction coefficient at the same time, and a new set of strength index is obtained; calculate this way for several times, until the slope reaches a critical failure state, and F_{ST} is used at this time, which is the safety factor of the slope (Quan et al., 2008):

$$F_{sr} = \frac{c}{c'} = \frac{\tan(\phi)}{\tan(\phi')} \tag{1}$$

In the formula, the original c and ϕ are cohesion and friction angle of slope. The reduced values are labelled c' and ϕ'. They are used when the slope reaches a critical damage at the level of Poe safety factor. It can be seen from the basic principles of strength reduction that the safety factor is obtained by the method clearly, and the method is simple enough to be applied into practical engineering.

The determination of instability criterion has also been a focus of discussion on the overall stability analysis of foundation. There are five parts proposed about the slope failure criterion currently (Quan et al., 2008): the convergence criterion; plastic zone generalized shear strain criterion or generalized plastic strain criterion; criterion dynamics criterion; displacement or displacement criterion; and mutation rate criterion.

Each criterion has its own scope of application here the fifth criterion is chosen as the criterion of slope failure, namely, the displacement of the feature point mutations can occur suddenly or displacement will increase quickly when the slope is close to its destruction. Because the displacement of the mutation will mean the beginning of part instability, displacement mutation selection criterion or displacement mutation rate criterion has the physical advantages, and the only problem lies in the selection of feature point. Theoretically speaking, only points

http://dx.doi.org/10.5772/66806

within slip surface can serve as the feature points in terms of selecting feature points, starting from the inside retaining wall engineering dot located near the surface of the soil excavation is more suited to the convergence point discriminating judgment.

3. Plaxis software and percolation theory

3.1. Plaxis software

Plaxis geotechnical engineering finite element analysis software is a software that is used to solve the problems of geotechnical engineering, such as deformation, stability, as well as groundwater seepage, and it has already become the world-renowned geotechnical engineering finite element analysis software. Compared with other similar types of geotechnical engineering software, Plaxis software has certain advantages on soil stability calculation or seepage calculation (Wei et al , 2011).

In the Plaxis software, groundwater percolation theory is mainly based on percolation theory of finite elements. Flow in porous media can be described by Darcy's law. Considering the vertical flow within the x–y plane:

$$q_x = -k_x \frac{\partial \phi}{\partial x}; q_y = -k_y \frac{\partial \phi}{\partial y} \tag{2}$$

In the formula, q is treated as the flow rate ratio, which is calculated by the permeability and groundwater head gradient obtained. Head is defined as:

$$\phi = y - \frac{p}{r_w} \tag{3}$$

In the formula, y is in a vertical position and represents the pore water pressure (negative pressure) and severe water. For steady-state flow, the continuous application conditions are:

$$\frac{\partial q_x}{\partial x} + \frac{\partial q_y}{\partial y} = 0 \tag{4}$$

Formula (4) represents the total amount of water flowing into the unit body total water per unit time and is equal to the outflow. After the simulation of the entire discrete objects, corresponding groundwater head in any place within the cell can be used to represent the node cell values:

$$\phi(\xi , \eta) = N \varphi^e \tag{5}$$

In the formula, N is the shape function, and ξ and η are the local coordinate units. In accordance with Eq. (2) that the flow rate is based on the gradient of the groundwater head, the gradient matrix can be determined. It is the spatial derivative of interpolating function. To describe saturated soil (less saturation line) and nonsaturated soil (phreatic line), reduction function is introduced in the Darcy's theorem:

$$q_x = -K^r k_x \frac{\partial \phi}{\partial x}; q_y = -k_y \frac{\partial \phi}{\partial y} \tag{6}$$

The value of reduction function below the saturation line is 1 (positive pore pressure), and the value above the phreatic line is less than 1 (negative pore pressure). In the transition zone above the phreatic line, the function value is reduced to 10^{-4}. In the transition zone, function logarithmic is in linear relationship:

$$\lg K^r = -\frac{4h}{h_k} \tag{7}$$

In the formula, h represents the head pressure and h_k represents the head pressure where the reduction function reduces to the head pressure of 10^{-4}. In Plaxis, the default is 0.7 m (with the selected unit of length has nothing to do). In the numerical analysis, the ratio of the flow is written as:

$$q = -K^r RB \phi^e \tag{8}$$

among them:

$$q = \begin{bmatrix} q_x \\ q_y \end{bmatrix}; R = \begin{bmatrix} k_x & 0 \\ 0 & k_y \end{bmatrix} \tag{9}$$

Flow from the node ratio can be obtained by integrating the node traffic:

$$Q^e = -\int B^T q dV \tag{10}$$

In the formula, B^T is the transposed matrix. The following equation is applied in the unit level:

$$Q^e = K^e \varphi^e; K^e = \int K^r B^T RBq dV_0 \tag{11}$$

On the global level, all units of contributions are superimposed, and boundary conditions are applied (groundwater flow and head loss), which is formed in n unknown quantities of n equations:

$$Q = K\phi \tag{12}$$

In the formula, K is a global traffic matrix and Q includes boundary conditions specified as flow losses. When the saturation line is unknown (unconfined water issues), the pickup (Picard) iterative method is used to solve the balance of the system. At this point the problem can be solved by the iterative process which can be written as:

$$K^{j-1} \delta \phi^j = Q - K^{j-1} \phi^{j-1}; \phi^j = \phi^{j-1} + \delta \phi^j \tag{13}$$

In the formula, j is the iteration number, which is an unbalanced vector. In each iteration of the groundwater heads, nodes of unbalanced flow should be computed, and effective head should be added. New distribution of the groundwater head should be computed again according to formula (8) recalculation of the traffic, and integrated into the node traffic. This process continues to standard imbalance vector, namely, the error node traffic is smaller than the allowable error.

3.2. Soil model

Constitutive model of soil is the premise of stability calculation. At present, constitutive model of soil can be roughly divided into three categories: Elastic class model (elastic model, Duncan-Chang (DC) model), elastic-ideal plastic class model (Mohr-Coulomb (MC) model, Drucker-Prager (DP) model), and strain hardening elastoplastic model class (Modified Cambridge (MCC) model, Plaxis hardening soil (HS) model). MC model is the most widely used, but MCC model and the HS model have greater applicability in the simulation of the nature of the soil (Zhonghua and Weidong, 2010; Feng and Po, 2011).

HS model is put forward by Schanz (Duncan, 1996), which is an isotropic hardening elastoplastic model. HS model can consider both the shear and compression hardening, and the use of Mohr-Coulomb failure criterion. The basic idea is to assume partial HS model and vertical drained triaxial stress test should remain in hyperbolic relationship. For HS elastoplastic model to express this relationship, HS model considers soil dilatancy and neutral loading. Ideal elastoplastic model is different, and HS model in stress space yield surface is not fixed and it varies with plastic strain and expansion. HS models can adapt to describe a variety of damage and deformation behavior of soil types and is suitable for various geotechnical engineering applications, such as embankment filling, foundation bearing capacity, slope stability analysis and excavation, and so on. The following numerical simulation of the following will adopt HS model.

4. Numerical simulation

4.1. Project overview

Specific examples of excavation adopt the foundation of Sheng (2008). Excavation width is 20 m, depth 10 m, with two 15 m deep and 0.35 m thick concrete diaphragm walls and two rows of anchors as shoring structure, where the first row of anchor length is 14.5 m with 33.7° inclination, and the second row bolt length is 10 m, with an angle of 45°. Considering the surrounding load factors, a load of 10 and 2.5 kN/m^2 is added around the pit. Related soils are filling (0–3 m), sand (3–15 m), and sand and mud (>15 m), and the underground water level in the initial state lies in 3 m below the surface.

Combing with the case background, a geometric mode with 80 m width and 20 m height is established by the Plaxis software, and the generated geometric model and network are shown in **Figure 1** and **Figure 2**. Parameters related to soil properties and structures are shown in **Table 1** and **Table 2**, by taking the default value of the software, which is no longer listed in **Table 1**. The excavation pit is divided into three stages, namely, the first excavation of the subsurface 3 m, then reexcavation of 4 m, and the last remaining excavation 3 m. Plaxis is divided into six steps of excavation.

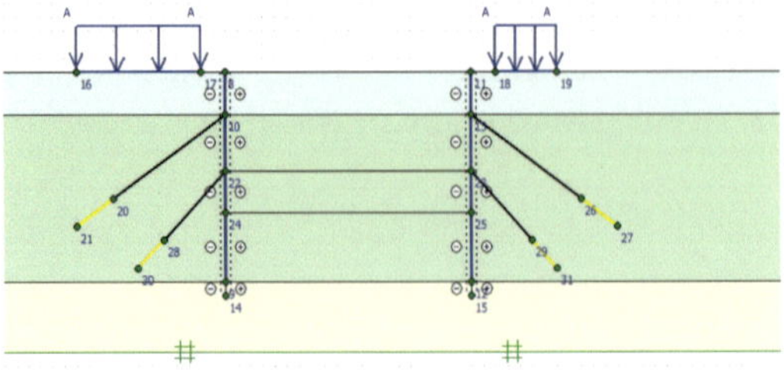

Figure 1. The geometric model of foundation.

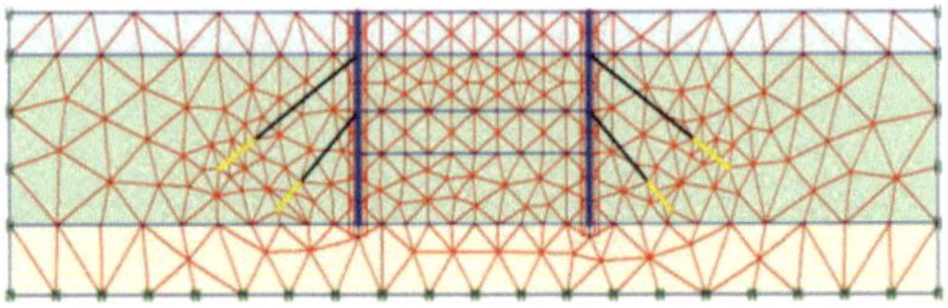

Figure 2. The grid division of foundation.

Parameter	Name	Filling	Sand	Sand and mud
Severe natural (kN/m³)	γ_{unsat}	16.00	17.00	17.00
Severe saturation (kN/m³)	γ_{sat}	20.00	20.00	19.00
Horizontal permeation coefficient (m/day)	k_x	1.000	0.500	0.100
Vertical permeability coefficient (m/day)	k_y	1.000	0.500	0.100
Test secant stiffness of triaxial test (kN/m²)	E^{ref}_{50}	22,000	40,000	20,000
The main tangent stiffness in the loading consolidation apparatus (kN/m²)	E^{ref}_{oed}	22,000	40,000	20,000
Unloading/reloading stiffness (kN/m²)	E^{ref}_{ur}	66,000	40,000	20,000
Power exponential function	m	0.50	0.50	0.60
Group cohesiveness (kN/m²)	c	1.00	1.00	8.00
Friction angle (angle)	ϕ	30.00	34.00	29.00
Dilation angle (degree)	ψ	0.00	4.00	0.00
Interface reduction factor	R_{inter}	0.65	0.70	1.00

Table 1. The soil parameters of foundation.

4.2. Simulation results

The simulation is divided into two cases: one is that considers the seepage, and the other is that does not consider the seepage and displacement. They are shown in **Figure 3**.

http://dx.doi.org/10.5772/66806

	Parameter	Name	Numerical value
Diaphragm wall panel trench	Axial rigidity (kN)	EA	1.2×10^7
	bending rigidity (kN. m)	EI	1.2×10^3
	Equivalent thickness (m)	d	0.346
	Severe (kN/m³)	w	8.3
	Poisson ratio	v	0.15
Bolting	Axial stiffness (kN)	EA	2×10^5
	horizontal spacing (m)	LS	2.5
Grout	Axial rigidity (kN)	EA	2×10^5

Table 2. Structural parameters.

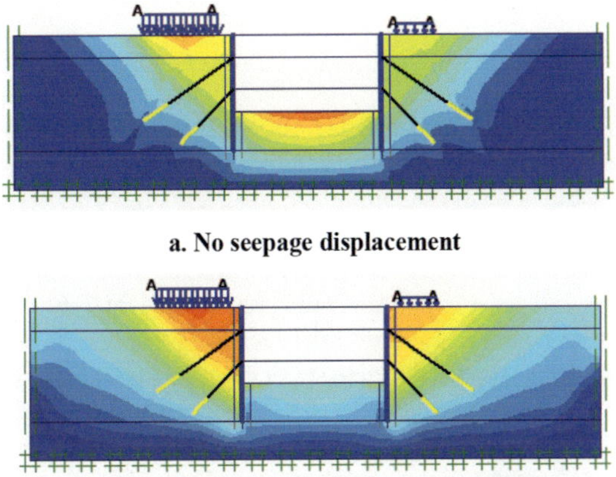

a. No seepage displacement

b. the seepage and displacement nephogram

Figure 3. Yaxi Expressway.At the end of excavation displacement contours. (a) No seepage displacement and (b) the seepage and displacement nephogram.

From the numerical simulation computed by finite element, it can be seen that when the seepage is not considered, the maximum displacement is 22 mm, which occurs in pit bottom. When considering seepage, the maximum displacement is 47 mm, which occurs in the soil layer with a load of 10 kN/m^2. By comparing the differences between the two, it can be seen that in most regions of the pit, soil displacement is greater than the case that the seepage is considered, it is not displaced when considering the seepage. So when seepage is not considered, foundation displacement calculation is a little dangerous.

When considering seepage (**Figure 4**), it can be seen that the boundary seepage pit is slightly arc-shaped, grout, and anchor near the foot of the slope location seepage velocity, and the maximum value will reach 387.73×10^{-3} m/day. By contrast, considering the seepage and displacement map and when not considering seepage, seepage velocity larger field position, displacement difference reaches 20 mm, which is slightly large, indicating the presence of seepage field, in terms of the pit, it will increase the displacement of its territories. If the flow is not considered, it is unreasonable to valuate soil excavation pit stabilization with the calculated displacement.

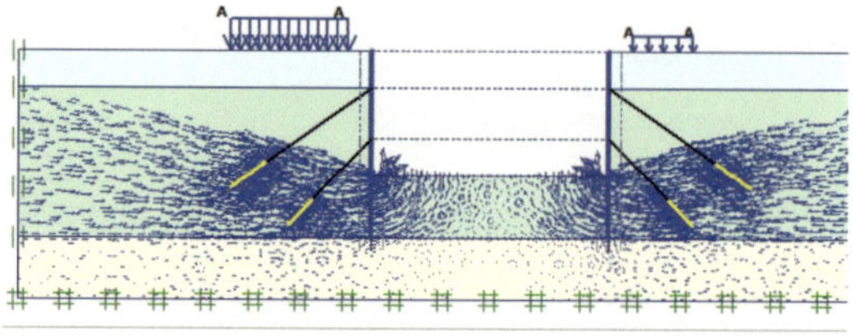

Figure 4. Seepage field with seepage.

When comparing the difference between displacement, it can be seen that there exists great difference between these two cases that with and without considering the seepage flow. Differences between the soil stress can be analyzed through **Figure 5** and **Figure 6**. **Figure 5** shows the effective stress diagram in which flow is not considered, because in the case without considering the water, effective stress and total stress is always equal, and the total stress diagram is no longer listed specially. **Figure 6** shows effective stress diagram and the total stress that considers seepage. It can be seen from **Figure 5**, the maximum effective stress occurs near the body part of bolting and grouting, the maximum effective stress is −363.91 kN/m^2. From **Figure 6**, it can be seen when considering the seepage, there are similarities between pit effective stress exhibited seepage pit and without considering the effective stress distribution. And all, the maximum occurs in the vicinity of bolting and grouting body soil. There are also great differences. When considering the seepage, the maximum effective stress reaches −407.11 kN/m^2. In terms of the entire distribution, when considering the seepage pit, the distribution of effective stress is not as intensive as shown in **Figure 5**, which is so concentrated in the vicinity of the distribution of grout; therefore, when considering effective stress of seepage pit, most of the soil area is larger than that does not consider effective stress of seepage time. By analyzing the total stress diagram when considering the seepage, it can be seen that the total maximum stress occurs when there is seepage near the foot of the slope anchor grouting and excavation position, which has certain pertinence with the maximum speed occurring at the seepage field. By analyzing **Figure 4** together with **Figure 6**, it can be found that the increase of the total stress mainly manifests in the area where there is seepage, and the greater speed the seepage becomes, the more obvious the seepage increases.

By comparing and considering the two cases, it can be found that when seepage is not taken into account, both the soil displacement and stress are small, so in this case, the calculated conditions will reduce the accuracy of numerical simulation.

http://dx.doi.org/10.5772/66806

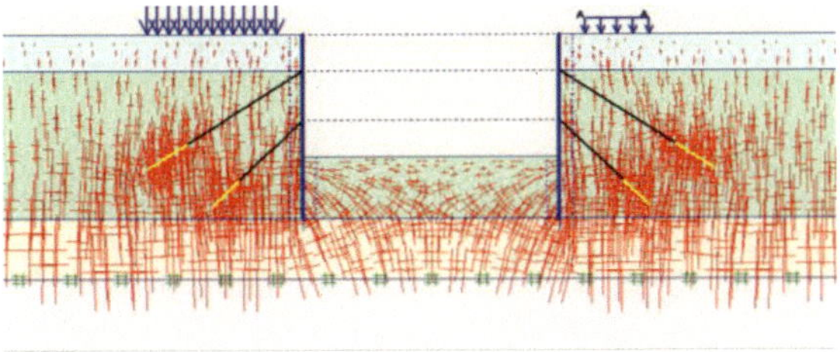

Figure 5. The absence of effective stress of seepage pit.

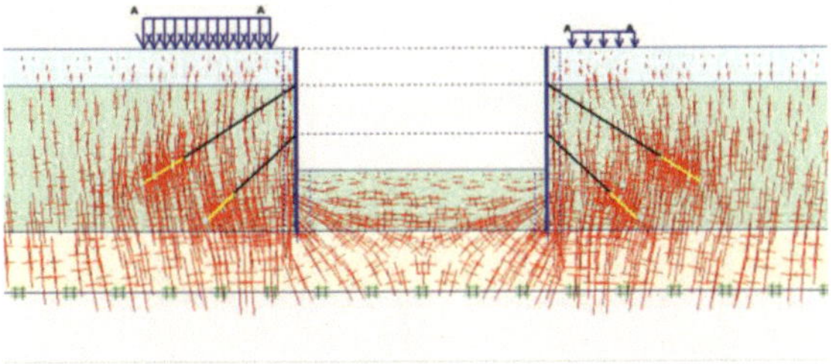

a. The effective stress seepage of foundation pit

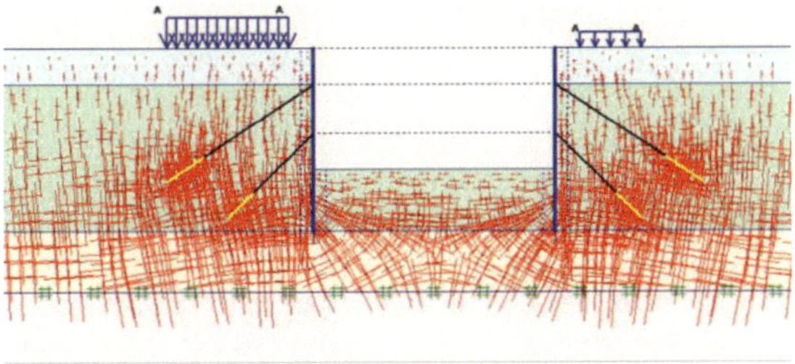

b. The total stress seepage of foundation pit

Figure 6. The stress of foundation with seepage pit.

4.3. Finite element strength reduction

To further study the impact of seepage pit stability in the original calculation step by considering the seepage and without considering, add a new step 7, reset displacement to zero, and conduct strength reduction operation. Select mutations displacement for instability criterion, select point A as the displacement point, as shown in **Figure 7**. Point A has a distance of 28 m to the left edge, and 7 m to the upper boundary. The displacement corresponding to A under different reduction coefficient is shown in **Figure 8**.

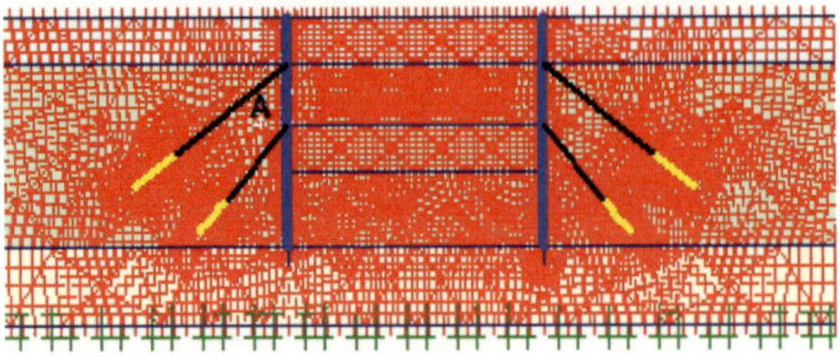

Figure 7. The position of displacement point A.

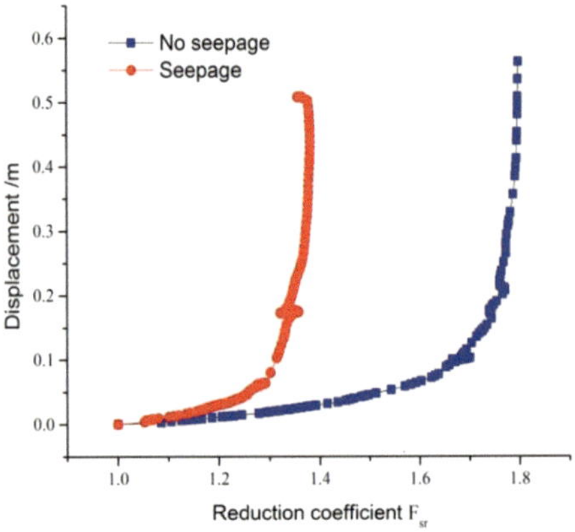

Figure 8. The corresponding values for the various displacement reduction factor.

By analyzing **Figure 8**, it can be seen that for the case when there is no seepage, point mutation displacement of A occurs between 1.6 and 1.7 reduction factor, and by combining the specific data, the stability factor of 1.67 is determined. For seepage cases, it can be seen that point A displacement occurs between 1.2 and 1.3 mutation, and the steady flow coefficient was 1.28 by combining with the specific data. By comparing these two cases, we find that the gap between large and stable coefficient is a little great, in which when seepage is not considered, the stability coefficient is 30% larger than the contrary case with more errors. The strength reduction operation further indicates that the results without considering seepage are rather dangerous. And the stability of the excavation cannot be assessed reasonably.

5. Conclusion

In this chapter, the process of numerical simulation program is excavated by applying Plaxis and combining the example. It analyzed the stability under seepage pit. The conclusions are as follows:

(1) By finite element numerical simulation, it can be seen that in most regions of the pit, soil displacement with considering seepage is larger than when not considered. And in greater flow velocity field position, displacement values are greater, too.

(2) The differences between the soil stresses can be analyzed through **Figure 5** and **Figure 6**. In effective stress when considering the seepage, there is distribution similarity in pit effective stress. With the contrary case, it has also shown great differences. When considering the seepage, most of the effective stress soil excavation area is greater than the effective stress without considering the seepage. By analyzing the total stress diagram, it can be seen that when considering the seepage, the total stress distribution has a certain relevance with flow field velocity distribution. And the increase of the total stress mainly manifests in the area with seepage, and the greater speed of the seepage is, the more obvious the increase is.

(3) Comparing these two cases through strength reduction method, it can be seen that the stability factor is larger, when seepage is not considered, and the stability coefficient is 30% larger than the contrary case with more errors.

(4) Through the overall analysis, calculated seepage is unreasonable without considering the case of seepage, which will reduce the practical significance of engineering. The impact of seepage should be taken into account when analyzing foundation stability.

Author details

Hu Qizhi, Liu Zhou*, Song Guihong and Zhuang Xinshan

*Address all correspondence to: 247366875@qq.com

Hubei University of Technology, Wuhan, China

References

Chang Hee, M., Jiqing, L. and Po, C., 2002. Reply to the discussion of "circular slip of earth slope under seepage action finite element calculation". Journal of Geotechnical Engineering, to discuss 3: 399–402.

Cheng, D., Zheng, Y. and Xiaosong, T., 2009. Using FEM strength reduction overall stability of foundation under seepage analysis. China Civil Engineering Journal, 42 (3): 105–110.

Duncan, J.M., 1996, State of the art: Limit equilibrium and finite element analysis of slopes. Journal of the Geotechnical Engineering, ASCE, 122 (7): 577–596.

Feng, H. and Po, C.,2011, Wave should be magnificent. Affect the overall stability of the excavation of soil constitutive model of strength reduction. Rock and Soil Mechanics, 32 (Suppl 2): 592–597.

Huangchun, E. and Xiaonan, G.X.L., 2001. Stability analysis considering seepage pit slope. China Civil Engineering Journal, 34 (4): 98–101.

Qiang, S., Branch, Z. and Lee, G., 2007. Calculation pit slope stability analysis considering the effect of water pressure. Journal of Engineering Geology, 15 (3): 403–406.

Quan, H. S., Jun, L. and Kong, X.B., 2008. DDA strength reduction method and its application in slope stability. Rock Mechanics and Engineering, 27 (1): 2799–2806.

Sheng, N. C., 2008. Excavation Seepage and Its Engineering Application. Wuhan: Wuhan Institute of Rock and Soil Mechanics Chinese Academy of Sciences.

Wei, S., Hang, L. and Wei, R., 2011. FLAC3D in geotechnical engineering. Beijing: China Water Power Press, 2011.

Zhonghua, X. and Weidong, W., 2010. Numerical analysis foundation environmentally sensitive choice Turkey constitutive model. Rock and Soil Mechanics, 31 (1): 258–264.

Zhuanzheng, T., Qiu, P. and Yue, W., 2012. Wuhan, a municipal channel excavation accident hazard analysis process and the lessons. Geotechnical Engineering, 34 (Suppl): 735–738.

Chapter 30

Experimental Testing of Punching Shear Resistance of Concrete Foundations

Martina Janulikova and Pavlina Mateckova

Additional information is available at the end of the chapter

http://dx.doi.org/10.5772/66807

Abstract

Foundation structures, their testing, and modeling are a wide area to research. A lot of different concrete elements are tested and modeled in the world. Analysis of interaction between the foundation structures and the subsoil has been developed for many years. For the determination of stress in foundation structure, it is needed to determine the influence of the stiffness, respectively, the pliability of subsoil to structural internal forces, and vice versa, how the stiffness of the foundation structure affects the resulting subsidence. This chapter deals with experimental tests of concrete foundation slabs. Tests are carried out at the steel test frame structure by dimension 2 × 2.5 × 5 m, which is placed open air at the Faculty of Civil Engineering in Ostrava. Tested slabs are by dimension 2 × 2 m and have different thickness between 100 and 200 mm. A lot of physical quantities are tested in those experiments and experiments are then multidisciplinary because geotechnical, acoustic, strain gauges, and deformation measurements are conducted. This chapter addresses especially with punching shear analysis and maximum punching resistance. A number of experimental tests of concrete foundation slabs were carried out. Slabs classically reinforced, prestressed, or FRC were tested, but slabs were not reinforced with shear reinforcement. During the experiment, the interaction between the concrete foundation and the subsoil was monitored. Most of the slabs were disrupted by punching shear. If the slab was disrupted by punching shear, dimension and shape of the punching failure were monitored and measured, and results were compared between them. Last but not the least, results from the experiment and results according to design methods used in EC2 are compared in this chapter. The maximum shear design force according to EC2 was lower than the one from the experiment.

Keywords: punching shear analysis, punching, shear resistance, soil–foundation interaction

1. Introduction

A series of experimental test are carried out at the Faculty of Civil Engineering through various project. Most of them concentrate on interaction between subsoil and foundation structures because it is very interesting and an important field of research in civil engineering. Foundation structure is the most important part of whole structure and their quality has an important effect on quality of buildings. Properly designed and carried out foundations can be used for a very long time. These foundations can influence durability of the building. On the contrary, wrongly designed and carried out foundation can cause a lot of problems. For the right design of foundation structure it is necessary to know the behavior of the concrete foundation on the subsoil. For this reason experimental test of foundation slab are performed. Slabs with dimension 2 × 2 m and with thickness 0.1–0.2 m are tested with concentrated load. Load is introduced through distributing plates with dimension 0.2 × 0.2 m or 0.4 × 0.4 m. Slabs are reinforced with classic reinforced bars, with prestressed bars, as FRC concrete or its combination. This chapter is focused on classically reinforced slabs.

2. Experimental tests

2.1. Steel test equipment

Aforementioned tests are performed on the steel frame test equipment (**Figures 1** and **2**) which is placed outside the premises of the Faculty of Civil Engineering in Ostrava (Czech Republic) (Cajka, 2014; Cajka et al., 2016a,b; Mynarcik et al., 2016; Buchta et al., 2016). The basic principle on this equipment is clear from **Figures 1** and **2**.

Steel frame is built on concrete strips which are anchored in the soil using micropiles to ensure the bearing capacity and to prevent the lifting of steel frame. Foundation slab is concreted under steel frame approximately in the center between foundation strips to prevent influencing of test results by eccentric placement of slab between foundation strips. Tested foundation slabs are loaded with vertical force which is introduced using system of steel attachment which can be changed according to the thickness of the tested slab. These steel attachments are placed on hydraulic press which makes vertical force on the foundation. Maximal force which can be developed is 1000 kN. Under the steel frame is the original subsoil which consists from clayey soil.

2.2. Basic principle and course of test

This chapter deals with slab of dimension 150 × 2000 × 2000 mm from the concrete C35/45 reinforced by hand-knotted mesh 8/100/100 (**Figure 3**). Average value of characteristic compressive strength was 47.6 MPa, which means that it is rather a concrete class C45/55. It is probably caused by the long time between concreting and testing (about 4 months). Loaded area of 400 × 400 mm large was chosen. The soil is going to creep during the loading of foundations so the load is introduced in steps. Steps 50 kN after 30 minutes were chosen in this test. In each step load was introduced and 30 minutes to keep calm because of creep. Because of the subsoil's behavior a longer period would be better but 30 minutes is a compromise with regard to the

http://dx.doi.org/10.5772/66807

feasibility of the test which has to be executed in one day. **Figure 3** shows the concrete slab used in the test.

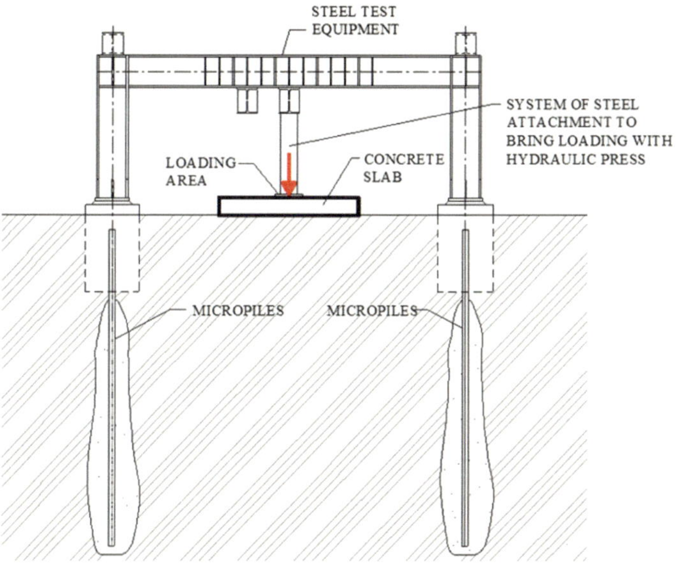

Figure 1. Scheme of steel test equipment.

Figure 2. Photo of steel test equipment.

However, the calculated value of bearing capacity was much lower than it was decided during the testing to test this slab, maximum 750 kN at first. But the slab was not corrupted at the first set of loading. Then it was decided to conduct a second test and in this test 1000 kN was to be applied, which is also the maximal bearing capacity of a steel testing equipment.

Figure 3. Scheme of tested slab.

3. Results

3.1. Deformation of concrete foundation

Deformations of the concrete slab are monitored using 16 sensors, see **Figure 3**. In the graph of **Figure 4** are shown deformations at the first set of test of this slab. It is clear from this figure that a great part of the deformation was returnable. It means that the majority of the test was performed in elastic area.

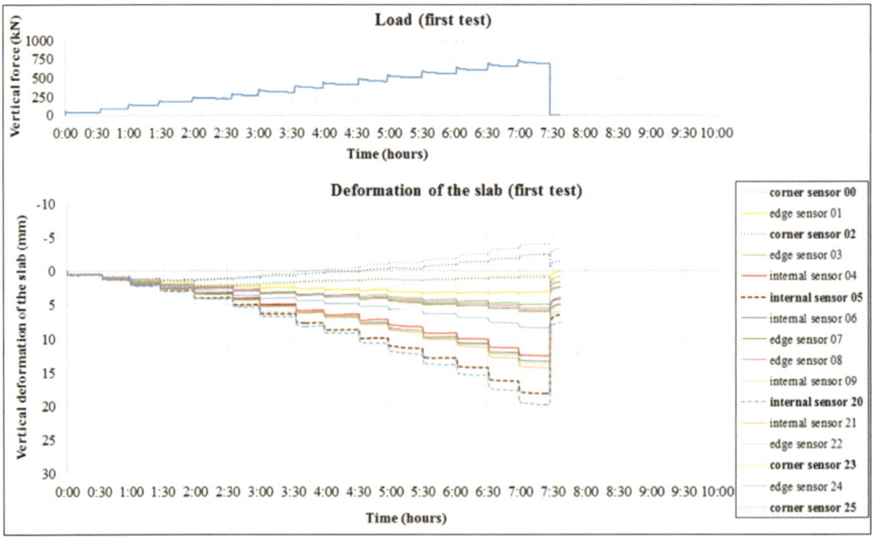

Figure 4. Deformation of the foundation slab (first set of measurement).

http://dx.doi.org/10.5772/66807

On this base a second test on the same slab was carried. In this test the slab was corrupted by punching shear at a force of 945 kN. Results from this test are shown in **Figure 5**.

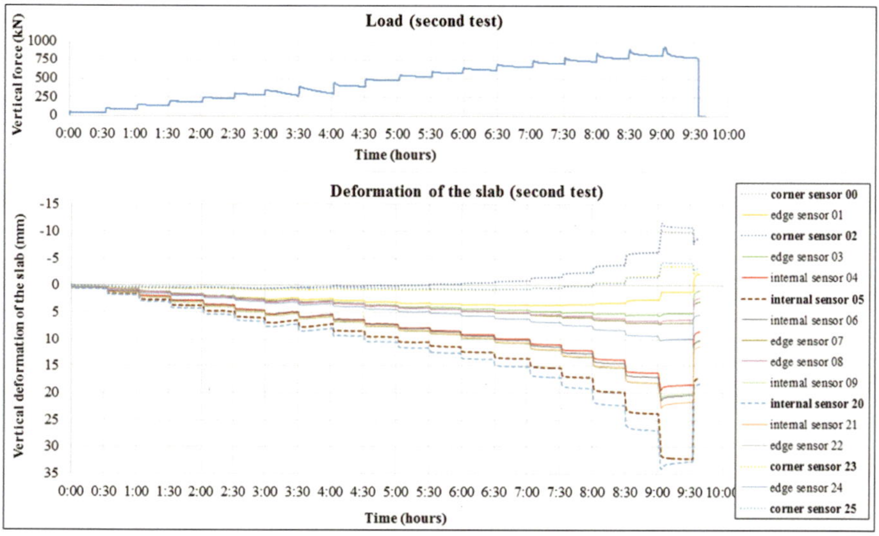

Figure 5. Deformation of the foundation slab (second set of measurement).

Results from the test are used for the analysis of other numerical modeling (Cajka et al., 2016a,b) or comparison with other computing methods (Labudkova and Cajka, 2015).

3.2. Punching shear failure

Because several slabs were failed by punching, shear attention is focused on punching shear analysis and punching shear failure monitoring which is a wide area to research (Hegger et al., 2007, Siburg and Hegger, 2014; Siburg et al., 2014). Punching capacity is compared through calculation according to Eurocode 2.

Slabs were not reinforced with shear reinforcing so bearing capacity were calculated according to the equation for element without shear reinforcement which is valid in the interval $\langle a; 2d \rangle$:

$$v_{Rd} = C_{Rd,c} \cdot k \cdot (100 \cdot \rho_1 f_{ck})^{\frac{1}{3}} \cdot 2d/a \geq v_{\min} 2d/a \qquad (1)$$

where f_{ck} is in MPa, and a is the distance from the periphery of the column to the control perimeter considered.

The maximal shear stress for the described slab according Eq. (1) is 0.999 MPa and from this value maximal possible applied force were calculated that causes the slab damage is 393 kN (with characteristic values) and 190 kN (according to EC including all safety coefficient). The achieved value was 945 kN at the second test. In **Figure 6** are shown cracks on the bottom surface of the slab and **Figures 7** and **8** show cuts on this slab.

Figure 6. Cracks on the bottom surface of the slab.

Figure 7. Lateral cuts of slab.

Figure 8. Diagonal cuts of slab.

4. Conclusion

The experimental test on the concrete foundation slab was introduced in this chapter. The calculated value for the punching shear resistance was 392.6 kN (characteristic value). This

http://dx.doi.org/10.5772/66807

slab was corrupted by force 945 kN. The real value of shear resistance is, as expected, larger as according to Eurocodes. That means that Eurocodes are of course on the safe side. In this case, the real resistance was more than two times higher than the calculated value.

Acknowledgements

This outcome has been achieved with the financial support of the project GACR No. 16-08937S "State of stress and strain of fiber reinforced composites in interaction with the soil environment." In this undertaking, theoretical results gained by conceptual development of research, development, and innovations for 2016 at the VŠB-Technical University of Ostrava (granted by the Ministry of Education, Youths and Sports of the Czech Republic) were partially exploited.

Author details

Martina Janulikova* and Pavlina Mateckova

*Address all correspondence to: martina.janulikova@vsb.cz

Faculty of Civil Engineering, VSB-Technical University of Ostrava, Ostrava, Poruba, Czech Republic

References

R. Cajka. Comparison of the calculated and experimentally measured values of settlement and stress state of concrete slab on subsoil. Applied Mechanics and Materials. 501–504, 867–876, 2014. DOI: 10.4028/www.scientific.net/AMM.501-504.867.

R. Cajka, P. Mynarcik, J. Labudkova.. Experimetal measurement of soil-prestressed foundation interaction. International Journal of GEOMATE. 10, 2101–2108, 2016a.

R. Cajka, J. Labudkova, P. Mynarcik. Numerical solution of soil-foundation interaction and comparison of results with experimetal measurements. International Journal of GEOMATE. 11, 2116–2122, 2016b.

P. Mynarcik, J. Labudkova, J. Koktan. Experimental and numerical analysis of interaction between subsoil and post-ensioned slab-on-ground. Jurnal Teknologi. 78, 23–27, 2016. DOI: 10.11113/jt.v78.8530.

V. Buchta, M. Janulikova, R. Fojtik. Experimental tests of reinforced concrete foundation slab. Procedia Engineering. 114, 530–537, 2016. DOI: 10.1016/j.proeng.2015.08.102.

R. Cajka, J. Labudkova. Numerical modeling of the subsoil-structure interaction. Key Engineering Materials. 691, 333–343, 2016. DOI: 10.4028/www.scientific.net/KEM.691.333.

J. Labudkova, R. Cajka. Comparison of the results from analysis of nonlinear homogeneous and nonlinear inhomogeneous half-space. Procedia Engineering. 114, 522–529, 2015. DOI: 10.1016/j.proeng.2015.08.101.

C. Siburg, J. Hegger. Experimental investigations on the punching behaviour of reinforced concrete footings with structural dimensions. Structural Concrete. 15, 331–339, 2014. DOI: 10.1002/suco.201300083.

J. Hegger, M. Ricker, B. Ulke, M. Ziegler. Investigations on the punching behaviour of reinforced concrete footings. Engineering Structures. 29, 2233–2241, 2007. DOI:10.1016/j.engstruct.2006.11.012.

J. Hegger, G. A. Sherif, M. Ricker. Experimental investigations on punching behavior of reinforced concrete footings. ACI Structural Journal. 103, 604–613, 2006.

C. Siburg, M. Ricker, J. Hegger. Punching shear design of footings: critical review of different code provisions. Structural Concrete. 15, 497–508, 2014. DOI: 10.1002/suco.201300092.

Chapter 31

Finite Element Analysis on Seismic Behavior of Ultra-High Toughness Cementitious Composites Reinforced Concrete Column

Jun Su and Jun Cai

Additional information is available at the end of the chapter

http://dx.doi.org/10.5772/66808

Abstract

In order to study the seismic behavior of ultra-high toughness cementitious composites reinforced concrete column, the concrete columns were simulated based on the finite element program OpenSees. The simulated hysteresis curves and skeleton curves were in good agreement with the test curves. The results well reflected the seismic performance of ultra-high toughness cementitious composite (UHTCC) reinforced concrete columns under earthquake and showed that the constitutive relation and the related parameters had good applicability for the simulation of fiber concrete columns. The UHTCC reinforced concrete column had higher bearing capacity and energy dissipation capacity.

Keywords: ultra-high toughness cementitious composites, concrete column, low cyclic loading, finite element analysis

1. Introduction

In the earthquake, reinforced concrete columns often lead to plastic hinge under compression, bending, and shear. Shear failure occurs frequently, such as protection layer spalling, reinforcement exposed, concrete crushing, deformation of steel bars, and even overall collapse. In the Seismic code GB50011 (2010), shear stirrups are allocation at the end of column and the diameter, spacing and reinforcement length are stipulated. But, due to the construction factors, the connection between confined concrete and protective layer cannot be guaranteed and leading to hidden danger.

Ultra-High toughness cementitious composite (UHTCC) is a kind of high performance cement matrix composite based on micromechanics and fracture mechanics (Xu and Li, 2008).

It shows the obvious characteristic of strain hardening and high toughness under tension and shearing force and it strengthens the softening performance of traditional cementitious material. Moreover, the characteristics of stable cracking effectively improved the durability and make the deformation coordinate with the steel bar. In view of this, based on the test research, this chapter uses OpenSees to simulate and analyze the seismic behavior of UHTCC reinforced concrete columns and provide reference for engineering application.

2. Numerical model of OpenSees

Concrete column adopted the flexibility-based fiber model in OpenSees program, for every root fibers only consider the axial constitutive relation, and every fiber can be defined different constitutive relations.

Concrete constitutive relation used the Concrete 02 model (Based on the Kent and Park (1971) uniaxial concrete constitutive model), which reflects the restriction of the stirrup by considering the peak stress, peak strain, and the softening curvature of compressive concrete.

A steel constitutive model adopted the Steel 02 material provided by OpenSees. This model considers the double broken line constitutive relation to reflect the Bauschinger effect and have good stability (Liu et al., 2012).

As the constraint effect of stirrups, the section is divided into cover concrete and core concrete according to the different stress-strain relations of the protective layer and confined concrete. **Figure 1** shows the column fiber section. The core concrete is divided into 10 × 10, a total of 100 grids, and adopted the confined concrete constitutive model.

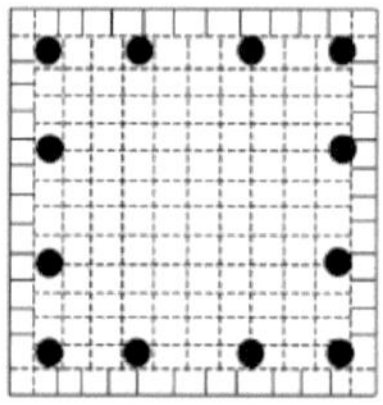

Figure 1. Sketch of column section fiber model.

3. Model validation

The quasistatic test of side and middle columns under low cyclic loading was provided in the study by Tang (2011). The column specimens were numbered Za1, Za2, Zb1, and Zb2: a represented the middle column and b represented the side. The remaining parameters are given in **Table 1**. Comparison of hysteretic curves and skeleton curves between the test and OpenSees is shown in **Figures 2** and **3**. **Table 2** shows the peak load.

Specimens	Dimension (mm²)	Length (mm)	Strength (MPa)	Longitudinal reinforcement	Hoop reinforcement
Za1	200 × 200	850	30.1	8Φ8	Φ6/70
Za2	200 × 200	850	30.1	8Φ8	Φ6/70
Zb1	200 × 200	850	30.1	4Φ10 + 4Φ8	Φ6/70
Zb2	200 × 200	850	30.1	4Φ10 + 4Φ8	Φ6/70

Table 1. Specimen size and reinforcement.

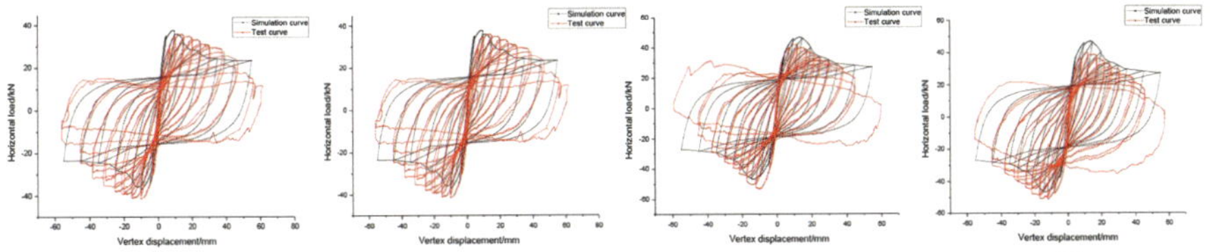

(a) (b) (c) (d)

Figure 2. Comparison of Hysteresis curve. (a) Za1, (b) Za2, (c) Zb1, (d) Zb2.

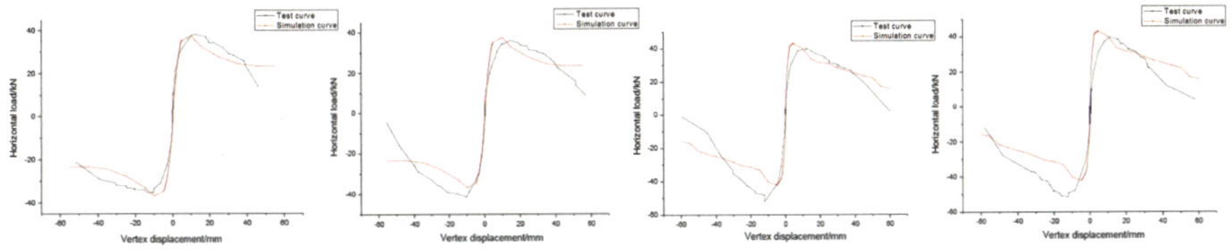

(a) (b) (c) (d)

Figure 3. Comparison of Skeleton curve. (a) Za1, (b) Za2, (c) Zb1, (d) Zb2.

Figure 2 shows that the hysteresis curves based on OpenSees were in good agreement with the test results in initial stiffness, pinch degree, and trend. The hysteresis curves showed linear change before Za, Zb yield, and the stiffness degradation was not obvious. But, in the processes of loading and unloading, the stiffness degradation increased gradually. Finite element simulation could well reflect the phenomenon of column crack open and close, develop, and pinch. There are some differences during the late period of the loading process, and the simulation curves were very full. The main reasons were as following: generally, the column specimens were loaded to 85% limit load, but this specimen was loaded to collapse. It is difficult to accurately reflect the stiffness degradation and cumulative damage in the later stage of loading. In the stress-strain model, the parameters of the descending segment and the strength coefficient of steel bars were not very reasonable.

Specimens	Test value (kN)		Simulation value (kN)		Error (%)	
	Positive	Negative	Positive	Negative	Positive	Negative
Za1	38.003	-35.446	37.600	-36.828	1.06	3.90
Za2	36.002	-41.465			4.44	11.18
Zb1	42.059	-53.081	46.650	-46.673	10.92	12.12
Zb2	39.613	-51.435			17.76	9.26

Table 2. Peak load of simulation and test.

4. Numerical simulation of UHTCC reinforced concrete column

4.1. Design of column specimen

According to the size and reinforcement shown in **Table 1**, the model of UHTCC reinforced column was established and numbered UZ1 and UZ2.

The characteristic value of cube strength took $f_{cu} = 40$ MPa, and the elastic modulus was $E_c = 17{,}000$ MPa, stiffness decreased to 0.1 E_c when unloading. As the confinement effects from hoop steels, the strength increasing coefficient K took 1.1. The longitudinal reinforcement used HRB335 and strength variable coefficient took $\delta = 0.07$, the modulus of elasticity was 2.00×10^5 N/mm^2 according to the specification. The specific parameters are summarized in **Tables 3** and **4**.

Areas	f_c/N/mm^2	ε_c	ε_u	f_t/N/mm^2	E_c/N/mm^2
Cover concrete	40	0.005	0.034	5.98	1.7×10^4
Core concrete	44	0.00517	0.0577	5.98	1.7×10^4

Note: ε_c, ε_u are peak strain and ultimate strain, respectively. f_t represents the ultimate tensile strength.

Table 3. UHTCC column parameters.

Diameter	f_y/N/mm^2	E_s/N/mm^2	b	R_0	cR_1	cR_2	a_1–a_4
6	441	2.00×10^5	0.01	12	0.800	0.13	0
8	582	2.00×10^5	0.01	12	0.800	0.13	0
10	481	2.00×10^5	0.01	12	0.800	0.13	0

Table 4. Reinforcement parameter.

4.2. Finite element simulation and analysis

The comparisons of UZ1 and UZ2 with ordinary column are shown in **Figure 4**.

Bearing capacity. The yield and ultimate load of UHTCC column were significantly higher than that of ordinary column. The average yield load of concrete column was 35.31 kN, the

http://dx.doi.org/10.5772/66808

average ultimate load was 40.25 kN, and the UHTCC column was 40.53 and 47.93 kN, respectively, which was increased by 14.8 and 19.1%.

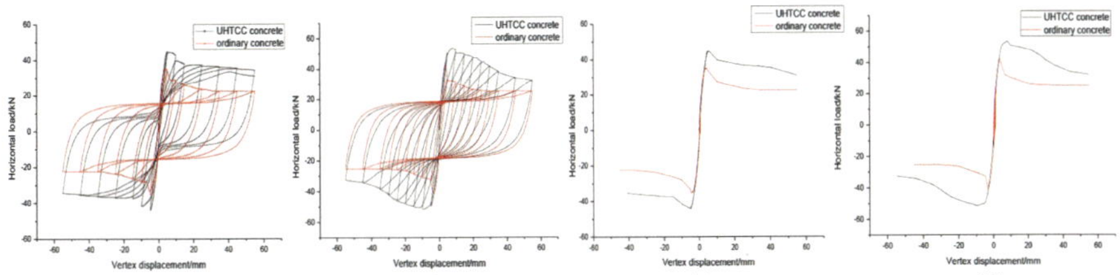

(a) (b) (c) (d)

Figure 4. Comparisons of hysteresis curves and skeleton curves. (a) Za1, UZ1. (b) Zb1, UZ2. (c) Za1, UZ1. (d) Zb1, UZ2.

Hysteresis curves and skeleton curves. The hysteresis loops of UHTCC columns (UZ1 and UZ2), in contrast to that of ordinary column, was fuller. Its linear stage was longer, the deformation of elastic stage was slight and the slope change was not obvious. Although the stiffness and strength in the late stage were gradually reduced, the degradation became flat.

Deformation and energy dissipation. The yield points P_y and Δ_y was calculated by the energy method, meanwhile, defined the vertex of curves as the ultimate load P_{max} and ultimate displacement Δ_{max}. The failure load P_u equaled 0.85 P_{max} and the corresponding displacement was Δ_u. The ductility coefficient was defined as the ratio of the ultimate displacement to the yield displacement; **Figure 5** shows the method. The equivalent viscous damping coefficient h_e was used to evaluate the capacity of energy dissipation. In **Figure 6**, $h_e = (S_{BEF} + S_{EDF})/2\pi(S_{AOB} + S_{COD})$. The simulation results are shown in **Table 5**.

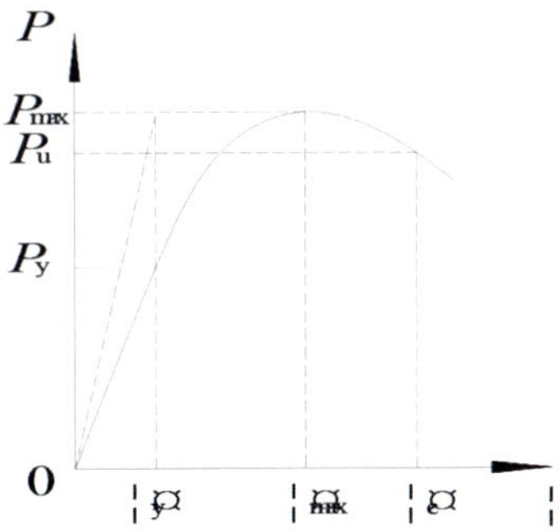

Figure 5. Determination of characteristic points damping.

As shown in **Table 5**, the ductility coefficient of UHTCC column is higher than that of ordinary column. With the cycle of load, the decline of the hysteresis curve of UHTCC columns occurred slowly. Compared with Za1 and Zb1, the viscous damping coefficients of UZ1 and

UZ2 were increased by 14.1 and 18%, respectively. It indicated that the high toughness of UHTCC can effectively improve the deformation and energy dissipation capacity of concrete columns.

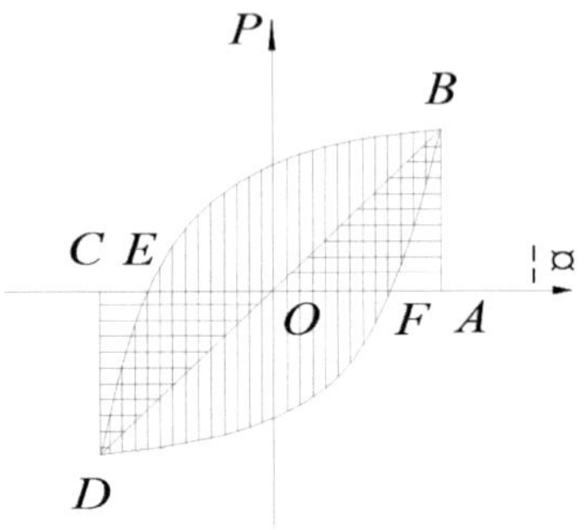

Figure 6. Determination of equivalent viscous coefficient.

Specimen	Δu Positive	Negative	Δy Positive	Negative	Δu/Δy Positive	Negative	h_e
Za1	11.4	-11.2	4.0	-4.1	2.85	2.73	7.34
UZ1	20.6	-19.3	4.8	-4.7	4.29	4.11	8.37
Zb1	10.6	-10.2	4.1	-3.9	2.59	2.62	7.21
UZ2	28.3	-29.1	5.1	-5.3	5.55	5.49	8.51

Table 5. The ductility coefficient and energy dissipation of each specimen.

4.3. Analysis of influence factor

Axial compression ratio. In the case of other factors remained constant, changed the vertical axial force, and calculated the bearing capacity of UHTCC columns under different axial compression ratios. The results are shown in **Figure 7**. In the range of axial compression ratio less than 0.7, the horizontal bearing capacity and ultimate displacement increased with the increase of the axial compression ratio.

Volume-stirrup ratio. If the axial compression ratio remains constant then it can be seen than calculated bearing capacity of UHTCC columns is influenced by changing volume-stirrup ratios. It can be seen that, in the case of other factors unchanged, the horizontal bearing capacity of the specimens increased little with the increase of the volume-stirrup ratio (**Figure 8**).

http://dx.doi.org/10.5772/66808

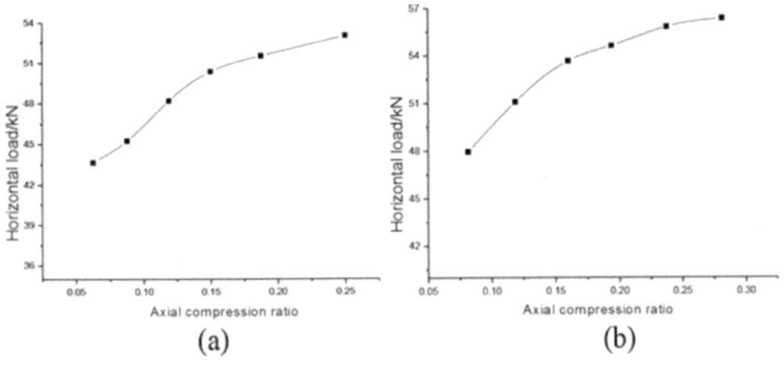

Figure 7. Relation curve of bearing capacity and axial compression ratio. (a) UZ1 and (b) UZ2.

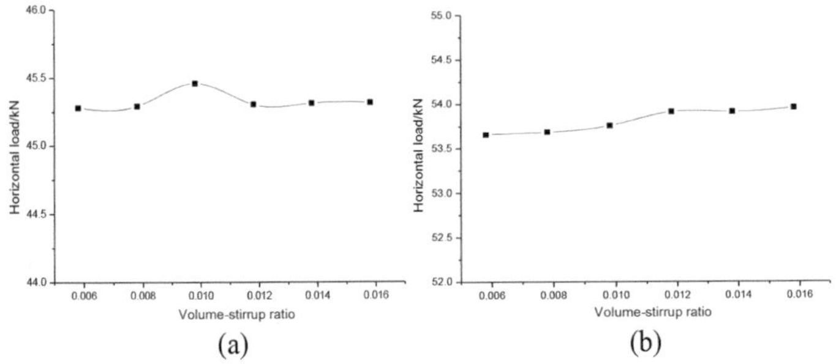

Figure 8. Relation curve of bearing capacity and volume-stirrup ratio. (a) UZ1 and (b) UZ2.

5. Main conclusions

The seismic behavior of UHTCC reinforced concrete column based on OpenSees finite element program was analyzed in this chapter, and the conclusions were as follows:

The flexibility-based fiber model, the Concrete 02 model, and the Steel 02 material can exactly simulate the hysteresis characteristic and energy dissipation of columns under low cyclic loading, and it verified the reliability of OpenSees.

The stiffness of hysteretic curves of the test degenerated obviously, but the simulated curves declined relatively flat at the later stage. On the one hand, it was because the model did not fully take the interaction of various parameters into consideration, on the other hand, although the Steel 02 material considered the Bauschinger effect, the fatigue effect of steel bar under low cyclic loading had not been embodied, so that the decline became flat.

Compared with the ordinary column, the UHTCC column had higher yield strength and ultimate strength, and the UHTCC can effectively improve the ductility. With the cycles increased, the stiffness degradation became flat. The higher viscous damping coefficient also indicates that its energy dissipation capacity was better than that of ordinary column.

The finite element simulation results of the lower axial compression ratio were closer to the test results. With the increase of the axial compression ratio, the horizontal-bearing capacity increased and the specimens under the lower axial compression ratio had better deformation performance. Under the same conditions, the bearing capacity has no significant change with the increase of the volume-stirrup ratio.

Acknowledgements

The research reported in this chapter was made possible by the financial support from the School of Civil Engineering and Architecture, Hubei University of Technology. The authors would like to express their gratitude to this organization and Mr. Su for the support.

Author details

Jun Su and Jun Cai*

*Address all correspondence to: 505742800@qq.com

Hubei University of Technology, Wuhan, China

References

GB50011-2010. Ministry of Construction of the People's Republic of China. Code for seismic design of buildings, Beijing: China Architecture & Building Press, 2010.

D. C. Kent, R. Park. Flexural members with confined concrete, ASCE, 97 (7): 1969–1990, 1971.

L. J. Liu, M. M. Jia, X. H. Yu. Study of constitutive relations for steel-hoop-confined concrete material. Industrial Construction, 42: 188–191, 2012.

D. Y. Tang. Experimental research and theoretical analysis on the seismic collapse resistance of equal span RC frame structures. Beijing: Tsinghua University, 2011.

S. L. Xu, H. D. Li. A review on the development of research and application of ultra high toughness cementitious composite. Civil Engineering Journal, 41 (6): 45–59, 2008.

Chapter 32

Load Transfer Coefficient of Transverse Cracks in Continuously Reinforced Concrete Pavements Using FRP Bars

Chunhua Hu and Liang Chen

Additional information is available at the end of the chapter

http://dx.doi.org/10.5772/66809

Abstract

Continuously reinforced concrete pavement (CRCP) with fiber reinforced polymer bars (FRP) shows several advantages over the conventional CRCP, such as no erosion, lightweight, and low modulus. And the load transfer performance of the transverse cracks affects its service life. As the traditional evaluation system of crack load transfer failed to eliminate the influence of pavement deflection, the simple deflection value ratios for the cracks of the load transfer coefficient are not accurate enough to evaluate the load transfer performance of transverse crack in FRP-CRCP. In order to further understand the effects and improve the current load transfer coefficient of transverse cracks in FRP-CRCP, the relationship between the deflection value and load transfer coefficient in FRP-CRCP was analyzed. Using this approach, a more accurate prediction equation has been put forward to express the load transfer capacity.

Keywords: continuously reinforced concrete pavements, transverse cracks, vehicle load, load transfer, deflection value

1. Introduction

The aim of this chapter is to modify the current load transfer coefficient of transverse cracks in FRP-CRCP and put forward an accurate formula. CRCP as a type of high performance concrete pavement has high strength, good evenness, comfortable driving, long service life, and low maintenance cost advantages. More and more people pay attention to the research and application of CRCP (Zhigang Zhou and Qisen Zhang, 2000; Changshun Hu et al., 2001). The spacing and width of transverse cracks in the pavement can influence the performance of CRCP. Tabatabaie (1978), Huang (1985), and Guo et al. (1995) developed the illi-slab,

KENSLABS, JSLAB, and other 2D finite element analysis programs to analyze the rigid pavement. According to Winkler foundation of the theory of elastic thin plates, they used beam elements and spring elements to simulate the load transfer function model. At present, there are two problems in the study of load transfer coefficient of transverse cracks in FRP-CRCP. On the one hand, simulation of a transverse crack model in finite element is easy to implement, but the related research which determines the load transfer ability of transverse cracks is relatively few. On the other hand, currently it is difficult to obtain the accurate shear stiffness of transverse cracks in FRP-CRCP through the existing calculation method, and thus it is impossible to accurately given crack prediction equation.

2. Load transfer mode of transverse crack in CRCP

The transfer of shear forces depends on parameters of the shear surface aggregate. The load transfer ability depends on the transverse crack width, the shape of the aggregate, the relative stiffness of the plate and foundation, and times of load action. Virtual equivalent filler simulation is in the cracks using an equivalent filler to simulate the packing, and cracking of concrete mechanical interlock action can be carried out by adjusting the equivalent sealing material stiffness to the realization of load transfer ability of the simulation (Jian-ming and Sheng-fei, 2009).

It is assumed that there are embedded transverse cracks in concrete on both sides of a short rod and there are small shear force and bending and toque moments. In addition to the plate and the foundation of the relative stiffness, factors that influence the load transfer ability includes gap widths and dowel bar parameters and construction factors, such as spacing, diameter, length, and the elastic modulus.

3. Evaluation system for load transfer efficiency of transverse crack in FRP-CRCP

For estimate of the load transfer capacity of transverse cracks in FRP-CRCP, the method is that the wheel load is applied at the side of plate, and measures capacity of load transfer from the direct-bearing plate to the nondirect-bearing plate. The rate of load values applied on plates on either hand of crack can show intuitively the load transfer efficiency (%), as in Eq. (1):

$$K_j = \frac{P_2}{P_1} \text{x} 100\% \tag{1}$$

where k_j is the coefficient of load transference for transverse crack in pavement structure; P_1 is the load applied on half of direct-bearing plate; P_2 is the load transmit from wheel pressure to nondirect bearing plate; $P = P_1 + P_2$, the sum of P_1 and P_2 is the total wheel pressure applied on the cracking location. It is impossible to accurately given crack prediction equation.

http://dx.doi.org/10.5772/66809

4. Experimental model of the FRP-CRCP

Three-dimension finite element analysis (FEA) is adopted to analyze the selected pavement structure. Running direction is along the x direction; pavement transverse section is along the y direction; and vertical upward direction of pavement is along the z direction. The calculating width of the x direction is 8 m, the calculating width of the y direction is 6 m, and the calculating width of the z direction is 2 m (exclusive of structure thickness), in order to decrease influence of boundary constraint on the stress ability of model structure, according to grounding wire size and character of perpendicular stress. Longitudinal reinforcement design for surface course in continuous reinforced concrete pavement requires that maximal crack width is less than 1 mm. Breadth of transverse fissure in this chapter is 1 mm.

In consequence of FRP bars set transversely and longitudinally in FRP-CRCP, shrinkage and temperature shrinkage in the process of hardening of concrete is contained. So, some fine transverse cracks are produced pavement in slab. The continuous reinforced concrete pavement becomes the pavement structure with cracks. So, when three-dimensional finite element model is established, influence of steel on serviceability of pavement structures must be considered. The number of model elements varies with the crack width. In order to improve computational accuracy and consider running speed, the model dimension of reinforced continuous reinforced concrete pavement plate is 0.1 m after much experimenting. The model dimension of base course, subbase course, and soil base course are 0.5 m. In this way, both computing speed and calculation accuracy are enhanced suitably. This chapter analyzes mainly the force characteristics of the structure and load transfer capacities in the same section in where sheet fissure occurs in FRP-CRCP with 0.5 m fracture interval. **Figure 1** shows the schematic diagram of pavement structure.

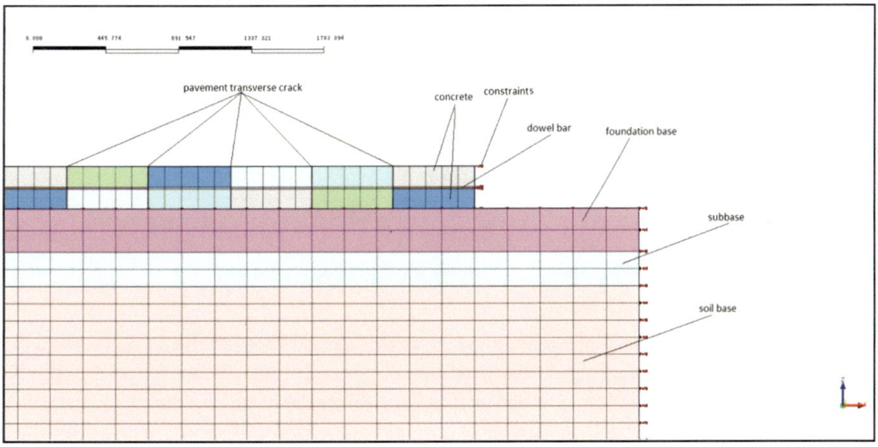

Figure 1. Schematic diagram of pavement structure.

As a numerical computation method, finite element method's computational accuracy is concerned with element size, model dimension, and so on. For pavement structure, the foundation is essentially semi-infinite elastic solid and its range is infinite in the horizontal and vertical direction. As a result of the incompatibility between the finite element method and elastic half space theory, the foundation size must be appointed. In the process of appointment for foundation size, on the assumption that the other parameters are invariable, foundation size will enlarge gradually, so stress in cement concrete pavement slab will also be stable gradually. The foundation size is the adopted value, when the result is convergent.

5. Results and discussion

To research foundation stiffness on the panel, ordinary cement was selected for the concrete pavement, load was applied in the middle of the pavement transverse crack, the pavement slab under the layer modulus was converted into an equivalent modulus and the foundation modulus was used, the selected foundation moduli were of 40, 100, 150, 200, 250, and 300 MPa. Calculation of top plate deflection variation degree difference and slab maximum principal stress with the foundation modulus, top plate deflection difference, and the maximum principal stress changes are shown in **Figures 2** and **3**.

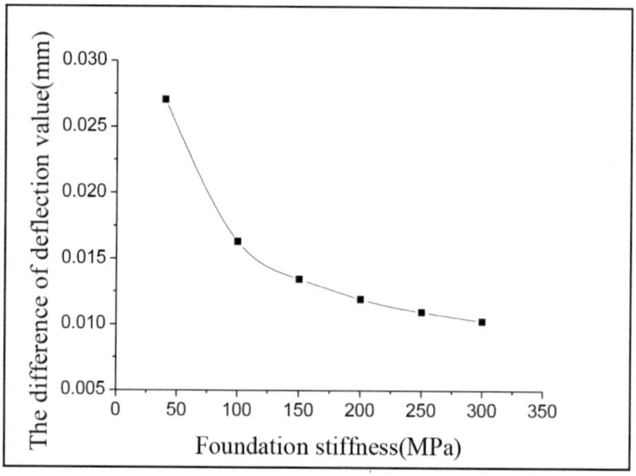

Figure 2. Relationship between the top deflection of the plate and the foundation stiffness.

Figure 2 shows that the top plate deflection differences decreases with: the increase of elastic modulus of the foundation. Foundation modulus of pavement has effect on load transfer to some extent. As the foundation modulus varies from 40 to 300 MPa, surface plate deflection difference decreases from 0.02706 to 0.01028 mm with reduction of 62%. However decreased degree of surface plate deflection difference gradually decreases with the increase of elastic modulus of the foundation. Even in the severe cracking, where there is no interlocking

http://dx.doi.org/10.5772/66809

concrete function of load transfer and without the dowel bar load condition, the results show that the deflection difference exists between the loading plate and unloading plate. The traditional evaluation method of pavement crack load transfer coefficient has been used with no consideration for the influence of base and foundation on load transfer capacity of cracks in the pavement structure.

Figure 3 shows with the increase of foundation modulus the bottom plate of the maximum principal should stress σ_{max} reduced. When the foundation modulus from 40 to 300 Mpa, the maximum principal should stress σ_{max} from 1.3543 fell to 0.7319 MPa with 45% of reduction, but the maximum principal stress decreased with the increase of elastic modulus of the foundation and gradually weakened. Obviously, the good foundation support condition has the very big function to reduce the CRCP plate bottom stress.

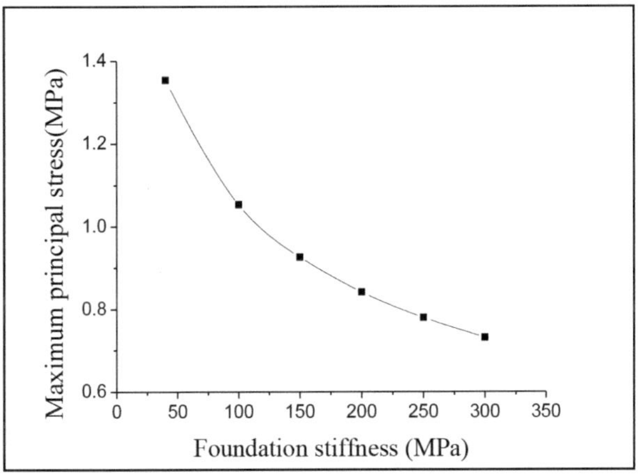

Figure 3. Relation curve of maximum principal stress and foundation stiffness.

6. Evaluation index of load transfer coefficient of transverse cracks in CRCP

For the convenience of the cracks on both sides of the bending relationship with transverse crack load transfer coefficient of deflection, so the definition of the difference of deflection value d_d and relative difference of deflection value d_r is shown in Eqs. (2) and (3):

$$d_d = d_2 - d_1 \tag{2}$$

$$d_r = \frac{d_d + 10^{-6}}{d_1} \times 100\% = \frac{(d_2 - d_1 + 10^{-6})}{d_1} \times 100\% \tag{3}$$

d_1: bending deflection value of the load plate on one side of the crack, mm;

d_2: the deflection of the load plate on the other side of the crack is not directly loaded, mm.

In the selected model to maintain constant computation parameters, single axle standard static load, tire pressure of 0.7 MPa, positions of the load for crack side plate, loose pavement without power transmission, pavement cracks in different degree of damage with different load transfer capacity and crack stiffness were used. There can be gradual changes in fracture load transfer coefficient values at different levels of damage, variation analysis deflection of CRCP pavement cracks at the time crack different stiffness and different load transfer capability. **Table 1** shows the results of simulated crack rigidity and cracks around the top plate deflection difference.

As can be seen from **Table 1**, the absolute deflection value of CRCP road surface decreases with the increase of the stiffness of the crack, and the transfer capacity can be increased. With more serious cracks and damage, crack spread load quickly decreases. When cracks lost the transmission capacity of load and dowel bar if loose, the cement panel bending heavy difference close to 0.006 mm, the continuous reinforced concrete pavement service life is greatly reduced.

Crack stiffness (%)	0	0.5	1	5	10	20	40	60	80	100
dd ($\times 10^{-3}$ mm)	3.35	2.32	1.79	0.62	0.42	0.18	0.11	0.08	0.05	0.05
dr ($\times 10^{-3}$)	2.58	1.82	1.41	0.49	0.28	0.14	0.09	0.06	0.04	0.039

Table 1. Load position when the second lower pass crack stiffness change on both sides of the plate deflection difference d_d relative deflection difference d_r.

As can be seen from **Figure 1**, simulation results of absolute deflection difference Si bearing capacity satisfy the exponential relationship, the correlation coefficient is 0.98501 that CRCP in this model of pavement structure crack bending heavy difference with crack load transfer ability with good linear relationship, the fitting formula is shown as Eqs. (4) and (5).

$$y = 0.00134\,x^{-1.09771} \tag{4}$$

$$k_C = 0.00134 \times \left(\frac{d_2 - d_1 + 10^{-6}}{d_1}\right)^{-1.09771} \times 100\% \tag{5}$$

k_C: continuously reinforced concrete pavement load transfer coefficient, %;

d_1: The deflection value of the load plate on one side of the crack directly, mm;

d_2: On the other side of the crack is not directly under the load plate deflection value, mm.

The fitting equation (5) eliminates the effects of subgrade layer by use of subtraction of deflections at the top and at the bottom, the traditional load transfer coefficient is calculated by

http://dx.doi.org/10.5772/66809

falling weight deflection experiment. It is defined as the ratio of deflection for loaded and unloaded plate. No consideration is taken into account for the offset effect of the soil base to the deflection, which is influenced by the simulation (**Figure 4**).

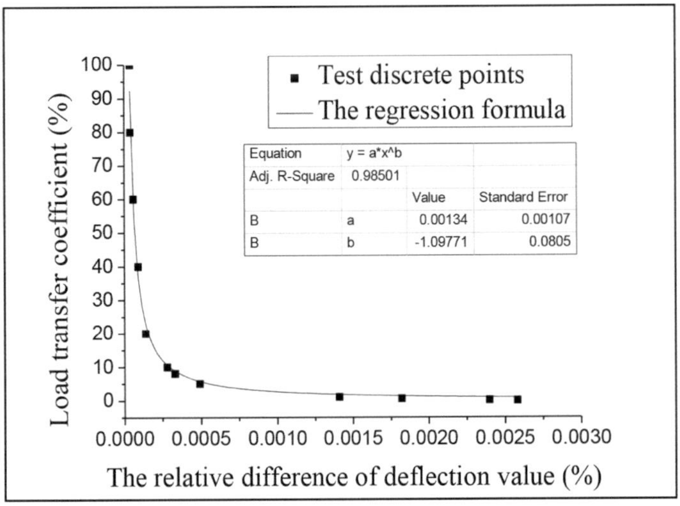

Figure 4. The relationship between the relative deflection difference and load transfer coefficient under the load.

Table 1 shows that there are large stiffness differences for the 1% prequel rod. If the material is reinforced then the transfer charge effect is very large, especially when the crack stiffness in 0% and the concrete interlocking function is lost completely. When the crack stiffness is larger, namely it there is the interlocking effect of the concrete then the force transmission is smaller compared to the interlocking concrete effect. For example, when the crack stiffness is 80% and the distribution reinforcement normal load transfer can be made on each side of the crack deflection, difference is reduced from 0.00, 006 to 0.00, 005 mm, reduced by 16%. This means at this time the dowel normal operation have less effect on load transfer coefficient of cracks compared to the concrete interlocking.

7. Conclusion

By using the basic principles of the finite element method continuously reinforced concrete pavement structural model parameters under selected conditions have been described, and the use of finite element method calculation model of continuously reinforced concrete pavement structure, the reliability of the model was validated. Considered in the model of the load transfer between the cracks of the pavement structure suffered constraints and other factors, the grass-roots and foundation factors from the model results and the impact of the difference between the deflection at the end of pavement on the maximum principal stress cracks, draw better evaluation methods to crack load transfer capabilities. The main conclusions are as follows:

(1) The basic level and the stiffness of the foundation have influences on the deflection of the pavement, top plate deflection difference between the bottom plate with the maximum principal stress and grassroots foundation stiffness increases significantly with the decrease in phenomenon. Even in severe cracks damage. There is no concrete interlocking effect on load transfer and do not set the dowel bar load transfer conditions when the load side plate crack. The results show that it is not directly affected by the load plate but directly affected by the Dutch deflection plate.

(2) When the damage degree of the transverse crack is serious, the effect of the load transfer function is relatively large. When the crack stiffness is 0%, there is complete loss of cracks of concrete interlocking action, the force transmission rod normal transmission bearing can make cracks on both sides of the curved, big difference decreases by 46.5%; crack damage to a lesser extent. The force transmission rod load transfer effect compared with interlocking concrete load transfer function produces a relatively small effect. When the crack stiffness is in 80%, the normal load transfer that can make the difference between the two sides of the crack is only reduced by 17%.

Acknowledgements

This research was supported by the Natural Science Foundation of China (51178167), the Natural Science Foundation of Hubei Province (2014CFB599), the key project of Science and Technology research projects Hubei Ministry of Education (D20131405), and the research project of Hubei University of Technology (BSQD12057).

Author details

Chunhua Hu* and Liang Chen

*Address all correspondence to: hu_chunhua@163.com

School of Civil Engineering and Architecture, Hubei University of Technology, Wuhan, China

References

Changshun, Hu. et al. 2001. Study on the test pavement of continuous reinforced concrete pavement, Highway, Chang'an University, no. 7, China.

Guo H., Sherwood J.A., Snyder, M., 1995. B1 component dowel bar model for load transfer systems in PCC pavements. Journal of Transportation Engineering, ASCE, 121(3): 289–298.

Huang Y.H., 1985. A computer package for structural analysis of concrete pavements. Third international conference on concrete pavement design and rehabilitation. Indiana: Purdue University, pp. 295–307.

http://dx.doi.org/10.5772/66809

Jian-ming Ling, Sheng-fei Guan, 2009. Analysis on load transfer efficiency of doweled joints in PCC pavement. Shanghai: Tongji University.

Tabatabaie A.M., 1978. Structural analysis of concrete pavement joints. Illinois: University of Illinois at Urbana-Champaign.

Zhigang Zhou, Qisen Zhang, 2000. A summary of the research on continuously reinforced concrete pavement, Journal of Changsha Communications University, Changsha Communications University, Changsha, China.

Chapter 33

Protection of Buildings at Areas Affected by Mining Activities

Pavlina Mateckova, Martina Janulikova and
David Litvan

Additional information is available at the end of the chapter

http://dx.doi.org/10.5772/66810

Abstract

Building structures in areas affected by underground mining demand specific treatment due to expected terrain deformation. In this chapter, expected horizontal terrain deformation and its effect especially on foundations structure is analyzed. Through the friction between subsoil and foundations, the foundation structure must resist significant normal forces. The idea of sliding joints between subsoil and foundation structure, which eliminates the friction in footing bottom, comes from the 1970s. The bitumen asphalt belt given rheological properties has been proven as an effective material for sliding joints. In the chapter, there are test results of shear resistance of currently used asphalt belts and example of sliding joint application in real structure.

Keywords: underground mining, protection of structures, sliding joint, asphalt belt

1. Introduction

Building structures in areas affected by underground mining demand specific treatment due to expected terrain deformation, special requirements mentioned also in CSN 730039 (2015). Terrain deformation comprises subsidence, declination, curvature, and horizontal deformation (**Figure 1**). The most demanding, and also most expensive, are requirements for terrain horizontal deformation. One of the reasons is that, through the friction between subsoil and foundations, the foundation structure must resist significant normal forces. The idea of sliding joints between subsoil and foundation structure, which eliminates the friction in footing bottom, comes from the 1970s (Balcarek and Bradac, 1982). In the beginning there were several materials considered (e.g., use of cardboard with ash, isinglass, graphite). Finally, the bitumen asphalt belt given its rheological properties has been proven as an effective material for sliding joints. When the terrain deformation velocity is low the shear resistance of bitumen asphalt belt is also low.

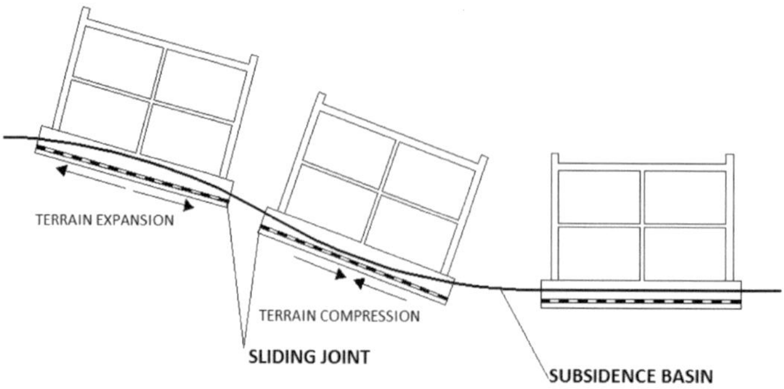

Figure 1. Horizontal terrain deformation on the margin of subsidence basin.

2. Sliding joint testing

2.1. Testing principle

When using asphalt belt as a sliding joint, the shear resistance is the main material characteristic. It was found out that the shear resistance of slide joints is primarily dependent on the deformation velocity. When the deformation velocity is slow the shear resistance of the bitumen sliding joint is low. The terrain deformation velocity could be estimated on the basis of exploitation plan.

However, testing of the shear resistance for particular deformation velocity is problematic. It was decided to appoint experimentally the deformation velocity for different shear stresses. Using linear regression, it is possible to appoint the shear resistance of a slide joint as a function of deformation rate.

Results of asphalt belt testing can be used also for parametrical subsoil model. On the basis of test results it is possible to appoint horizontal resistance with parameters C_{1x}, C_{1y} analogically to vertical resistance C_{1z} defined in Winkler one parametrical model (Cajka[a,b], 2013).

2.2. Testing equipment

At VSB–Technical University of Ostrava unique equipment was designed for shear resistance testing (**Figure 2**). Experiments on testing equipment started in 2008 and have been in process continuously.

Asphalt belt specimens are placed in between concrete blocks with a dimension of 300 × 300 × 100 mm. Specimens are exposed to a vertical load. A horizontal load is applied after a one-day delay. Displacement of the middle concrete block is measured for 6 days, and sometimes also for more days.

The specimens were exposed to vertical load of 100 and 500 kPa, values that correspond with the expected stress in the footing bottom. The horizontal load is between 1.0 and 2.0 kN, such that the deformation velocity corresponds with expected terrain deformation velocity.

http://dx.doi.org/10.5772/66810

Figure 2. Testing equipment.

2.3. Testing the temperature dependence

One of the important factors that affect the rheological shear resistance of a sliding joint is temperature. For that reason, selected materials have been tested by dependence on temperature. The testing equipment was placed in a temperature-controlled room with limit from −20 to +40°C (**Figure 3**). The aim is to determine the slide joint shear resistance for temperatures expected in a footing bottom, more detailed description is presented in the paper of Cajka and Mateckova (2011).

Figure 3. Testing equipment in temperature controlled room.

3. Selected test results

3.1. Different types of asphalt belt

Primitive asphalt is refined with oxidization or modified with an admixture of polymers. Depending on the type of admixture polymer there are asphalt belts modified with rubber, usually styrene-butadiene-styrene (SBS asphalt) and thermoplastics, mostly amorphous polypropylene (APP asphalt). Oxidized and modified asphalts possess different temperature sensitivity, elasticity, and plasticity or adhesiveness also in correlation with the amount of admixture. Consequently, the asphalt belts show different rheological shear characteristics for the group of oxidized bitumen asphalt belts, SBS asphalt belts, and APP asphalt belts.

3.2. Oxidized and modified asphalt belt

Until now 14 types of different trademark bitumen asphalt belts have been tested within new testing since the year 2008. The thickness of asphalt belts is between 3 and 5 mm, and they are predominantly covered with mineral gritting. Four types were oxidized, nine types were SBS modified, and one type was APP modified. Specimens of APP modified asphalt belt were not available. Six types of asphalt belt were tested with dependence on temperature. Particular experiment results were published in few papers, such as Cajka et al. (2012), Cajka et al. (2011), Cajka and Manasek (2007), and Janulikova and Stara (2013).

Generally, tested SBS asphalt belts show higher deformation than oxidized asphalt belts. This finding is demonstrated in **Figure 4**, with two asphalt belts: an oxidized asphalt belt with a thickness of 3.5 mm with fine-grained mineral gritting, and a modified SBS asphalt belt with a thickness of 4.7 mm with a slate gritting. Nearly the same shape of chart is for the relative displacement in relation with the specimen thickness.

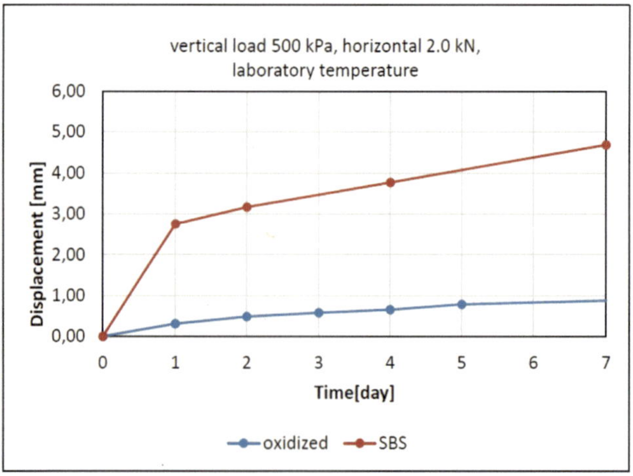

Figure 4. Experiment results, different types of asphalt belt.

http://dx.doi.org/10.5772/66810

3.3. Temperature dependence

In **Figure 5**, test results of specimens made of oxidized asphalt belt exposed to various temperatures are presented. Higher temperature leads to higher deformation, both for the group of oxidized and SBS modified asphalt belts.

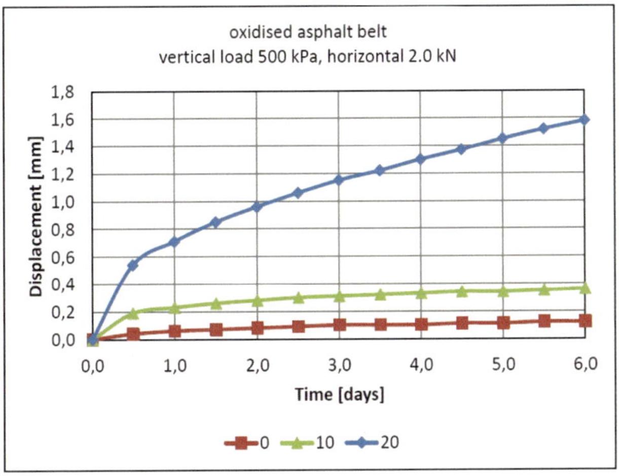

Figure 5. Experiment results, temperature dependence.

4. Sliding joint application

4.1. Complex of buildings: University of Ostrava

The new buildings of the Faculty of Science, University of Ostrava (**Figure 6**) are situated in an area with extremely unstable subsoil. The area is intersected with a tectonic fault activated by underground mining activity. Though the effects of undermining on the surface are gradually subsiding, in combination with tectonic fault and quicks and it could lead to significant subsoil deformation.

The entire design concept is adapted to extreme foundation conditions. The substructure is rigid slab-wall structure realized on a sliding joint made of SBS modified asphalt belt with a thickness 3.5 mm, reinforced with composite polyester fleece, without surface coating. Asphalt belt used at this building foundation was tested at VSB–Technical University of Ostrava.

4.2. Golf club building

A natural golf course was created in Czech Republic in the Moravian-Silesian Region (**Figure 7**). The territory is a protected coal deposit territory where mining is still in process and related phenomena occur actively on the surface. A load bearing structure is made of steel and masonry. The entire design concept is adapted to undermining effects, e.g., building is divided into deformation units and rectification is possible. The structure is realized on a sliding joint consisting of two layers of SBS modified asphalt belt. More details are in the study of Cajka et al. (2014).

Figure 6. University of Ostrava Building, application of sliding joint.

Figure 7. Golf club building at undermined territory.

http://dx.doi.org/10.5772/66810

5. Conclusion

The aim of the authors was to present possible protection of buildings affected with underground mining. Friction in footing bottom and consequent internal forces in foundation structure resulting from horizontal terrain deformation is possible to reduce with sliding joint made of asphalt belt.

Testing of asphalt belt sliding joint is presented in this chapter. Shear resistance test results for different types of asphalt belts are listed, and dependence on temperature is mentioned. Testing of sliding joints at Faculty of Civil Engineering, VSB–Technical University of Ostrava indicates better shear characteristics of SBS modified asphalt belts then oxidized asphalt belts.

Though the bitumen sliding joint was successfully applied in few buildings, sliding joints have not been widely used yet. Experiments should contribute to a wider utilization of bitumen sliding joint and thus enable design of more durable and sustainable building structures.

Acknowledgments

The paper has been supported by the project of "Conceptual development of science and research activities 2016" on the Faculty of Civil Engineering, VSB – Technical University of Ostrava.

Author details

Pavlina Mateckova*, Martina Janulikova and David Litvan

*Address all correspondence to: pavlina.mateckova@vsb.cz

Faculty of Civil Engineering, VSB-Technical University of Ostrava, Ostrava, Poruba, Czech Republic

References

V. Balcarek, J. Bradac: "Utilization of bitumen insulating stripes as sliding joints for buildings on undermined area," Civil Engineering Journal, vol. 2, 1982, in Czech.

CSN 730039, 2015. Design of constructions on the mining subsidence areas. Czech code, UNMZ. In Czech

Cajka, R.[a], 2013. Horizontal friction parameters in soil–structure interaction tasks. Advanced Materials Research, 818, 197–205. DOI:10.4028/www.scientific.net/AMR.818.197

Cajka, R.[b], 2013. Analytical derivation of friction parameters for FEM calculation of the state of stress in foundation structures on undermined territories. Acta Montanistica Slovaca 18.4, 254–261. WOS: 000343184100006

Cajka, R., Mateckova, P., Janulikova, M., 2012. Bitumen sliding joints for friction elimination in footing bottom. Applied Mechanics and Materials, 188, 247–252. DOI: 10.4028/www.scientific.net/AMM.188.247

Cajka, R., Janulikova, M., Mateckova, P., Stara, M., 2011. Modelling of foundation structures with slide joints of temperature dependent characteristics. Proceedings of the Thirteenth International Conference on Civil, Structural and Environmental Engineering Computing, Crete, Greece. DOI:10.4203/ccp.96.208

Cajka, R., Labudek, P., Burkovic, K., Cajka, M., 2014. Construction of a green golf club buildings on undermined area. 4th International Conference on Green Building, Materials and Civil Engineering, 861–867.

Cajka, R., Manasek, P., 2007. Finite element analysis of a structure with a sliding joint affected by deformation loading. Proceedings of the Eleventh International Conference on Civil, Structural and Environmental Engineering Computing, St. Julians, Malta. DOI:10.4203/ccp.86.18

Cajka, R., Mateckova, P., 2011. Temperature distribution of slide joint in reinforced concrete foundation structures. 17th International Conference on Engineering Mechanics. Svratka, Czech Republic. WOS: 000313492700017

Janulikova, M., Stara, M., 2013. Multi-layer rheological sliding joint in the foundation structures. Transactions of the VSB–Technical University of Ostrava, Civil Engineering Series, 13(2), 41–46. DOI: 10.2478/tvsb-2013-0008

Chapter 34

High-Speed Railway Tunnel Hood: Seismic Dynamic Characteristic Analysis

Wang Ying-xue, He Jun, Jian Ming,
Chang Qiao-lei and Ren Wen-qiang

Additional information is available at the end of the chapter

http://dx.doi.org/10.5772/66811

Abstract

When a high-speed train is passing through a tunnel, micro-compression wave may be created at the tunnel exit, which will affect the environment around the railway line. Setting hood at tunnel entrance is one of the efficacious ways for solving this problem. While in an earthquake region, in addition to consideration of controlling micro-compression wave, the seismic safety of hood structure must not overlook the factor. In this chapter, using finite difference method, several types of hood seismic dynamic characteristic were analyzed, and their seismic dynamic respond stress curves were drawn out. As a result, the recommended hood type was determined, which is helpful for hood design in high intensity earthquake zone.

Keywords: tunnel, hood, seismic dynamic characteristic, finite difference method

1. Introduction

In China, high-speed railway technology got quick development and more and more high-speed railway tunnels were built in high earthquake intensity zones. Large amount of post-earthquake investigation shows that tunnel entrance is liable to earthquake effect and result in damage. However, for solving train-tunnel aerodynamic effect, tunnel hood has become an indispensable accessory of tunnel structure. So how to improve tunnel hood seismic characteristic is one of the hot point on tunnel seismic research.

To the tunnel entrance seismic character, much amount of research has been made and some important conclusions have been drawn out.

Gao et al. (2009) did a thorough tunnel damage field investigation along the Dujiangyan-Wenchuan highway after the Wenchuan earthquake. The inspection results showed that the disaster induced relatively serious damage to the tunnel entrance.

Through collecting and analyzing information of seismic damage of the tunnel portals in Wenchuan earthquake, Wang et al. (2012) summed up the major factors that affected seismic damage extension, and pointed out using fuzzy synthetic evaluation method for the estimation of seismic risk level of mountain tunnel portals. Zheng (2007), using 3D distinct element method, analyzed the tunnel entrance seismic response. Cui (2010), using finite difference software FLAC-3D and experiment method, researched the seismic design calculation method of tunnel shallow-buried portal.

From previous discussion, it can be seen that many researchers have done study on tunnel entrance seismic character, while few topic pointed to the high-speed tunnel hood. In fact, for relief of aerodynamic effect, the hood structure usually set opening on top or side district. So in high seismic area, the tunnel hood will be more fragile compared with common tunnel portal.

In this chapter, using finite element software, Abaqus, the seismic character of two opening hood and nonopening hood was calculated and compared, and the recommended hood structure was given out. The types of hood are side-strip and top combination opening hood, two-side-strip opening hood and two seam opening hood.

2. Numerical simulation parameter

2.1. Numerical model

In this chapter, finite element software Abaqus was used for simulating the tunnel hood dynamic response under seismic wave. The dimension of the model is shown in **Table 1**.

Whole model	*x/m*	*y/m*	*z/m*	**Explanation**
	134.7	**100**	**77.8**	Yang slope degree is 45°, and vault buried depth is 30 m.
Tunnel hood	14.7	20	12.28	Cross-section area is 100 m^2; the thickness of the lining is 0.7 m.

Table 1. Model size and tunnel hood structure size.

For obtaining a good relief microcompression effect, the hood opening parameter must be determined through a large amount of analysis and calculation. Jiang (2014) gave out 20 m length of three types of hood optimum parameter, which is shown in **Figure 1**. The numerical model meshes of the whole model and hood structure are shown in **Figures 2** and **3**. For absorption of the seismic reflection wave, the cycle sides of the model is set as infinite element, and the bottom of the model is set as viscous-elastic boundary.

http://dx.doi.org/10.5772/66811

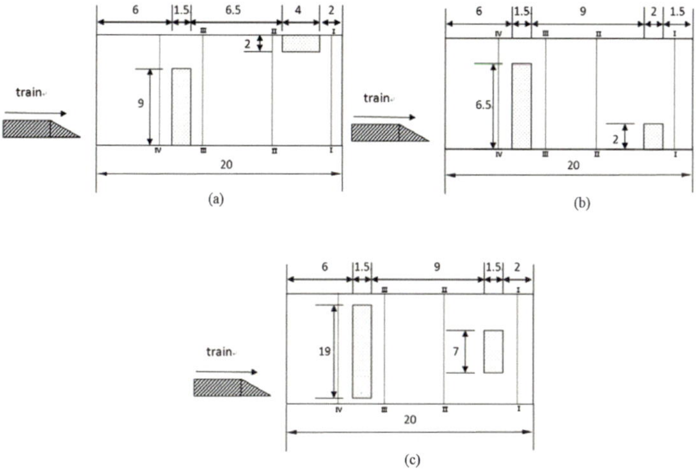

Figure 1. The diagram of the calculation model. (a) Side-strip and top combination opening hood (opening ratio = 45%). (b) Two-side-strip opening (opening ratio = 27.5%). (c) Two seam opening hood (opening ratio = 39%).

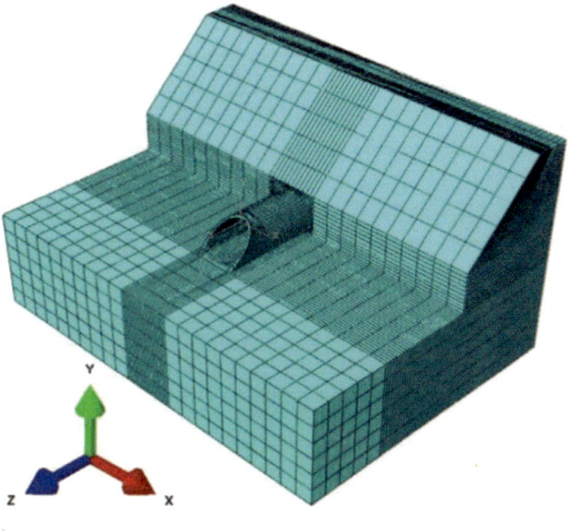

Figure 2. The arrangement plan of hood structure (units: m; bilateral symmetry).

The lining of the tunnel is C35 concrete and the surrounding rock is grade IV. Their mechanical parameters are shown in **Table 2**.

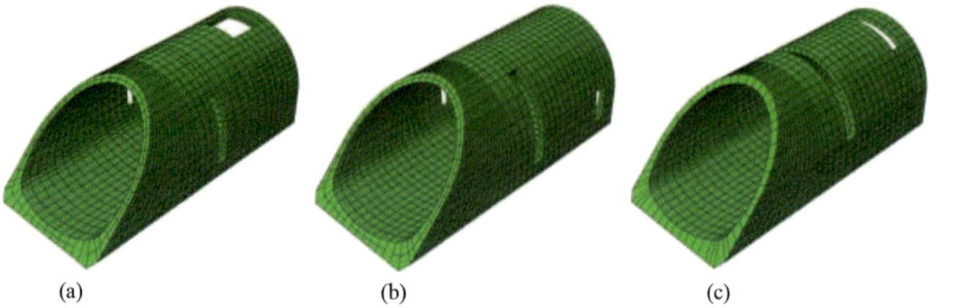

Figure 3. The model diagram of different hood structure. (a) Side-strip and top combination opening hood. (b) Two-side-strip opening hood. (c) Two seam opening hood.

Type of material	Density (kg/m³)	Young modulus (GPa)	Poisson ratio	Cohesion force (MPa)	Friction angle (°)
Surrounding rock	2100	8	0.31	0.6	33
C35 concrete	2500	31.5	0.2	—	—

Table 2. Material mechanics parameters.

The input seismic wave is scaled Wolong wave and is shown in **Figure 4**, the acceleration pick of which was 0.62 m/s² and was set on the bottom of the model for simulating the seismic effect in vertical direction.

The calculation was divided into two steps. First, only setting gravity load, the initial stress condition of the whole model would be obtained. Second, using dynamic implicit model and setting seismic load at the bottom of the model, the dynamic response of the lining was simulated. In the tunnel cross-section, there are nine monitor points on the lining of the tunnel, as shown in **Figure 5**, for recording the dynamic response in different position. In the tunnel axis direction, the hood was set with four monitor in cross-section. The position to tunnel and hood crossing surface is 1 m (surface I-I), 7 m (surface II-II), 11.5 m (surface III-III), and 15 m (surface IV-IV), respectively, and is shown in **Figure 1**.

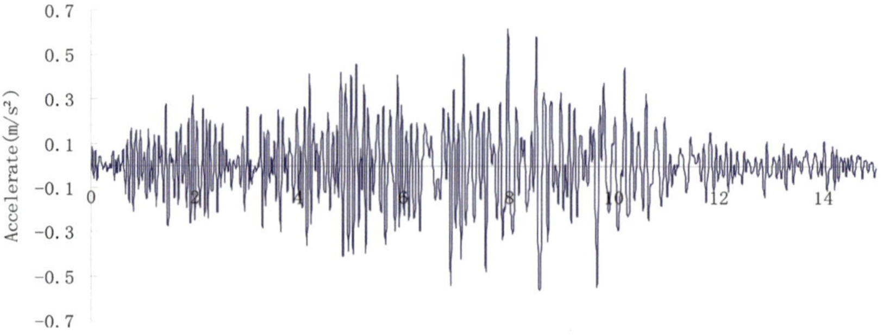

Figure 4. The accelerate curve of scaled Wolong wave.

http://dx.doi.org/10.5772/66811

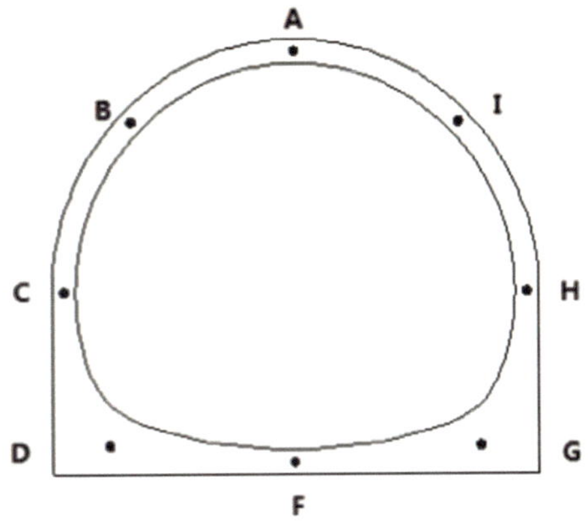

Figure 5. The schematic diagram of monitoring points.

3. Calculation result

3.1. Initial stress condition

Under dead weight of the surrounding rock and lining, the stress status of structure is almost the same to different calculation condition. So only the maximum principal stress contour graphic of side strip and top open combination opening hood was provided (**Figure 6**). It can be seen that the peak value of the maximum principal stress is 0.67 Mpa.

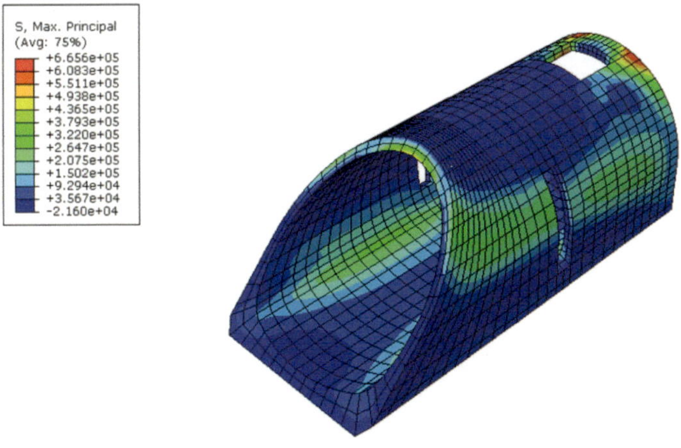

Figure 6. The max principal stress contour of side-strip and top combination opening hood structure.

3.2. Stress condition under seismic load

3.2.1. Stress contour condition analysis

The maximum principal stress conditions of the hood structures at peak period of seismic wave ($t = 8.92$) were shown in **Figure 7** and **Table 3**.

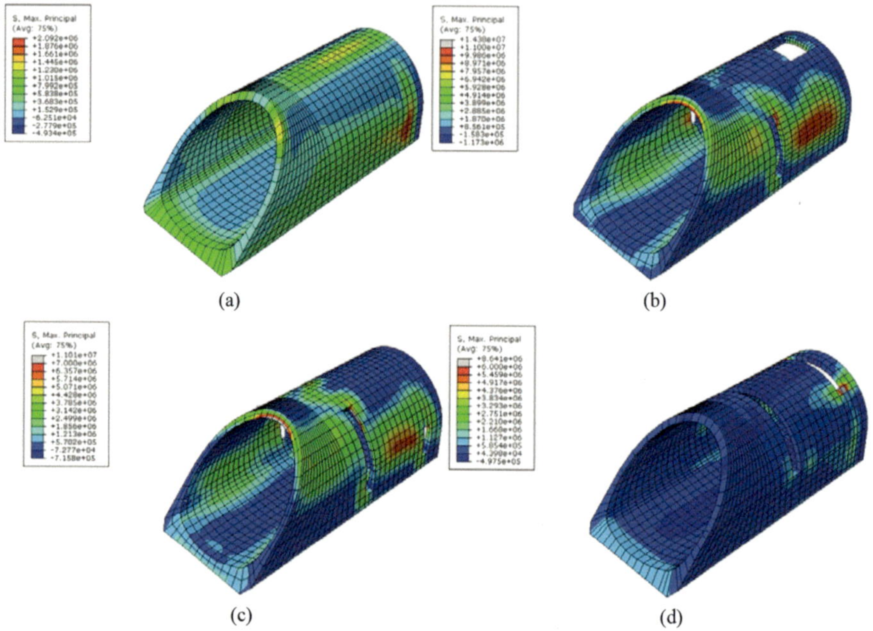

Figure 7. The max principal stress contour of different hood structure ($T = 8.92$ S). (a) No open. (b) Side-strip and top combination opening hood. (c) Two-side-strip opening hood. (d) Two seam opening hood.

Type of hood	**No opening hood**	**Side-strip and top combination opening hood**	**Two-side-strip opening hood**	**Two seam opening hood**
Value of maximum principal stress (MPa)	2.09	14.38	11.01	8.64

Table 3. The max principal stress of different hood structure.

The simulation results showed that:

1. Under seismic load, the peak value of maximum principal stress is more than 2 MPa. It can be checked out that the tensile strength of C35 is 1.57 MPa, which means that no matter of the setting opening, the seismic load can result in damage to the hood structure.

2. Differences of setting opening position can affect the peak value of maximum principal stress. Peak value of maximum principal stress to side-strip and top combination opening hood is up to 14.38 MPa, while the value to two seam opening hood is only 8.64 MPa. So choosing appropriate type of hood is very helpful for promoting structural safety in the earthquake region.

3.2.2. The stress condition discrepancy analysis

The discrepancy at hood cross-section and axial direction is shown in **Tables 4–7** and **Figures 8–11**.

Type of hood	**Maximum principal stress/MPa**							
	A	B	C	D	F	G	H	I
No open	0.87	-0.01	0.08	-0.11	0.01	-0.1	0.08	-0.01
Side-strip and top combination opening hood	2.17	0.02	2.93	1.82	0.02	1.81	2.82	0.03
Two-side-strip opening hood	0.01	0.05	2.89	-0.08	-0.03	-0.08	2.87	0.05
Two seam opening hood	0.11	2	0.04	-0.17	-0.002	-0.17	0.04	1.9

Table 4. The maximum principal stress of different hood structure at surface I-I.

Type of hood	**Maximum principal stress/MPa**							
	A	B	C	D	F	G	H	I
None opening hood	1.46	0.02	1.2	-0.06	0.1	-0.05	1.18	0.02
Side-strip and top combination opening hood	2.92	2.02	6.4	1.75	-0.01	1.75	6.43	2.02
Two-side-strip opening hood	0.98	0.71	4.1	1.26	-0.08	1.26	4.1	0.71
Two seam opening hood	-0.01	0.59	0.69	-0.04	0.12	-0.04	0.73	0.51

Table 5. The maximum principal stress of different hood structure at surface II-II.

Type of hood	**Maximum principal stress/MPa**							
	A	B	C	D	F	G	H	I
None opening hood	1.17	0.04	0.93	-0.07	0.11	-0.06	0.95	0.03
Side-strip and top combination opening hood	0.46	0.86	3.05	1.28	-0.01	1.27	3.06	0.83
Two-side-strip opening hood	0.05	0.13	1.18	0.85	0.01	0.85	1.18	0.12
Two seam opening hood	0.004	0.23	0.91	0.003	0.13	0.009	1.03	0.24

Table 6. The maximum principal stress of different hood structure at surface III-III.

Type of hood	Maximum principal stress/MPa							
	A	B	C	D	F	G	H	I
None opening hood	0.76	0.13	0.77	-0.08	0.12	-0.07	0.76	0.16
Side-strip and top combination opening hood	0.46	0.86	3.05	1.28	-0.01	1.27	3.06	0.83
Two-side-strip opening hood	2.08	1.5	3.2	0.58	-0.01	0.57	3.2	1.49
Two seam opening hood	0.06	0.12	1	0.05	0.13	0.05	1.16	0.12

Table 7. The max principal stress of different hood structure's surface IV-IV.

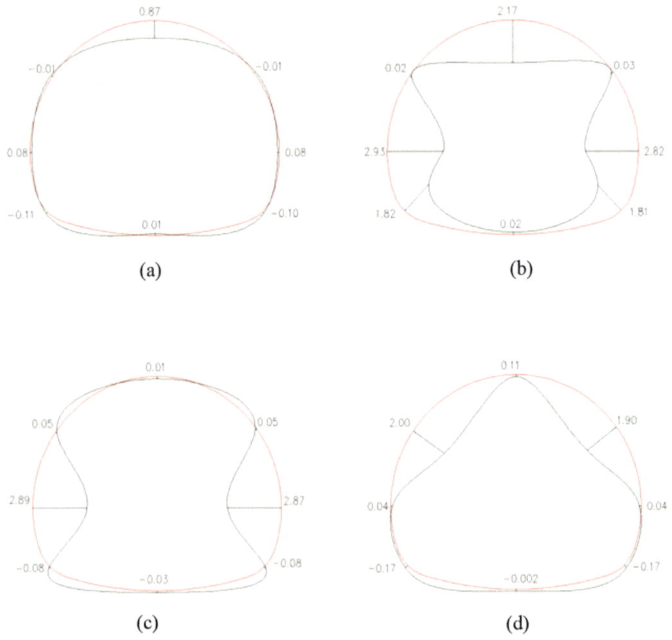

Figure 8. The max principal stress curve of different hood structure's surface I-I (units: Mpa). (a) None opening hood. (b) Side-strip and top combination opening hood. (c) Two-side-strip opening hood. (d) Two seam opening hood.

The results showed that:

1. Although there are some variations because of the differences in the type of hood, the stress maximum region is almost the same, near to the waist of the hood structure.

2. Except the two seam opening hood, in the hood axial direction, the peak value of max principal stress appeared at the section II-II, which is 7 m to the hood and tunnel cross point.

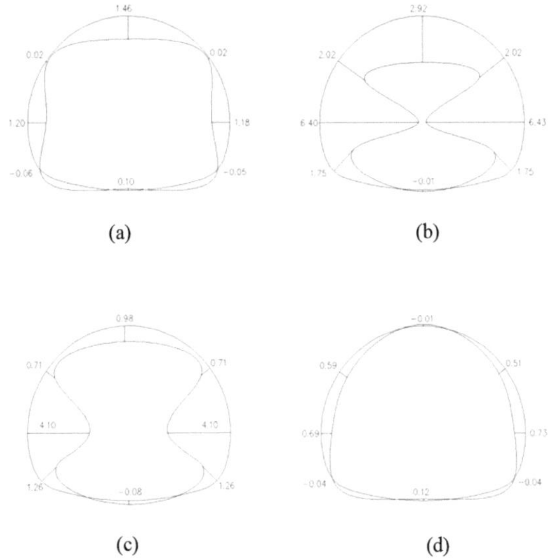

Figure 9. The max principal stress curve of different hood structure on surface II-II (units: Mpa). (a) None opening hood. (b) Side-strip and top combination opening hood. (c) Two-side-strip opening hood. (d) Two seam opening hood.

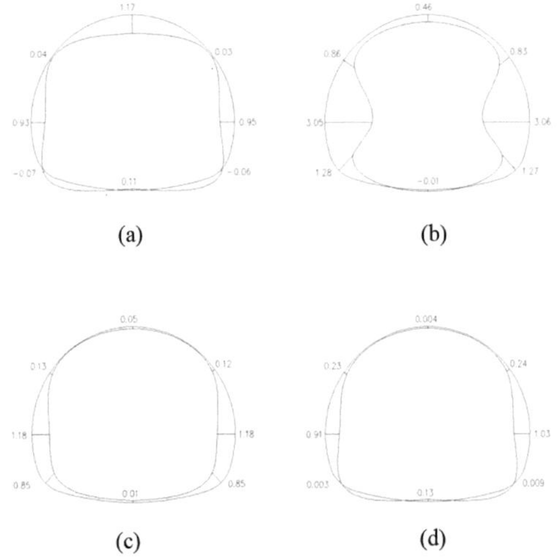

Figure 10. The max principal stress curve of different hood structure at surface III-III (units: Mpa). (a) None opening hood. (b) Side-strip and top combination opening hood. (c). Two-side-strip opening hood. (d) Two seam opening hood.

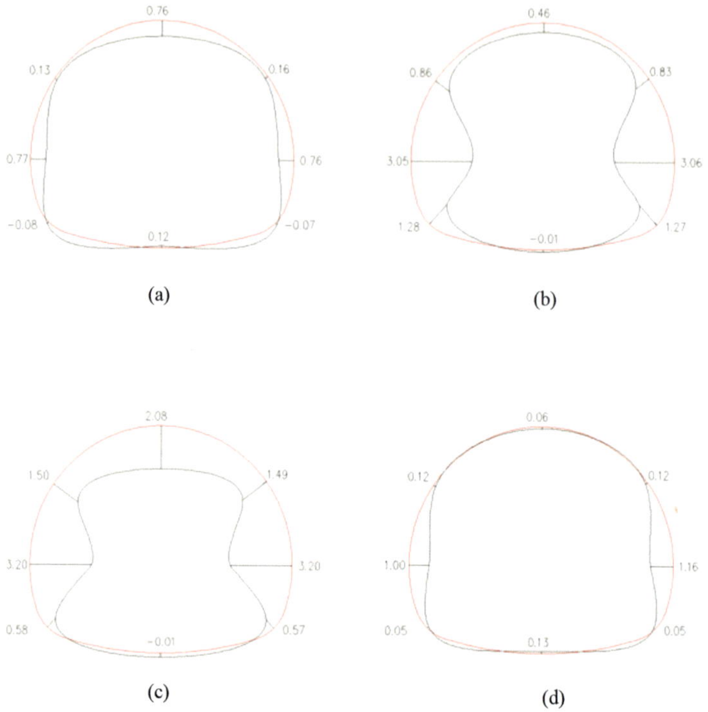

Figure 11. The max principal stress curve of different hood structure at surface IV-IV (units: Mpa). (a) No open. (b) Side-strip and top combination opening hood. (c) Two-side-strip opening hood. (d) Two seam opening hood.

3. Among these hoods, the peak value on two seam opening hood is the lowest. In some checking point, the max principal stress on its structure is even lower than that on no opening hood as shown in **Figure 12** and **Table 8**.

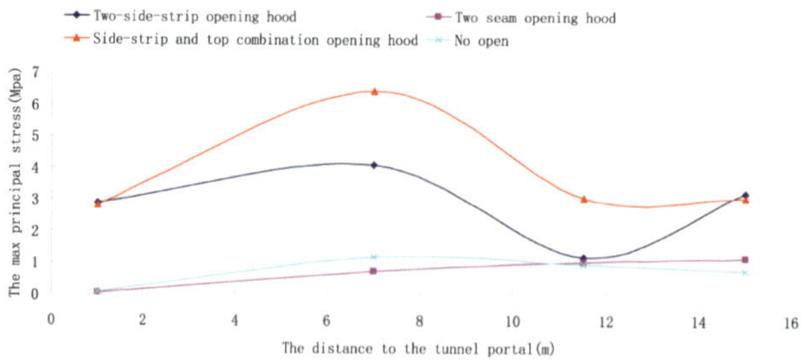

Figure 12. The max principal stress of different monitor cross-section at the waist of the hood structure.

Type of hood	Maximum principal stress value/Mpa			
	Section I-I	Section II-II	Section III-III	Section IV-IV
None opening hood	0.08	1.18	0.95	0.76
Side-strip and top combination opening hood	2.82	6.43	3.06	3.06
Two-side-strip opening hood	2.87	4.1	1.18	3.2
Two seam opening hood	0.04	0.73	1.03	1.16

Table 8. The maximum principal stress of different monitor cross-section at the waist of the hood structure.

4. Conclusion

The chapter discussed several types of hood structures' safety and dynamic response under Wolong seismic wave, the acceleration peak of which was 0.6 m/s^2, using finite element method. The analysis results showed that:

1. Under seismic load, the peak value of maximum principal stress may exceed the tensile strength of structure material (C35). No matter setting opening or not, the seismic load can result in damage to the hood structure.
2. Differences of setting opening position can affect the peak value of maximum principal stress.
3. Although there is some variation because of the differences in types of hood, the stress maximum region is almost same, near to the waist of the hood structure.
4. Except the two seam opening hood, in the hood axial direction, the peak value of max principal stress appeared at the section II-II, which is 7 m to the hood and tunnel cross point.
5. Among these hoods, the peak of maximum principal stress value on two seam opening hood is the lowest. So this hood structure is recommended in tunnel entrance for high-speed railway line.

Acknowledgment

This chapter is supported by the National High Technology Research and Development Program ("863" Program) of China (grant No. 2011AA11A103-3-3-2).

Author details

Wang Ying-xue*, He Jun, Jian Ming, Chang Qiao-lei and Ren Wen-qiang

*Address all correspondence to: wangyingxue@home.swjtu.edu.cn

Key Laboratory of Transportation Tunnel Engineering, Ministry of Education, Southwest Jiaotong University, Chengdu, China

References

G. Cui. The seismic design calculation method and test study of tunnel shallow-buried portal and rupture stick-slipping section, Southwest Jiaotong University Doctor Degree Dissertation, 2010.

B. Gao, Z. Wang, S. Yuan, Y. Shen. Lessons learnt from damage of highway tunnels in Wenchuan earthquake, Journal of Southwest Jiaotong University, Vol. 44, No. 3, pp. 336–341, 2009.

Q. Jiang, Two-stage aerodynamic characteristics and parameter analysis of buffer structure at high-speed railway tunnel entrance, Southwest Jiaotong University Master Degree Dissertation, 2014.

Z. Wang, Z. Zhang, B. Gao, Y. Shen. Factors of seismic damage and fuzzy synthetic evaluation on seismic risk of mountain tunnel portals, Journal of Central South University (Science and Technology), Vol. 43, No. 3, pp. 1122–1130, 2012.

Z. Zheng. Analysis of mountain tunnel seismic damage study on dynamic response of tunnel entrance, Southwest Jiaotong University Master Degree Dissertation, 2007.

Chapter 35

Theoretical Solution for Tunneling-Induced Stress Field of Subdeep Buried Tunnel

Qinghua Xiao, Jianguo Liu, Shenxiang Lei, Yu Mao,
Bo Gao, Meng Wang and Xiangyu Han

Additional information is available at the end of the chapter

http://dx.doi.org/10.5772/66812

Abstract

In the traditional Kirsch solution of stress field induced by tunneling in rock mass, the body force was not taken into consideration, and therefore the Kirsch solution is limited to demonstrate stress redistribution of deep-buried tunnel. In order to consider the effect of body force on the stress redistribution induced by tunneling, a new secondary stress field solution for tunnel between shallow and deep tunnel (called subdeep tunnel) is developed with elastic mechanics and complex function employed. The stress field from theoretical solution is verified by numerical models, and the results showed good agreements with each other. This solution can be the basic theory in the analysis of the stress and field of subdeep tunnel, which have not been valuated through theoretical study yet.

Keywords: subdeep tunnel, stress field, theoretical analysis, complex function, elastic mechanics

1. Introduction

Tunnels are always classified as shallow and deep in tradition, and the classifying standard is the buried depth of tunnel, also known as h_q, which is the limit height that a pressure arch would form in the surrounding rock, and h_q is dependent on the quality of rock mass and tunnel height (Xu et al., 2000). This classification standard is just an empirical approach and is not based on mechanical behavior of rock mass. When tunneling in intact rock mass near the ground surface, the rock mass have partial self-loading capacity which on the other hand is not enough to reach stability for rock mass. And in the presented research, a new type of tunnel classified as subdeep tunnel is described as: primary stress field of tunnel varies along the depth while the redistribution of secondary stress field of tunnel does not reach the ground surface or just have limited influence.

In general, the tunnel support design follows two kinds of strategies based on the tunnel classification of depth, and essentially, the support design strategy is a response to the stress redistribution and rock failure modes. For shallow tunnel, the surround rock has a very limited capacity to form a pressure arch and is very easy to collapse (Yang and Yang, 2008), and therefore the support system are designed to bear the whole weight of loosen zone above the tunnel (Peng and Liu, 2009; Terzaghi, 1943), and the loosen zone is determined by rock properties and tunnel section shapes (Wang et al., 2014). However, in this design approach self-loading capacity of surround rock is not considered in subdeep tunnel, especially in intact rock mass, which would cause a significant waste of materials and always brings uneconomic results. On the other hand, according to the definition of subdeep tunnel, pressure arch which requires a significant burying depth cannot form and thus self-stability cannot be expected, which makes researches on the pressure arch (Li, 2006; Poulsen, 2010; Sansone and Silva, 1998) and support design not suitable for subdeep tunnel.

For a better understanding of mechanical behavior of surround rock for subdeep tunnel, which is the basis for a reasonable support design, a theoretical solution of secondary stress field is put forward with elastic mechanics and complex function adopted. And the derivation process is described in detail as well as the results. The stress field including radial, tangential, and shear stresses are described and discussed. For an intuitive comparison, a numerical model is also built and the results from both theoretical and numerical models are presented, which is a powerful verification for the theoretical solution. Besides, the reasonable depth of subdeep tunnel is suggested by the theoretical results. And this solution for subdeep tunnel can be a very useful theoretical basis for safe and economic tunneling.

2. Theoretical model and primary stresses for subdeep tunnel

2.1. Subdeep tunnel model

In the stress analysis of subdeep tunnel, the following assumptions are made:

- The rock mass is elastic, homogeneous, and intact;
- Only self-weight induced stress field exists in the tunnel site;
- The whole section of the tunnel is excavated in one step; and
- The tunnel is long enough and the model can be treated as planer strain.

As shown in **Figure 1**, a circular tunnel with a radius of a is excavated at depth of h_t. The vertical stress σ_z at the bottom boundary is caused by the self-weight of rock mass and $\sigma_z = 0$ at the ground surface; horizontal stress σ_x varies along z. And this stress field model can be decomposed into two parts, one is the primary stress field as shown in **Figure 1(b)**, and the other is induced by stress-released at tunnel boundary, as shown in **Figure 1(c)**.

The primary stress field in **Figure 1(b)** can be described as follows according to Heim's hypothesis (Fegert, 2013) and Gold Nick's hypothesis (Dessler, 1982):

$$\begin{cases} \sigma_z = \gamma(h_t - z) \\ \sigma_x = k_0\sigma_z = k_0\gamma(h_t - z) \\ \tau_{xz} = 0 \end{cases} \tag{1}$$

where z is the coordinate from tunnel center; k_0 is the lateral stress coefficient; h_t is the depth from the tunnel center to ground; and $k_0 = \frac{\nu}{1-\nu}$.

2.2. Analysis of the stress to release at tunnel boundary

The stress in polar coordinates in **Figure 1** can be converted from Eq. (1) in rectangular coordinates as follows:

$$\begin{cases} \sigma_r = \sigma_x \cos^2\theta + \sigma_z \sin^2\theta + \tau_{xz}\sin 2\theta = \dfrac{\sigma_z + \sigma_x}{2} - \dfrac{\sigma_z - \sigma_x}{2}\cos 2\theta + \tau_{xz}\sin 2\theta \\ \sigma_\theta = \sigma_x \sin^2\theta + \sigma_z \cos^2\theta - \tau_{xz}\sin 2\theta = \dfrac{\sigma_z + \sigma_x}{2} + \dfrac{\sigma_z - \sigma_x}{2}\cos 2\theta - \tau_{xz}\sin 2\theta \\ \tau_{r\theta} = \dfrac{\sigma_z - \sigma_x}{2}\sin 2\theta + \tau_{xz}\cos 2\theta \end{cases} \tag{2}$$

Where σ_r = radial stress, σ_θ= tangential stress, and $\tau_{r\theta}$= shear stress.

By submitting Eq. (1) into Eq. (2), the primary stress field can be expressed as:

$$\begin{cases} \sigma_r = \dfrac{1+k_0}{2}\gamma(h_t - z) - \dfrac{1-k_0}{2}\gamma(h_t - z)\cos 2\theta \\ \sigma_\theta = \dfrac{1+k_0}{2}\gamma(h_t - z) + \dfrac{1-k_0}{2}\gamma(h_t - z)\cos 2\theta \\ \tau_{r\theta} = \dfrac{1-k_0}{2}\gamma(h_t - z)\sin 2\theta \end{cases} \tag{3}$$

where $z = r\sin\theta$.

When $r = a$, which means at the tunnel boundary:

$$\begin{cases} \sigma^0_{ra} = \dfrac{1+k_0}{2}\gamma(h_t - a\sin\theta) - \dfrac{1-k_0}{2}\gamma(h_t - a\sin\theta)\cos 2\theta \\ \sigma^0_{\theta a} = \dfrac{1+k_0}{2}\gamma(h_t - a\sin\theta) + \dfrac{1-k_0}{2}\gamma(h_t - a\sin\theta)\cos 2\theta \\ \tau^0_{r\theta a} = \dfrac{1-k_0}{2}\gamma(h_t - a\sin\theta)\sin 2\theta \end{cases} \tag{4}$$

Then, Eq. (4) is the stress to release in the following steps shown in **Figure 1**.

2.3. Complex function for elastic mechanics

With complex function, stress components in **Figure 1** in polar coordinate can be expressed as

$$z: \begin{cases} \sigma_\theta + \sigma_r = 4\mathrm{Re}\chi'_1(z) \\ \sigma_\theta - \sigma_r + 2\mathrm{i}\tau_{r\theta} = 2[\bar{z}\chi''_1(z) + \psi'_1(z)]\mathrm{e}^{2\mathrm{i}\lambda} \end{cases} \tag{5}$$

where Re is the real part of complex function, $\bar{z}$ is the conjugate expression of z, $\chi_1(z)$ and $\psi_1(z)$

are complex potential function, $\chi_1'(z)$ and $\chi_1''(z)$ are the first and second derivatives of $\chi_1(z)$, respectively, and $\psi_1'(z)$ is the first derivative of $\psi_1(z)$.

2.4. Stress release

Similar to the solving process of an infinite plate with a hole (Wang, 2008), as the stress field approximates to zero, the solving process are described as follows:

1. Submitting Eq. (4) into the Fourier coefficient equations;
2. Through the obtained equations, solving the constant coefficients; and
3. As the problem has been assumed to be a planar strain work, which means $k_0 = \frac{v}{1-v} = v'$, the following equations can be obtained based on steps (1) and (2):

$$\left\{\begin{array}{l}
\chi'_1(z) = \frac{a_1}{z} + \frac{a_2}{z^2} + \frac{a_3}{z^3} = \frac{(1+v')\gamma a^2}{8}\frac{\mathrm{i}}{z} + \frac{(1-k_0)\gamma h_t a^2}{2}\frac{1}{z^2} - \frac{(1-k_0)\gamma a^4}{4}\frac{\mathrm{i}}{z^3} \\
\overline{\chi'_1(z)} = \frac{\bar{a}_1}{\bar{z}} + \frac{\bar{a}_2}{\bar{z}^2} + \frac{\bar{a}_3}{\bar{z}^3} = -\frac{(1+v')\gamma a^2}{8}\mathrm{i}\bar{z} + \frac{(1-k_0)\gamma h_t a^2}{2}\frac{1}{\bar{z}^2} + \frac{(1-k_0)\gamma a^4}{4}\frac{\mathrm{i}}{\bar{z}^3} \\
\chi''_1(z) = \frac{-a_1}{z^2} + \frac{-2a_2}{z^3} + \frac{-3a_3}{z^4} = -\frac{(1+v')\gamma a^2}{8}\frac{\mathrm{i}}{z^2} - \frac{2(1-k_0)\gamma h_t a^2}{2}\frac{1}{z^3} + \frac{3(1-k_0)\gamma a^4}{4}\frac{\mathrm{i}}{z^4} \\
\psi'_1(z) = \frac{b_1}{z} + \frac{b_2}{z^2} + \frac{b_3}{z^3} + \frac{b_4}{z^4} + \frac{b_5}{z^5} = \frac{(3-v')\gamma a^2}{8}\frac{\mathrm{i}}{z} + \frac{(1+k_0)\gamma h_t a^2}{2}\frac{1}{z^2} + \frac{(v'-k_0)\gamma a^4}{4}\frac{\mathrm{i}}{z^3} \\
\qquad + \frac{3(1-k_0)\gamma h_t a^4}{2}\frac{1}{z^4} - (1-k_0)\gamma a^6\frac{\mathrm{i}}{z^5} \\
\chi_1(z) = \frac{(1+v')\gamma a^2}{8}\mathrm{i}\ln z - \frac{(1-k_0)\gamma h_t a^2}{2}\frac{1}{z} + \frac{(1-k_0)\gamma a^4}{8}\frac{\mathrm{i}}{z^2} + c_1 \\
\psi_1(z) = \frac{(3-v')\gamma a^2}{8}\mathrm{i}\ln z - \frac{(1+k_0)\gamma h_t a^2}{2}\frac{1}{z} - \frac{(v'-k_0)\gamma a^4}{8}\frac{\mathrm{i}}{z^2} - \frac{(1-k_0)\gamma h_t a^4}{2}\frac{1}{z^3} + \frac{(1-k_0)\gamma a^6}{4}\frac{\mathrm{i}}{z^4} + c_2 \\
\overline{\psi_1(z)} = \frac{(3-v')\gamma a^2}{8}\mathrm{i}\ln\bar{z} - \frac{(1+k_0)\gamma h_t a^2}{2}1\bar{z} + \frac{(v'-k_0)\gamma a^4}{8}\frac{\mathrm{i}}{\bar{z}^2} - \frac{(1-k_0)\gamma h_t a^4}{2}\frac{1}{\bar{z}^3} - \frac{(1-k_0)\gamma a^6}{4}\frac{\mathrm{i}}{\bar{z}^4} + \bar{c}_2
\end{array}\right. \quad (6)$$

By submitting Eq. (6) into Eq. (5), the relation between stress components can be expressed by:

$$\left\{\begin{array}{l}
\sigma_\theta + \sigma_r = 2\mathrm{Re}\left(\frac{(1+v')\gamma a^2}{4}\frac{\mathrm{i}}{z} + (1-k_0)\gamma h_t a^2\frac{1}{z^2} - \frac{(1-k_0)\gamma a^4}{2}\frac{\mathrm{i}}{z^3}\right) \\
\sigma_\theta - \sigma_r + 2\mathrm{i}\tau_{r\theta} = 2[\bar{z}\chi''_1(z) + \psi'_1(z)]\mathrm{e}^{2\mathrm{i}\theta} \\
= 2\left[-\frac{(1+v')\gamma a^2}{8}\frac{\bar{z}\mathrm{i}}{z^2} - \frac{2(1-k_0)\gamma h_t a^2}{2}\frac{\bar{z}}{z^3} + \frac{3(1-k_0)\gamma a^4}{4}\frac{\bar{z}\mathrm{i}}{z^4} + \frac{(3-v')\gamma a^2}{8}\frac{\mathrm{i}}{z}\right.
\end{array}\right.$$

$$\left. + \frac{(1+k_0)\gamma h_t a^2}{2}\frac{1}{z^2} + \frac{(v'-k_0)\gamma a^4}{4}\frac{\mathrm{i}}{z^3} + \frac{3(1-k_0)\gamma h_t a^4}{2}\frac{1}{z^4} - (1-k_0)\gamma a^6\frac{\mathrm{i}}{z^5}\right]\mathrm{e}^{2\mathrm{i}\theta} \quad (7)$$

where z, which is a complex valuable, can be expressed as:

$$
\begin{cases}
\dfrac{\mathrm{i}}{z}=\dfrac{\mathrm{i}}{r\mathrm{e}^{\mathrm{i}\theta}}=\dfrac{\mathrm{i}\mathrm{e}^{-\mathrm{i}\theta}}{r}=\dfrac{\sin\theta+\mathrm{i}\cos\theta}{r}\\
\dfrac{1}{z^2}=\dfrac{1}{r^2\mathrm{e}^{\mathrm{i}2\theta}}=\dfrac{\mathrm{e}^{-\mathrm{i}2\theta}}{r^2}=\dfrac{\cos2\theta-\mathrm{i}\sin2\theta}{r^2}\\
\dfrac{\mathrm{i}}{z^3}=\dfrac{\mathrm{i}}{r^3\mathrm{e}^{\mathrm{i}3\theta}}=\dfrac{\mathrm{i}\mathrm{e}^{-\mathrm{i}3\theta}}{r^3}=\dfrac{\sin3\theta+\mathrm{i}\cos3\theta}{r^3}
\end{cases}
\tag{8}
$$

By submitting Eq. (8) into the Eq. (7), a more simple relation between stress components are obtained, and by decomposing the imaginary part in the obtained equations, the subsidiary stress induced by the stress release process in **Figure 1** is obtained. The obtained subsidiary stress is based on the infinite plate assumption, and it is an approximate value. For a more accurate solution, the subsidiary stress at ground surface should be released again, which is a very complicated process. By solving the obtained subsidiary stress components with Eq. (3), the stress field solution for subdeep tunnel can be expressed as:

$$
\begin{cases}
\sigma_r=\dfrac{1+k_0}{2}\gamma(h_t-r\sin\theta)-\dfrac{1-k_0}{2}\gamma(h_t-r\sin\theta)\cos2\theta+\gamma\Bigg\{-\dfrac{(1+k_0)h_ta^2}{2r^2}\\
+\left[\dfrac{(3+v')a^2}{4r}+\dfrac{(v'-k_0)a^4}{4r^3}\right]\sin\theta+(1-k_0)h_t\left(\dfrac{2a^2}{r^2}-\dfrac{3a^4}{2r^4}\right)\cos2\theta-(1-k_0)a\left(\dfrac{5a^3}{4r^3}-\dfrac{a^5}{r^5}\right)\sin3\theta\Bigg\}
\end{cases}
$$

$$
\sigma_\theta=\frac{1+k_0}{2}\gamma(h_t-r\sin\theta)+\frac{1-k_0}{2}\gamma(h_t-r\sin\theta)\cos2\theta+\gamma\Bigg\{\frac{(1+k_0)h_ta^2}{2r^2}
$$

$$
-\left[\frac{(1-v')a^2}{4r}-\frac{(v'-k_0)a^4}{4r^3}\right]\sin\theta+(1-k_0)h_t\frac{3a^4}{2r^4}\cos2\theta+(1-k_0)a\left(\frac{a^3}{4r^3}-\frac{a^5}{r^5}\right)\sin3\theta\Bigg\}
$$

$$
\tau_{r\theta}=\frac{1-k_0}{2}\gamma(h_t-r\sin\theta)\sin2\theta
$$

$$
+\gamma\left[\frac{(1-v')a^2}{4r}+\frac{(v'-k_0)a^4}{4r^3}\right]\cos\theta+(1-k_0)h_t\left(\frac{a^2}{r^2}-\frac{3a^4}{2r^4}\right)\sin2\theta+(1-k_0)a\left(\frac{3a^3}{4r^3}-\frac{a^5}{r^5}\right)\cos3\theta\Big\}
\tag{9}
$$

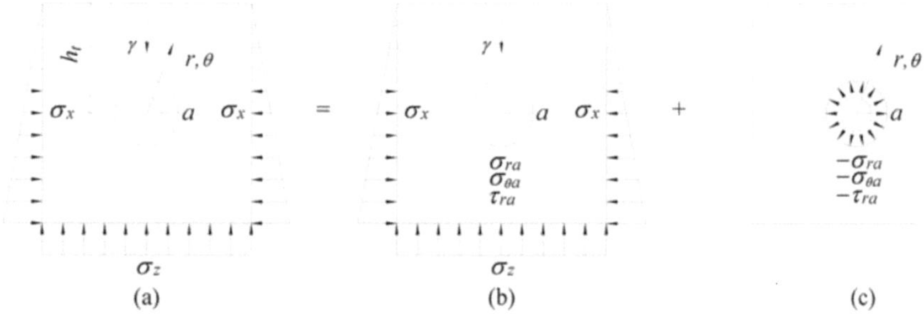

Figure 1. Mechanical model and decomposition of the secondary stress field of subdeep tunnel. (a) Secondary stress field. (b) Primary stress field. (c) Stress released.

3. Distribution of secondary stress field solution for subdeep tunnel

The secondary stress field of subdeep tunnel calculated from Eq. (9), and the v, and h_t are set at 0.3 and $6a$, respectively. The results are shown in **Figure 2**.

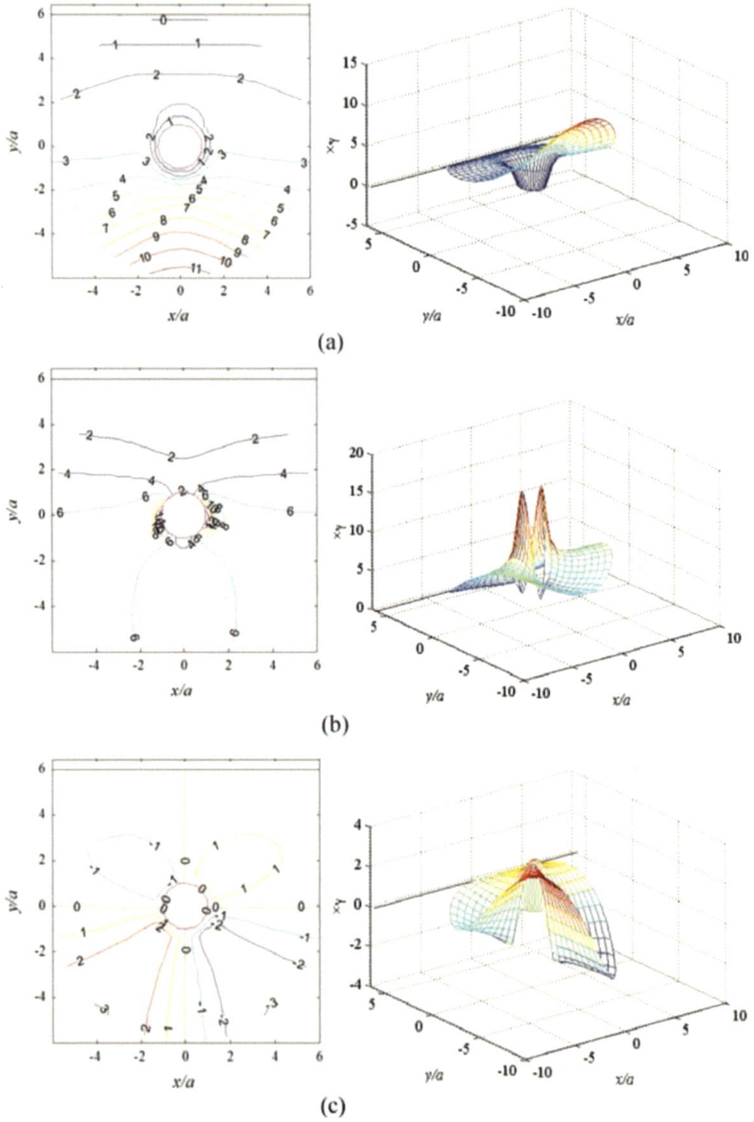

Figure 2. Second stress distribution in subdeep tunnels. (a) Distribution of σ_r. (b) Distribution of σ_θ. (c) Distribution of $\tau_{r\theta}$.

http://dx.doi.org/10.5772/66812

Stress distribution along r different directions in polar coordinate are also illustrated in **Figure 3** in a normalized form.

From **Figures 2** and **3**, the distributions of secondary stress field induced by tunneling for subdeep tunnel are different from the deep-buried tunnels, which has the following characteristics of its own:

1. The distribution of secondary subdeep tunnel just shows axial symmetry, not like the deep-buried tunnel, whose stress field shows bidirectional symmetry. This difference is caused by the variation of σ_z along z in the primary field.

2. The tangential stress near the tunnel boundary shows significant concentration at the side wall of the tunnel, to which close attention should be paid.

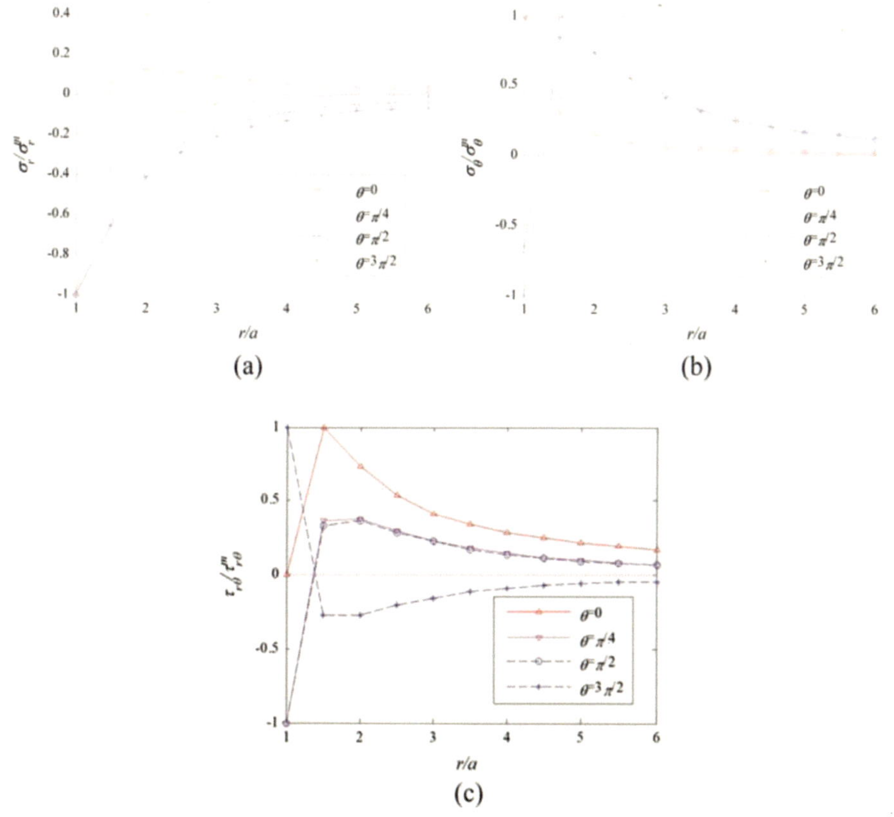

Figure 3. Stress distribution along r different directions. (a) Distribution of σ_r along r. (b) Distribution of σ_θ along r. (c) Distribution of $\tau_{r\theta}$ along r.

3. **Figure 3** shows the variation of stress induced by tunneling along polar radius, and it is quite clear that, the stresses are approximate to 0, which proves the simplification of zero stress boundary to be reasonable and correct. Besides, subdeep tunnel can be determined as tunnels buried at depth of more than 2.5 D (D is the diameter of tunnel), and less than the depth of deep tunnel.

From the stress analysis, secondary stress field for subdeep tunnel is demonstrated and the limit depth to distinguish shallow tunnel and subdeep tunnel is obtained, which can be the guideline for tunnel support in a more reasonable, economic, and safer way.

4. Numerical solution and its comparison with theoretical results

To verify the theoretical solution, a numerical model is developed. The parameters for theoretical and numerical simulation are shown in **Table 1**. And a numerical model is developed in Flac3D as shown in **Figure 4**.

Theoretical solution of horizontal stresses (σ_{xx}) of surrounding rock in subdeep tunnel is shown in **Figure 5(a)** in a rectangular coordinate system, and its numerical result is shown in **Figure 5(b)**.

Theoretical solution of vertical stresses (σ_{zz}) of surrounding rock in subdeep tunnel is shown in **Figure 6(a)**, and the numerical result is shown in **Figure 6(b)**.

Theoretical solution of shear stresses (τ_{xz}) of surrounding rock in subdeep tunnel is shown in **Figure 7(a)**, and its numerical result is shown in **Figure 7(b)**.

Contours of horizontal and vertical stress field from theoretical and numerical solution show good agreement with each other as shown in **Figures 5** and **6**, which proves the theoretical analysis reasonable and correct. On the other hand, shear stress from theoretical and numerical solution shows difference to some extent. In **Figure 7**, shear stress, from numerical solution, has a much smaller distribution area than that from numerical model, a probable explanation is that, the media in theoretical model is elastic while that in numerical model is elastic-plastic, and stress concentration near the tunnel boundary may redistribute after rock failure, leading to a smoother distribution of shear stress in numerical model.

Parameters of theoretical solution				Parameters of numerical simulation			
Item	**Symbol**	**Value**	**Unit**	**Item**	**Symbol**	**Value**	**Unit**
Tunnel's radius	a	3	m	Tunnel's radius	a	3	m
Depth of tunnel center	h_t	6 a	m	Depth of tunnel	d	15	m
Volume weight	γ	2.10E+04	N/m^3	Density	ρ	2.10E+03	kg/m^3
Elasticity modulus	E	9.00E+07	Pa	Bulk modulus	K	7.50E+07	Pa
Poisson's ratio	ν	0.30	-	Shear modulus	G	3.46E+07	Pa

Table 1. Parameters for theoretical and numerical simulation.

http://dx.doi.org/10.5772/66812

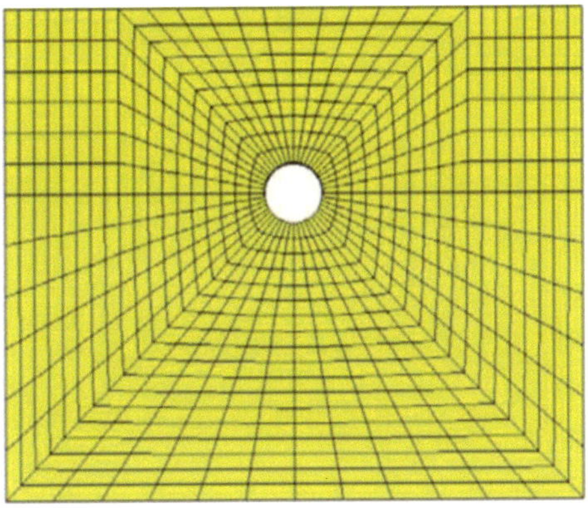

Figure 4. Numerical model of subdeep tunnels.

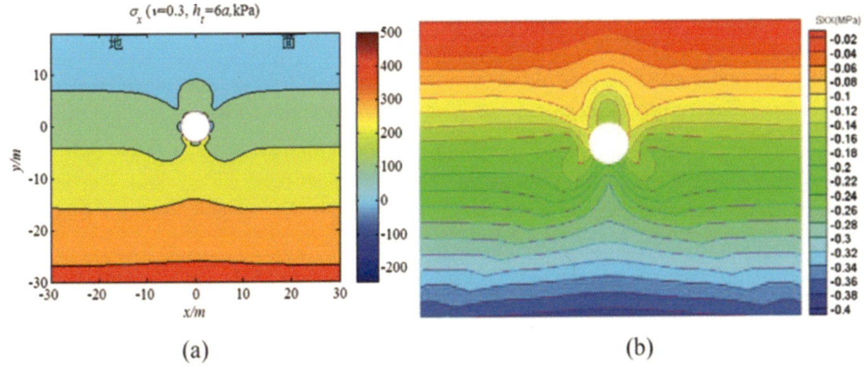

Figure 5. Theoretical and numerical solution of horizontal stress. (a) Theoretical solution. (b) Numerical solution.

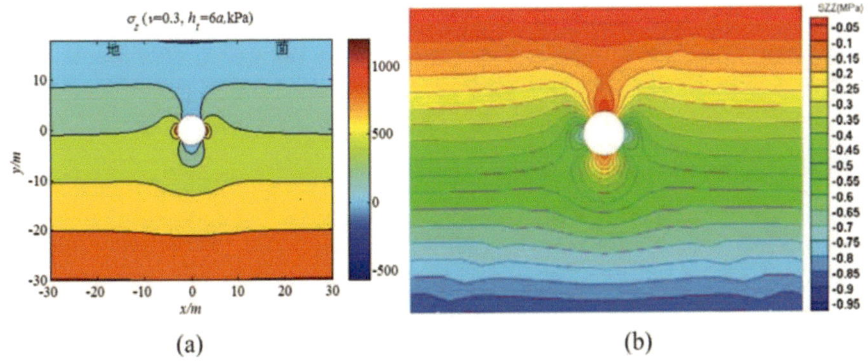

Figure 6. Theoretical solutions and numerical value of vertical stress. (a) Theoretical solution. (b) Numerical solution.

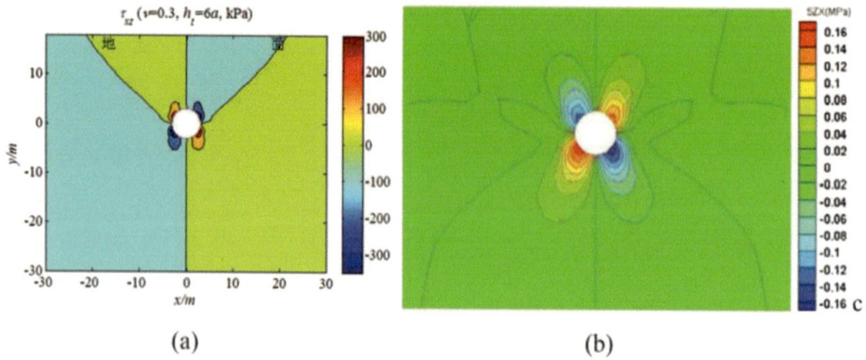

Figure 7. Theoretical and numerical solution of shear stress. (a) Theoretical solution. (b) Numerical solution.

5. Conclusions

In the presented research, the stress field for tunnel buried at depth between deep tunnel and shallow tunnel is analyzed with elastic mechanics and complex function, and conclusions can be drawn as:

1. The subdeep tunnel is proposed and through theoretical analysis, and stress fields show that essential difference exists between the mechanical behavior of deep tunnel and subdeep tunnel, and the depth to distinguish shallow tunnel and subdeep tunnel is suggested as 2.5 times of the tunnel diameter.
2. By numerical analysis, the theoretical solution is proved to be reasonable and with high accuracy. And this theoretical solution can be a very good guideline for the support design for subdeep tunnel with economy and safety.
3. This theoretical solution also has its limitations, and if used for deep tunnels, the vertical stress in rock would be incorrect totally. Besides, this solution is not suitable for tunnels buried in depth less than 2.5 D, which does not satisfy the boundary stress condition.
4. Through the comparison of theoretical analysis and numerical model, the theoretical results are proved to be effective in the determination of subdeep tunnel, which can be very helpful in the design of subtunnel support for economy and safety purpose.

Author details

Qinghua Xiao[1]*, Jianguo Liu[1], Shenxiang Lei[1,2], Yu Mao[1], Bo Gao[1], Meng Wang[1] and Xiangyu Han[1]

*Address all correspondence to: xqhbp@home.swjtu.edu.cn

1 School of Civil Engineering, Southwest Jiaotong University, Chengdu, China

2 China Railway 20th Bureau, Xi'an, China

http://dx.doi.org/10.5772/66812

References

A. J. Dessler. Gold's hypothesis and the energetics of the Jovian magnetosphere. Cosmology & Astrophysics Essays in Honor of Thomas Gold Ithaca, 137–143,1982 http://adsabs.harvard.edu/abs/1982coas.conf.137D.

M. D. Fegert. The insufficiency of S. Mark Heim's more pluralistic hypothesis. Theology Today 69, 497–510, 2013.

C. C. Li. Rock support design based on the concept of pressure arch. International Journal of Rock Mechanics & Mining Sciences 43, 1083–1090, 2006.

L. M. Peng, X. B. Liu. Tunneling Engineering. Central South University Press, Changsha, 2009.

B. A. Poulsen. Coal pillar load calculation by pressure arch theory and near field extraction ratio. International Journal of Rock Mechanics & Mining Sciences 47, 1158–1165, 2010.

E. C. Sansone, L. A. A. D. Silva. Numerical modeling of the pressure arch in underground mines. International Journal of Rock Mechanics & Mining Sciences 35, 436, 1989.

K. Terzaghi. Theoretical Soil Mechanics. John Wiley & Sons, New York, 1943.

G. Wang. Elastic Mechanics. China Railway Publishing House, Beijing, 2008.

Z. Wang, C. Qiao, C. Song, J. Xu. Upper bound limit analysis of support pressures of shallow tunnels in layered jointed rock strata. Tunnelling and Underground Space Technology 43, 171–183, 2014.

Z. Xu., R. Huang, S. Wang. Tunnel classifying in light of depth (i.e. thickness of overburden). The Chinese Journal of Geological Hazard and Control 11, 5–10, 2000.

F. Yang, J. Yang.. Limit analysis method for determination of earth pressure on shallow tunnel. Engineering Mechanics 25, 6, 2008.

Chapter 36

Rate Assessment of Slope Soil Movement from Tree Trunk Distortion

Karel Vojtasik, Pavel Dvorak and Milan Chiodacki

Additional information is available at the end of the chapter

http://dx.doi.org/10.5772/66813

Abstract

The chapter presents the methods for determining the rate movement of the slope, which is based on an evaluation of a distortion of the tree trunk. The distortion develops during the growth period of the tree, and its conditions are trajectory and speed of material movement, which takes away the root system of a tree. The distortion of the tree trunk explains kinematics of the root system movement. The distortion curve is constructed from the results of measurements carried out on several horizontal levels of the tree trunk. The speed of movement of the slope is calculated from the age of the tree and the length of the path of movement of the tree, which is derived from the distortion curve of the tree trunk. The length of the track of tree movement is drawn from horizontal position of the gravity point of the curve corresponding with the longitudinal axis of the tree trunk. This method is documented in one example. The method is appropriate to quantify the movement of the latent long slope.

Keywords: landslides, latent movement, rate of movement

1. Introduction

Soil movement on slope is a common phenomenon that deserves attention in particular when on the slope there is an intention for an extensive development that brings about earthworks as its consequence. The soil movement can have many causes. Gravity and transport media as water, snow, ice, or wind are inherent and cannot be bypassed. Other causes are conditioned with anthropogenic activity and those can be only mitigated. The soil movement differs in many respects. Landslides are divided according to value of the speed of movement of slope materials into several categories (Abramson et al., 1996). Most usually attention is paid to the influence of vegetation on slope stability (Morgan and Rickson, 1995), but the idea that some kind of long growing plants, e.g., the trees in specific way depict as well soil movement that has not yet been exploited. Determination of the rate of a slope soil material movement from trunk tree distortion growing on the slope is possible only for slow latent slope soil material movement. Latent movement escapes most observers and the slope is considered stable, although its degree of stability fluctuates around a value equal to the marginal state of equilibrium. Latent soil material movement can be exposed only over many years of continuous measurement, which is never done. This movement is typically indicated only with characteristic development of the trunk of tree. The distortion of a tree trunk can therefore be regarded as a direct form of continuous measurements of the slope soil material movement within an area that a root system of the tree develops in. With the age of the tree known the rate of a slope soil material movement can be set up.

2. Tree kinematics

The tree structure consists of two parts: underground root system that is manipulated by the creeping slope soil material and an above the ground part of tree trunk, a crown. Both the parts are linked together with stiff bond that transmits the principal compounds of soil movement, i.e., translation and rotation, on the above ground part of tree. The vital peak of a tree trunk executes a distortion of the tree trunk. It strives to keep the vertical direction of the growth of a tree and in this way it smooths out the tilt and shift of the root system of the tree being drifted through slope soil material. The tree trunk distortion determines type and size of the slope soil material movement. In general, there are two elementary shapes of the tree trunk distortion developments (see **Figure 1**):

- the vital peak and upper part of a tree trunk keep the vertical direction and distortion exhibits only the bottom irrespective of the middle tree trunk sections. This shape development refers to the latent soil material movement;
- excessive distortion of the entire tree trunk along and an apparent deviation of upper part of a tree trunk from the vertical direction refer to an advanced state of slope soil material movement.

Figure 1. Elementary shapes of the tree trunk distortion developments.

The apparent deviation of the upper part of a tree trunk from the vertical direction causes a load of a bending moment due to the weight of the above ground part of tree trunk a crown. This bending moment load opposes the root system together with soil environment. When the load exceeds a restrain capacity of the tree root system then the tree trunk inclines heavily

http://dx.doi.org/10.5772/66813

and finally tilts completely. This situation is inappropriate to set up the rate of a slope soil material movement, since for this situation it is not possible to separate which portion of the slope soil material movement belongs to the gravitation and to the soil ground yielding due to the bending moment.

The slope soil material movement describes exactly two vectors $\vec{u}_A(\vec{u}_{Ah}, \vec{u}_{Av})$ and $\vec{u}_B(\vec{u}_{Bh}, \vec{u}_{Bv})$, which state a displacement at two slope's points A and B. Both points are located in places where there are expected extreme values of displacements. They are located on the intersection line of a vertical plane of symmetry through the axis of the tree trunk with a slope's plane at antipodal points of tree root system extend (**Figure 2**).

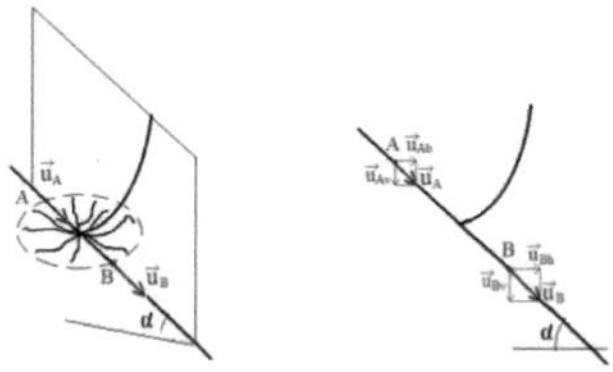

Figure 2. Kinematics of tree movement on slope.

Intersection line indicates the direction in which the maximum displacement occurs. Differentials of displacements between points A and B ($d_{uh} = \vec{u}_{Ah} - \vec{u}_{Bh}$; $d_{uv} = -\vec{u}_{Av} - \vec{u}_{Bv}$) induce simultaneous change of the distance length between A and B (ΔsAB) and change of inclination of slope section between points A and B ($\Delta\alpha$AB). When the distance between A and B extends then the inclination becomes flatter and vice versa, and when the distance between A and B contracts then the inclination becomes steeper.

A distinctive distortion of the tree trunk arises due to incremental change of the inclination ($\Delta\alpha$AB). **Figure 3** shows in sequence of the evolution for two forms of the tree trunk distortions one for the decreasing and second for increasing trend of inclination development of slope section.

In case when both the differentials d_{uh} and d_{uv} equal zero the change of inclination $\Delta\alpha$AB equals zero too and the tree root system does not rotate but translates so the tree trunk does not exhibit distortion. For this circumstance a rate movement of the slope soil material cannot be evaluated.

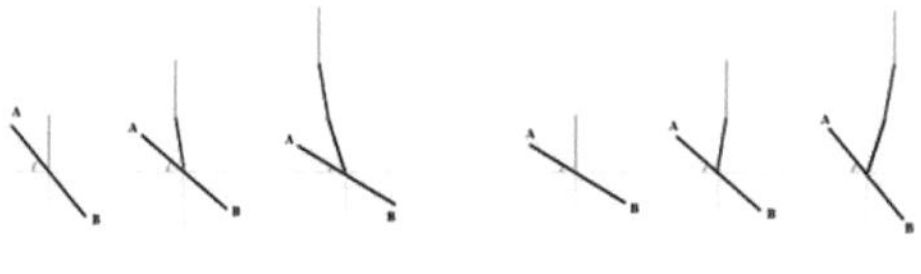

Figure 3. Phases of trunk distortion development of a growing tree during slope soil movement.

3. Evaluation of rate movement

The following assumptions are supposed to be taken into account:

- gravity is the only cause of material movement on slope;
- long term low rate of material movement on slope-latent movement;
- vital peak part of a tree trunk rectifies entirely the rotation of a tree root;
- soil yielding due a bending moment load from weight of the above ground part of tree is negligible and only moving slope materials drift the tree root system;
- the length of the path of the slope soil material movement corresponds to the trajectory of a tree trunk gravity point; consequently a horizontal distance between gravity point and heel point of the tree trunk (*s*) designates the value of the slope soil material movement in the horizontal direction for a tree lifetime (see **Figure 4**);
- tree lifetime is known (*t*);
- other factors affecting growth of the tree, such as climate and weather conditions, vegetation conditions, such as water and nutrients, site conditions like shading, or nearby standing objects are excluded.

For known values of the time of tree vegetation (*t*) and the value of the slope soil material movement in the horizontal direction (*s*) a simple calculation formula provides the value of average rate (*v*) of the slope soil material movement in the horizontal direction for the tree lifetime:

$$v = s/t \tag{1}$$

The value of the slope soil material movement in the horizontal direction (*s*) is solved analytically using approximation function that substitutes distortion line of tree trunk axis and a formula for calculating the position of gravity point of the distortion line. The analytic approximation of the distortion line of tree trunk axis is expressed as an exponential function with two coefficients:

$$y = a \times (1 - e^{-b \times x}) \tag{2}$$

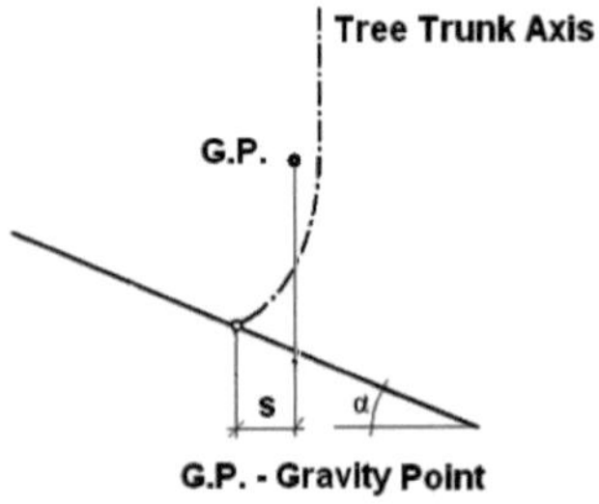

Figure 4. Scheme of tree trunk axis.

http://dx.doi.org/10.5772/66813

Determination of the coefficients "*a*" and "*b*" needs data from measurements made on the tree trunk. The measuring points are located along the vertical line on a peripheral surface of the tree trunk. This is necessary to localize four points minimally to implement the exponential function to state the distortion line of tree trunk axis satisfactory. The first point is placed at the lowest level over the terrain from which the shape of the peripheral surface of a tree trunk is not influenced by the development of the root system. The last is placed at the level from which the trunk is straight and vertical here above. Measured parameters are: the diameter of the tree trunk; vertical distance measured to the lowest point horizon; and horizontal distance measured to the highest point horizon. These parameters are worked out to set up the distortion line of tree trunk axis. The coefficients "*a*" and "*b*" are derived by applying the least square method from data that describe the distortion line of tree trunk axis. The horizontal coordinate of the gravity point of the distortion line of tree trunk axis is set up analytically. The tree lifetime can be established as sum of the tree herbaceous from core sample that was taken by Pressler's auger. Rough estimate of the tree's lifetime consists in measuring the girth and empirical formulas derived for the species of tree and geoclimatic conditions of the site that implement the measured girth. The second estimate is sufficient as the error of a few years does not matter in particular when the tree's lifetime is more than 50 years. In turn, the development of a tree root system influences the gravitational slope soil material movement too. The tree root system shore up the movement of slope materials and counteracts the effects of gravity. The rate of movement for slope covered with tree vegetation will not be constant. With the development of the root system the rate should decrease dependently on density and extension of the root system. The gradual decrease of the rate of slope movement can be set up under the following assumptions: a downward trend of the rate; sequencing calculation for several time periods; and determination of the age state for all points on horizons of measurement for the distortion geometry of the tree trunk. The last assumption requires to conduct the exact evaluation for particular section of the tree trunk.

4. Example

Documentation example shows determination of the rate of slope soil movement on the Flysch landslide area at Syrakov hill near Jasenna village in the Zlin region below the first class road number 69 (I / 69). The tree is located on coordinates 49.2743558N, 17.8982797E and is a beech, whose age is estimated at 32 years (see **Figure 5**).

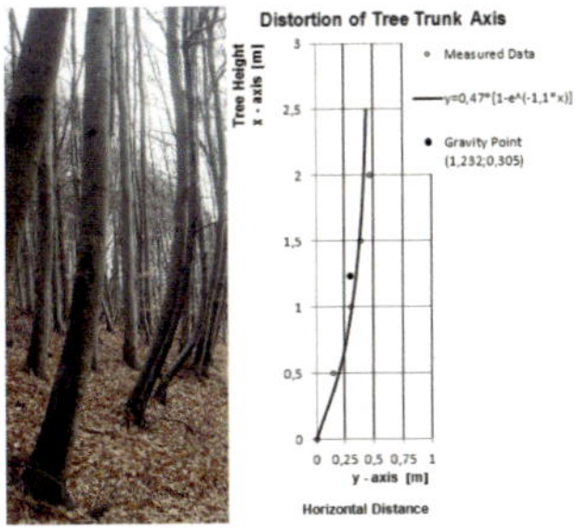

Figure 5. Illustrative example of evaluation of the rate of slope soil movement.

The graph in **Figure 5** displaces measured data, graph analytic function of the tree trunk axis, and the localization of the axis gravity point. The length of slope soil movement is to be about 9.5 mm per year ($s = 0.305$ m; $t = 32$ years).

5. Summary

Presented method allows easily with minimal instrumentation, costs, and only single measurement to determine the approximate value of the long-term rate of slope soil movement which other methods cannot ever achieve.

The method can be applied on all the sites that have not yet been investigated and which exhibit the latent slope soil movement.

The method can be applied in assessing the ground conditions for construction of many kind of engineering works such as roads, pipelines, power lines, as well as in foundations of civil engineering works on areas affected with long-term latent soil movement.

Acknowledgement

The chapter was prepared with the support of the Competence Centres of the Technology Agency of the Czech Republic (TAČR) within the project Centre for Effective and sustainable transport infrastructure (CESTI), project number TE01020168.

Author details

Karel Vojtasik[1]*, Pavel Dvorak[2] and Milan Chiodacki[2]

*Address all correspondence to: karel.vojtasik@vsb.cz

1 Department of Geotechnics and Underground Engineering, VŠB-Technical University of Ostrava, Ostrava-Poruba, Czech Republic

2 Minova Bohemia s.r.o., Lihovarska, Ostrava-Radvanice, Czech Republic

References

L.W. Abramson, T.S. Lee, S. Sharma and G.M. Boyce. *Slope Stability and Stabilization Methods.* New York: Wiley, 1996.

R.P.C. Morgan and R.J. Rickson. *Slope Stabilization and Erosion Control a Bioengineering Approach.* London: E&FN Spon, 1995.

Made in the USA
Las Vegas, NV
03 November 2026

57424103R00259